Lecture Notes in Control and Information Sciences

Edited by M. Thoma and A. Wyner

Vol. 117: K.J. Hunt
Stochastic Optimal Control Theory
with Application in Self-Tuning Control
X, 308 pages, 1989.

Vol. 118: L. Dai
Singular Control Systems
IX, 332 pages, 1989

Vol. 119: T. Başar, P. Bernhard
Differential Games and Applications
VII, 201 pages, 1989

Vol. 120: L. Trave, A. Titli, A. M. Tarras
Large Scale Systems:
Decentralization, Structure Constraints
and Fixed Modes
XIV, 384 pages, 1989

Vol. 121: A. Blaquière (Editor)
Modeling and Control of Systems
in Engineering, Quantum Mechanics,
Economics and Biosciences
Proceedings of the Bellman Continuum
Workshop 1988, June 13–14, Sophia Antipolis, France
XXVI, 519 pages, 1989

Vol. 122: J. Descusse, M. Fliess, A. Isidori,
D. Leborgne (Eds.)
New Trends in Nonlinear Control Theory
Proceedings of an International
Conference on Nonlinear Systems,
Nantes, France, June 13–17, 1988
VIII, 528 pages, 1989

Vol. 123: C. W. de Silva, A. G. J. MacFarlane
Knowledge-Based Control with
Application to Robots
X, 196 pages, 1989

Vol. 124: A. A. Bahnasawi, M. S. Mahmoud
Control of Partially-Known Dynamical Systems
XI, 228 pages, 1989

Vol. 125: J. Simon (Ed.)
Control of Boundaries and Stabilization
Proceedings of the IFIP WG 7.2 Conference
Clermont Ferrand, France, June 20–23, 1988
IX, 266 pages, 1989

Vol. 126: N. Christopeit, K. Helmes
M. Kohlmann (Eds.)
Stochastic Differential Systems
Proceedings of the 4th Bad Honnef Conference
June 20–24, 1988
IX, 342 pages, 1989

Vol.127: C. Heij
Deterministic Identification
of Dynamical Systems
VI, 292 pages, 1989

Vol. 128: G. Einarsson, T. Ericson,
I. Ingemarsson, R. Johannesson,
K. Zigangirov, C.-E. Sundberg
Topics in Coding Theory
VII, 176 pages, 1989

Vol. 129: W. A.Porter, S. C. Kak (Eds.)
Advances in Communications and
Signal Processing
VI, 376 pages, 1989.

Vol. 130: W. A. Porter, S. C. Kak, J. L. Aravena (E
Advances in Computing and Control
VI, 367 pages, 1989

Vol. 131: S. M. Joshi
Control of Large Flexible Space Structures
IX, 196 pages, 1989.

Vol. 132: W.-Y. Ng
Interactive Multi-Objective Programming
as a Framework for Computer-Aided Control
System Design
XV, 182 pages, 1989.

Vol. 133: R. P. Leland
Stochastic Models for Laser Propagation
in Atmospheric Turbulence
VII, 145 pages, 1989.

Vol. 134: X. J. Zhang
Auxiliary Signal Design in Fault
Detection and Diagnosis
XII, 213 pages, 1989

Vol. 135: H. Nijmeijer, J. M. Schumacher (Eds.)
Three Decades of Mathematical System Theor
A Collection of Surveys at the Occasion of the
50th Birthday of Jan C. Willems
VI, 562 pages, 1989

Vol. 136: J. Zabczyk (Ed.)
Stochastic Systems and Optimization
Proceedings of the 6th IFIP WG 7.1
Working Conference,
Warsaw, Poland, September 12–16, 1988
VI, 374 pages. 1989

For information about Vols. 1–116 please contact your bookseller or Springer-Verlag

Lecture Notes
in Control and Information Sciences 183

Editors: M. Thoma and W. Wyner

S. Hosoe (Ed.)

Robust Control

Proceedings of a Workshop
held in Tokyo, Japan, June 23 - 24, 1991

Springer-Verlag Berlin Heidelberg GmbH

Editor

Prof. Shigeyuki Hosoe
Nagoya University
Dept. of Information Engineering
Furo-cho, Chikusa-ku
Nagoya 464-01
JAPAN

ISBN 978-3-540-55961-0 ISBN 978-3-540-47320-6 (eBook)
DOI 10.1007/978-3-540-47320-6

Typesetting: Camera-ready by authors

61/3020 5 4 3 2 1 0 Printed on acid-free paper

PREFACE

The workshop on Robust Control was held in Tokyo, Japan on July 24-25, 1991. From 10 countries, more than 70 reseachers and engineers gathered together, and 33 general talks and 3 tutorial ones, all invited, were presented. The success of the workshop depended on their high level of scientific and engineering expertise.

This book collects almost all the papers devoted to the meeting. The topics covered include: H_∞ control, parametric/structured approach to robust control, robust adaptive control, sampled-data control systems, and their applications.

All the arrangement for the workshop was executed by a organizing committee formed in the technical committee for control theory of the Society of Instrument and Control Engineers in Japan.

For financial support and continuing cooperation, we are grateful to Casio Science Promotion Foundation, and many Japanese companies.

Shigeyuki Hosoe

CONTENTS

(J, J')-LOSSLESS FACTORIZATION USING CONJUGATIONS OF ZERO AND POLE EXTRACTIONS

Hidenori Kimura
Department of Mechanical Engineering for
Computer-Controlled Machinery, Osaka University

1. Introduction

A matrix $G(s)$ in L^∞ is said to have a *(J, J')-lossless factorization*, if it is represented as a product $G(s)=\Theta(s)\Pi(s)$, where $\Theta(s)$ is (J, J')-lossless [11] and $\Pi(s)$ is unimodular in H^∞. The notion of (J, J')-lossless factorization was first introduced by Ball and Helton [2] in geometrical context. It is a generalization of the well-known inner-outer factorization of matrices in H^∞ to those in L^∞. It also includes the spectral factorization for positive matrices as a special case. Furthermore, it turned out that the (J, J')-lossless factorization plays a central role in H^∞ control theory [1][4][7]. Actually, it gives a simple and unified framework of H^∞ control theory from the viewpoint of classical network theory. Thus, the (J, J')-lossless factorization is of great importance in system theory.

The (J, J')-lossless factorization is usually treated as a *(J, J')-spectral factorization* problem which is to find a unimodular $\Pi(s)$ such that $G^{\sim}(s)JG(s) = \Pi^{\sim}(s)J\Pi(s)$. Ball and Ran [4] first derived a state-space representation of the (J, J')-spectral factorization to solve the Nehari extension problem based on the result of Bart et.al. [5]. It was generalized in [1] to more general model-matching problems. Recently, Green et.al. [8] gave a simpler state-space characterization of the (J, J')-spectral factorization for a stable $G(s)$. Ball et al. [3] discussed some connections between the chain-scattering formulation of H^∞ and (J, J')-lossless factorization.

In this paper, we give a simple derivation of the (J, J')-lossless factorization for general unstable $G(s)$ based on the *theory of conjugation* developed in [9]. The conjugation is a simple operation of replacing some poles of a given rational matrix by their mirror images with respect to the origin by multiplication of a certain matrix. It is a state-space representation of the Nevanlinna-Pick interpolation theory [11] and the Darlington synthesis method of classical circuit theory. The (J, J')-lossless factorization for unstable matrix, which is first established in this paper, enables us to solve the general H^∞ control problem directly without any recourse to Youla parameterization [12].

In Section 2, the theory of conjugation is briefly reviewed. The (J, J')-lossless conjugation, a special class of conjugations by (J, J')-lossless matrix, is developed in Section 3. Section 4 gives a frequency domain characterization of the (J, J')-lossless factorization. Section 5 is devoted to the state space representation of the (J, J')-lossless factorization. The existence condition for the (J, J')-lossless factorization is represented in terms of two Riccati equations, one of which is degenerated in the sense that it contains no constant term.

Notations :

$$C(sI-A)^{-1}B+D := \left[\begin{array}{c|c} A & B \\ \hline C & D \end{array}\right], \qquad R^{\sim}(s) = R^T(-s), \qquad R^*(s) = R^T(\bar{s}),$$

$\sigma(A)$; The set of eigenvalues of a matrix A.

$RL^\infty_{m\times r}$; The set of proper rational matrices of size m×r without poles on the imaginary axis,

$RH^\infty_{m\times r}$; The set of proper stable rational matrices of size m×r.

2. Conjugation

The notion of conjugation was first introduced in [9]. It represents a simple operation which replaces the poles of a transfer function by their conjugates. The conjugation gives a unified framework for treating the problems associated with interpolation theory in state space [10]. In this section, the notion of conjugation is generalized to represent the relevant operations more naturally.

To define the conjugation precisely, let

$$G(s) = \left[\begin{array}{c|c} A & B \\ \hline C & D \end{array}\right], \qquad A \varepsilon R^{n\times n},\ D \varepsilon R^{m\times r} \tag{2.1}$$

be a minimal realization of a transfer function G(s).

Definition 2.1 Let $\sigma(A) = \Lambda_1 \cup \Lambda_2$ be a partition of $\sigma(A)$ into two disjoint sets of k complex numbers Λ_1 and n-k complex numbers Λ_2. A matrix $\Theta(s)$ is said to be a *Λ_1-conjugator* of G(s), if the poles of the product $G(s)\Theta(s)$ are equal to $\{-\Lambda_1\} \cup \Lambda_2$ and for any constant vector $\xi \neq 0$,

$$\xi(sI - A)^{-1}B\Theta(s) \neq 0 \tag{2.2}$$

for some s, where $-\Lambda_1$ is a set composed of the elements of Λ_1 with reversed sign.

As a special case, we can choose Λ_1 composed of the RHP elements of $\sigma(A)$. If A has no eigenvalue on the $j\omega$ - axis, Λ_1 conjugator of G(s) stabilizes the product $G(s)\Theta(s)$. In that case, Λ_1-conjugator is called a *stabilizing conjugator.* In the same way, we can define an *anti-stabilizing conjugator.*

Remark. The condition (2.2) is imposed to rule out the trivial case like $\Theta(s) \equiv 0$. It is automatically satisfied for the case where $\Theta(s)$ is right invertible.

THEOREM 2.2 $\Theta(s)$ is a Λ_1-conjugator of G(s) given by (2.4), if and only if $\Theta(s)$ is given by

$$\Theta(s) = \left[\begin{array}{c|c} -A^T & XB \\ \hline C_c & I \end{array}\right] D_c \tag{2.3}$$

where C_c, D_c, X, are such that X is a solution of the Riccati equation

$$XA + A^TX + XBC_cX = 0 \tag{2.4}$$

such that $A_c := A + BC_cX$ satisfies $\sigma(A_c) = \{-\Lambda_1\} \cup \Lambda_2$ and (A_c , BD_c) is controllable. In that case, the product $G(s)\Theta(s)$ is given by

$$G(s)\,\Lambda(s) = \left[\begin{array}{c|c} A + B\,C_c\,X & B \\ \hline C + D\,C_c\,X & D \end{array}\right] D_c \tag{2.5}$$

(*Proof*) From the definition, the A-matrix of the product $G(s)\Theta(s)$ has eigenvalues in $\sigma(A) \cup \sigma(-A^T)$. Hence, it can be assumed, without loss of generality, that the A-matrix of $\Theta(s)$ is equal to $-A^T$. Hence, we put

$$\Theta(s) = \left[\begin{array}{c|c} -A^T & B_c \\ \hline C_c & D_c \end{array}\right] \tag{2.6}$$

for some B_c C_c and D_c . Then, the product rule yields

$$G(s)\Theta(s) = \left[\begin{array}{cc|c} A & BC_c & BD_c \\ 0 & -A^T & B_c \\ \hline C & DC_c & DD_c \end{array}\right] \tag{2.7}$$

Since $-\Lambda_1 \subset \sigma(-A^T)$ and $\Lambda_2 \subset \sigma(A)$, we have

$$\begin{bmatrix} A & BC_c \\ 0 & -A^T \end{bmatrix}\begin{bmatrix} M_1 & M_2 \\ N_1 & N_2 \end{bmatrix} = \begin{bmatrix} M_1 & M_2 \\ N_1 & N_2 \end{bmatrix}\begin{bmatrix} A_1 & 0 \\ 0 & -A_1^T \end{bmatrix} \tag{2.8}$$

for some M_1 , M_2 , N_1 and N_2 such that $\sigma(A_1) = \{-\Lambda_1\} \cup \Lambda_2$.

Assume that $\Theta(s)$ is a Λ_1-conjugator. Since the modes associated with $-A_1^T$ are uncontrollable in (2.10), we have

$$\begin{bmatrix} BD_c \\ B_c \end{bmatrix} = \begin{bmatrix} M_1 \\ N_1 \end{bmatrix} B_1 \tag{2.9}$$

for some B_1. From (2.8), it follows that

$$AM_1 + BC_cN_1 = M_1A_1, \qquad -A^TN_1 = N_1A_1. \tag{2.10}$$

Hence, we have

$$(sI - A)\,M_1 - BC_cN_1 = M_1\,(sI - A_1) \tag{2.11}$$

Premultiplication of (2.11) by $(sI - A)^{-1}$ and postmultiplication by $(sI - A)^{-1}B_1$ yield

$$M_1 (sI - A_1)^{-1} B_1 = (sI - A)^{-1} B\Theta(s). \qquad (2.12)$$

Here, we used the relation $N_1 (sI - A_1)^{-1} B_1 = (sI + A^T)^{-1} B_c$, which is due to (2.9) and the second relation of (2.10). The condition (2.5) implies that M_1 is non-singular and (A_1, B_1) is controllable.

Taking $X=N_1M_1^{-1}$ in (2.10) yields (2.4). Since $M_1^{-1}A_CM_1=A_1$, A_C should be stable. It is clear that $(A_c, BD_c)=(M_1^{-1}A_1M_1, M_1B_1)$ is controllable because (A_1, B_1) is controllable.

Due to (2.9), we have $B_c=N_1B_1=N_1M_1^{-1}BD_C=XBD_C$, from which (2.3) follows. The derivation of (2.5) is straightforward.

Remark. Theorem 2.2 implies that the conjugator of G(s) in (2.4) depends only on (A, B). Therefore, we sometimes call it *a conjugator of (A, B)*.

Remark It is clear, from (2.7), that the order of $\Theta(s)$ given in (2.6) is equal to the rank of X which is equal to the number of elements in L_1.

3. (J, J')-Lossless Conjugators

A matrix $\Theta(s) \in \mathbf{RL}^{\infty}_{(m+r)x(p+q)}$ is said to be *(J, J')-unitary*, if

$$\Theta^{\sim}(s)\, J\, \Theta(s) = J' \qquad (3.1)$$

for each s. A (J, J')-unitary matrix $\Theta(s)$ is said to be *(J, J')-lossless*, if

$$\Theta^{*}(s)\, J\, \Theta(s) \le J', \qquad (3.2)$$

for each Re[s]≥0, where

$$J = \begin{bmatrix} I_m & 0 \\ 0 & -I_r \end{bmatrix}, \quad J' = \begin{bmatrix} I_p & 0 \\ 0 & -I_q \end{bmatrix}, \quad m \ge p, \quad r \ge q \qquad (3.3)$$

A (J, J')-lossless matrix is called simply a *J-lossless matrix.*

A conjugator of G(s) which is (J, J')-lossless is called a *(J, J')-lossless conjugator* of G(s). The existence condition of a J-lossless conjugator, which is of primary importance in what follows, is now stated.

THEOREM 3.1. Assume that (A, B) is controllable and A has no eigenvalue on the jω-axis. There exists a (J, J')-lossless stabilizing (anti-stabilizing) conjugator of (A, B), if and only if there exists a non-negative solution P of the Riccati equation

$$PA + A^TP - PBJB^TP = 0 \qquad (3.4)$$

which stabilizes (antistabilizes) $A_c := A - BJB^TP$. In that case, the desired (J, J')-lossless conjugator of G(s) and the conjugated system are given respectively by

$$\Theta(s) = \left[\begin{array}{c|c} -A^T & PB \\ \hline -JB^T & I \end{array}\right] D_c, \quad G(s)\Theta(s) = \left[\begin{array}{c|c} A - BJB^TP & B \\ \hline C - DJB^TP & D \end{array}\right] D_c \qquad (3.5)$$

where D_c is any matrix satisfying $D_c^T J D_c = J'$.

(Proof) Since (2.4) implies $X(sI + A^T) = (sI - A_C)X$, $\Theta(s)$ in (2.3) is represented as

$$\Theta(s) = \Theta_0(s) D_c,$$

$$\Theta_0(s) = I + C_c X (sI - A_c)^{-1} B, \quad A_c = A + B C_c X, \qquad (3.6)$$

Assume that P is a solution of the Lyapunov equation

$$P A_c + A_c^T P + X^T C_c^T J C_c X = 0. \qquad (3.7)$$

Then, it follow that

$$\Theta_0^{\sim}(s) J \Theta_0(s) = J + B^T (-sI - A_c^T)^{-1} (X^T C_c^T J + P B) + (J C_c X + B^T P) (sI - A_c)^{-1} B.$$

Since A_C is stable and (A_c, B) is controllable from the assumption, $\Theta(s)$ is J-unitary if and only if

$$J C_c X + B^T P = 0. \qquad (3.8)$$

In that case, we have

$$\Theta_0^*(s)J\Theta_0(s) = J - (s + \bar{s})B^T(\bar{s}I - A_c^T)^{-1}P(sI - A_c)^{-1}B$$

Hence, in order that $\Theta_0(s)$ is J-lossless, P should be non-negative. From (3.8), (3.7) is written as (3.4) and $\Theta_0(s)$ in (3.6) can be written as $\Theta_0(s) = I - JB^T(sI + A^T)^{-1}PB$, from which the representation (3.5) follows.

Now, we shall show a cascade decomposition of the (J, J')-lossless conjugation of (A, B) according to the modal decomposition of the matrix A. Assume that the pair (A, B) is of the form

$$(A, B) = (\begin{bmatrix} A_{11} & 0 \\ A_{21} & A_{22} \end{bmatrix}, \begin{bmatrix} B_1 \\ B_2 \end{bmatrix}). \tag{3.9}$$

The solution P of (3.4) is represented in conformable with the partition (3.9) as

$$P = \begin{bmatrix} P_{11} & P_{12} \\ P_{12}^T & P_{22} \end{bmatrix}. \tag{3.10}$$

LEMMA 3.2. If the pair (A, B) in (3.9) has a (J, J')-lossless conjugator $\Theta(s)$ of the form (3.5), then $\Theta(s)$ is represented as the product

$$\Theta(s) = \Theta_1(s)\,\Theta_2(s) \tag{3.11}$$

of a J-lossless conjugator $\Theta_1(s)$ of (A_{11}, B_1) and a (J, J')-lossless conjugator $\Theta_2(s)$ of $(A_{22}, \bar{B}_2)$ with $\bar{B}_2 = B_2 + P_{12}P_{22}^{\dagger}B_1$, where $P_{22}^{\dagger}$ is a peudo inverse of P_{22} in (3.10). Actually, $\Theta_1(s)$ and $\Theta_2(s)$ are given respectively as

$$\Theta_1(s) = \left[\begin{array}{c|c} -A_{11}^T & \Delta B_1 \\ \hline -J\,B_1^T & I \end{array}\right], \quad \Theta_2(s) = \left[\begin{array}{c|c} -A_{22}^T & P_{22}\bar{B}_2 \\ \hline -J\,\bar{B}_2^T & I \end{array}\right] D \tag{3.12}$$

where $\Delta = P_{11} - P_{12}P_{22}^{\dagger}P_{12}^T$ and D is any constant (J, J')-unitary matrix.

(Proof) Since $P \geq 0$, we have $P_{12}(I - P_{22}^{\dagger}P_{22}) = 0$. Therefore,

$$UPU^T = \begin{bmatrix} \Delta & 0 \\ 0 & P_{22} \end{bmatrix}, \quad (U^{-1})^T B = \begin{bmatrix} B_1 \\ \bar{B}_2 \end{bmatrix}, \quad U = \begin{bmatrix} I & -P_{12}P_{22}^{\dagger} \\ 0 & I \end{bmatrix}$$

From this relation and (3.4), it follows that

$$\begin{bmatrix} \Delta & 0 \\ 0 & P_{22} \end{bmatrix}\begin{bmatrix} A_{11} & 0 \\ \bar{A}_{21} & A_{22} \end{bmatrix} + \begin{bmatrix} A_{11}^T & \bar{A}_{21}^T \\ 0 & A_{22}^T \end{bmatrix}\begin{bmatrix} \Delta & 0 \\ 0 & P_{22} \end{bmatrix} - \begin{bmatrix} \Delta & 0 \\ 0 & P_{22} \end{bmatrix}\begin{bmatrix} B_1 \\ \bar{B}_2 \end{bmatrix} J \begin{bmatrix} B_1^T & \bar{B}_2^T \end{bmatrix}\begin{bmatrix} \Delta & 0 \\ 0 & P_{22} \end{bmatrix} = 0, \tag{3.13}$$

where $\bar{A}_{21} = A_{21} + P_{22}^{\dagger}P_{12}^TA_{11} - A_{22}P_{22}^{\dagger}P_{12}^T$. Hence, $\Delta A_{11} + A_{11}^T\Delta - \Delta B_1JB_1^T\Delta = 0$. Since $\Delta \geq 0$, $\Theta_1(s)$ in (3.12) is a J-lossless conjugator of (A_{11}, B_1). Also, it is clear that $\Theta_2(s)$ given in (3.12) is a (J, J')-lossless conjugator of $(A_{22}, \bar{B}_2)$. Since (3.11) implies that $\bar{A}_{21} = \bar{B}_2JB_1^T\Delta$, we have

$$\Theta_1(s)\Theta_2(s) = \left[\begin{array}{cc|c} -A_{11}^T & -\Delta B_1J\bar{B}_2^T & \Delta B_1 \\ 0 & -A_{22}^T & P_{22}\bar{B}_2 \\ \hline -JB_1^T & -J\bar{B}_2^T & I \end{array}\right] D = \left[\begin{array}{c|c} -UA^TU^{-1} & UPB \\ \hline -JB^TU^{-1} & I \end{array}\right] D = \Theta(s).$$

The proof is now complete.

4. (J, J')-Lossless Factorizations

A matrix $G(s) \in \mathbf{RL}^{\infty}_{(m+r)\times(p+q)}$ is said to have a *(J, J')-lossless factorization*, if it can be represented as

$$G(s) = \Theta(s)\,\Pi(s), \tag{4.1}$$

where $\Theta(s) \in \mathbf{RL}^{\infty}_{(m+r)\times(p+q)}$ is (J, J')-lossless and $\Pi(s)$ is unimodular in H^∞, i.e., both $\Pi(s)$ and $\Pi(s)^{-1}$ are in $\mathbf{RH}^{\infty}_{(p+q)\times(p+q)}$. It is a generalization of the inner-outer factorization of H^∞ matrices to L^∞ matrices.

Since $\Pi(s)$ is unimodular, it has no unstable zeros nor poles. Therefore, $\Theta(s)$ should include all the unstable zeros and the poles of G(s). The (J, J')-lossless conjugation introduced in the previous section gives a suitable machinery to carry out these procedures of "zero extraction" and "pole extraction."

If (4.1) holds, it follows that

$$\Pi(s)^{-1} J' \Theta^{\sim}(s) J G(s) = I_{p+q}. \tag{4.2}$$

This implies that G(s) should be left invertible. Hence, there exists a complement G'(s) of G(s) such that

$$[\, G(s) \;\; G'(s) \,]^{-1} = \begin{bmatrix} G^{+}(s) \\ G^{\perp}(s) \end{bmatrix} \tag{4.3}$$

exists. Obviously, $G^{+}(s)$ is a left inverse and $G^{\perp}(s)$ is a left annihilator of G(s), i.e.,

$$G^{+}(s)\, G(s) = I_{p+q}, \quad G^{\perp}(s)\, G(s) = 0 \tag{4.4}$$

The following result gives a method of carrying out a (J, J')-lossless factorization based on sequential conjugation of zeros and poles extraction.

THEOREM 4.1. (J, J')-lossless factorization of G(s) exists, if and only if the left inverse $G^{+}(s)$ of G(s) has a J-lossless stabilizing conjugator $\Theta_1(s)$ such that $G^{\sim}(s)J\Theta(s)$ has a (J, J')-lossless antistabilizing conjugator $\Theta_2(s)$ such that

$$G^{\perp}(s)\Theta_1(s)\Theta_2(s)=0 \tag{4.5}$$

(Proof) To prove the necessity, assume that G(s) allows a (J, J')-lossless factorization (4.1). Clearly, we have

$$\begin{bmatrix} G^{+}(s) \\ G^{\sim}(s)\, J \end{bmatrix} \Theta(s) = \begin{bmatrix} \Pi(s)^{-1} \\ \Pi^{\sim}(s) J' \end{bmatrix}.$$

Therefore, $\Theta(s)$ is a (J, J')-lossless stabilizing conjugator of $G^{+}(s)$, as well as a (J, J')-lossless anti-stabilizing conjugator of $G^{\sim}(s)J$. Due to Lemma 3.2, $\Theta(s)$ can be factored as (3.11) where $\Theta_1(s)$ is stable J-lossless matrix and $\Theta_2(s)$ is an anti-stable (J, J')-lossless matrix. From the well-known property of (J, J')-lossless matrix, $\Theta_2(s)$ has no unstable zero. From

$$G(s)=\Theta_1(s)\Theta_2(s)\Pi(s), \tag{4.6}$$

$G^{+}(s)\Theta_1(s) = \Pi_1(s)$ satisfies $\Pi_1(s)\Theta_2(s)\Pi(s) = I$. Since $\Theta_2(s)\Pi(s)$ has no unstable zeros, $\Pi_1(s)$ is stable. Therefore $\Theta_1(s)$ is a stabilizing conjugator of $G^{+}(s)$. Since $G^{\sim}(s)J\Theta_1(s)\Theta_2(s) = \Pi^{\sim}(s)J'$, and $\Pi^{\sim}(s)J$ is anti-stable, $\Theta_2(s)$ is a (J, J')-lossless anti-stabilizing conjugator of $G^{\sim}(s)J\Theta_1(s)$. The relation (4.5) is obvious from (4.6). Now, the necessity of the theorem has been proven.

Assume that there exist $\Theta_1(s)$ and $\Theta_2(s)$ satisfying the conditions of the theorem. Let $\Pi_1(s)=G^{+}(s)\Theta_1(s)$ and $\Pi_2(s) = G^{\sim}(s)J\Theta_1(s)\Theta_2(s)$. From the assumption, $\Pi_1(s)$ is stable and $\Pi_2(s)$ is anti-stable. Since $\Theta_1(s)$ is J-lossless, $\Theta_1(s)J\Theta^{\sim}{}_1(s)=J$ (see[6]). Therefore, we have $\Pi_1(s)J\Theta_1{}^{\sim}(s)JG(s) = I$. Since $\Pi_1(s)$ is stable, $G^{\sim}(s)J\Theta_1(s)$ has no stable zero. Since conjugator does not create zero, $\Pi_2(s)$ has no stable zero. Therefore, $\Pi_2{}^{\sim}(s)$ is unimodular.

Finally, we shall show that the relation (4.1) holds with $\Theta(s)=\Theta_1(s)\Theta_2(s)$ and $\Pi(s)=J'\Pi_2(s)$. Due to (4.5), $\Theta_1(s)\Theta_2(s)=G(s)U(s)$ for some U(s). Since $\Pi^{\sim}(s) = \Theta_2{}^{\sim}(s)\Theta_1{}^{\sim}(s)JG(s)$, we have $\Pi_2{}^{\sim}(s)U(s) = J'$. The assertion follows immediately.

Based on the above theorem and its proof, we can state an algorithm of (J, J')-lossless factorization.

[Algorithm of (J, J')-lossless Factorization]

Step 1. Find a J-lossless stabilizing conjugator $\Theta_1(s)$ of $G^{+}(s)$.

Step 2. Find a (J, J')-lossless anti-stabilizing conjugator $\Theta_2(s)$ of $G^{\sim}(s)J\Theta_1(s)$ satisfying (4.5).

Step 3. Put $\Theta(s)=\Theta_1(s)\Theta_2(s)$ and $\Pi(s)=J'\Theta^{\sim}(s)JG(s)$, which give a (J, J')-lossless factorization.

In the above algorithm, Θ_1(s absorbs all the unstable zeros of G(s). Hence, Step 1 is regarded as a procedure of *zero extraction*. Similarly, $\Theta_2(s)$ in Step 2 absorbs all the unstable poles of G(s), and hence, Step 2 is regarded as *pole extraction*.

5. State-Space Theory of (J, J')-Lossless Factorizations

In this section, we derive a state-space forms of (J, J')-lossless factorizations. Let

$$G(s) = \left[\begin{array}{c|c} A & B \\ \hline C & D \end{array}\right] \qquad A \in R^{n\times n},\ D \in R^{(m+r)\times(p+q)} \tag{5.1}$$

be a state-space form of G(s). The following assumptions are made :

(A_1) The matrix A has no eigenvalues on the jω-axis.

(A_2) The matrix D is of full column rank.

(A_3) $G^{\sim}(s)JG(s)$ admits no unstable pole-zero cancellation.

Due to the assumption (A_2), there exists a column complement D' of D such that

$$[\,D \quad D'\,]^{-1} = \left[\begin{array}{c} D^{+} \\ D^{\perp} \end{array}\right] \tag{5.2}$$

exists. Obviously,

$$D^{+}D = I_{p+q} \qquad D^{\perp}D = 0. \tag{5.3}$$

Now, we assume that G(s) has a (J, J')-lossless factorization (4.1). Since $G^{\sim}(s)JG(s)=\Pi^{\sim}(s)J\Pi(s)$, we have

$$D^{T} J D = D_{\pi}^{T} J' D_{\pi} \tag{5.4}$$

where $D_{\pi} = \Pi(\infty) \in R^{(p+q)\times(p+q)}$. Since $\Pi(s)$ is unimodular, D_{π} is non-singular. Hence, $D^{T}JD$ must be non-singular.

The left inverse $G^{+}(s)$ and a left annihilator $G^{\perp}(s)$ of G(s) satisfying (4.6) are represented in the state space as

$$\left[\begin{array}{c} G^{+}(s) \\ G^{\perp}(s) \end{array}\right] = \left[\begin{array}{c|c} A-LC & -L \\ \hline D^{+}C & D^{+} \\ D^{\perp}C & D^{\perp} \end{array}\right], \qquad L = BD^{+} + UD^{\perp}, \tag{5.5}$$

where U is an arbitrary matrix which is determined later.

Now, we shall carry out the (J, J')-lossless factorization of G(s) based on the algorithm derived in Section 4.

Step 1 (Zero Extraction)

Due to Theorem 3.1, a J-lossless stabilizing conjugator $\Theta_1(s)$ of $G^{+}(s)$ in (5.5) is calculated to be

$$\Theta_1(s) = \left[\begin{array}{c|c} -(A-LC)^{T} & -XL \\ \hline JL^{T} & I \end{array}\right], \tag{5.6}$$

where X is a non-negative solution of the Riccati equation

$$X(A - LC) + (A - LC)^{T}X - XL^{T}JL^{T}X = 0 \tag{5.7}$$

which stabilizes

$$A_1 := A - L(C + JL^{T}X) \tag{5.8}$$

Step 2 (Pole Extraction)

Straightforward computation using LD = B yields

$$G^{\sim}(s)J\Theta_1(s) = \left[\begin{array}{c|c} -A^{T} & C^{T}J \\ \hline -B^{T} & D^{T}J \end{array}\right]\Theta_1(s) = \left[\begin{array}{c|c} -A^{T} & C^{T}J + XL \\ \hline -B^{T} & D^{T}J \end{array}\right], \tag{5.9}$$

Due to Theorem 3.1, a (J, J')-lossless anti-stabilizing conjugator of (5.9) is given by

$$\Theta_2(s) = \left[\begin{array}{c|c} A & Y(C^{T}J + XL) \\ \hline -C - JL^{T}X & I \end{array}\right] DD_{\pi}^{-1} \tag{5.10}$$

where D_π is any matrix satisfying (5.4) and Y is a non-negative solution of the Riccati equation

$$YA^T + AY + Y(C^T + XLJ)\,J\,(C + JL^TX)Y = 0 \tag{5.11}$$

which stabilizes

$$A_2 := A + Y(C^T + XLJ)\,J\,(C + JL^TX), \tag{5.12}$$

Now, concatenation rule yields

$$\Theta(s) = \Theta_1(s)\Theta_2(s) = \left[\begin{array}{cc|c} -(A-LC)^T & XL(C+JL^TX) & -XB \\ 0 & A & Y(C^TJD+XB) \\ \hline JL^T & -C-JL^TX) & D \end{array}\right] D\pi \tag{5.13}$$

We use the free parameter U in (5.5) to satisfy (4.5). From (5.5), it follows that

$$G^\perp(s)\Theta(s) = \left[\begin{array}{c|c} A_1 & (I+YX)B+YC^TJD \\ \hline -D^\perp(C+JL^TX) & 0 \end{array}\right].$$

Assumption (A_3) implies that (A_1, $(I+YX)B+YC^TJD$) is controllable. Hence, (4.5) holds, if and only if

$$D^\perp(C+JL^TX) = 0. \tag{5.14}$$

Thus, the free parameter U in (5.5) must be determined such that (5.14) holds.

The relation (5.14) holds, if and only if

$$L^TX = -J\,(DF + C) \tag{5.15}$$

for some matrix F. Since LD=B for any choice of U, we have

$$F = -\,(D^TJD)^{-1}(B^TX + D^TJC). \tag{5.16}$$

Straightforward computation using (5.15) and (5.16) verifies that the Riccati equation (5.7) is written as

$$XA + A^TX - (C^TJD+XB)(D^TJD)^{-1}(D^TJC+B^TX)+C^TJC = 0. \tag{5.17}$$

The matrix A_1 given in (5.8) is calculated to be

$$A_1 = A + BF. \tag{5.18}$$

Also, the Riccati equation (5.11) and A_2 in (5.12) are represented respectively, as

$$YA^T+AY+YF^TD^TJDFY = 0\,, \qquad A_2 = A + YF^TD^TJDF \tag{5.19}$$

Step 3. Using (5.7), we can write the state-space form of $\Theta(s)=\Theta_1(s)\Theta_2(s)$ in (5.13) as

$$\Theta(s) = \left[\begin{array}{cc|c} A_1 & -BF & -B \\ 0 & A & -YF^TD^TJD \\ \hline -(C+DF) & DF & D \end{array}\right] D_\pi^{-1}, \tag{5.20}$$

where $D\pi$ is any matrix satisfying (5.4).

It remains to compute the unimodular factor $\Pi(s) = J'\Theta^\sim(s)JG(s)$. The concatenation rule yields

$$\Pi(s) = J'D_\pi^{-T}\left[\begin{array}{ccc|c} -A_1^T & 0 & -(C+DF)^TJC & -(C+DF)^TJC \\ F^TB^T & -A^T & F^TD^TJC & F^TD^TJD \\ 0 & 0 & A & B \\ \hline B^T & D^TJDFY & D^TJC & D^TJD \end{array}\right]$$

Carrying out the similarity transformation with the transformation matrix

$$T = \begin{bmatrix} I & 0 & 0 \\ 0 & I & 0 \\ 0 & -Y & I \end{bmatrix}\begin{bmatrix} I & 0 & -X \\ 0 & I & 0 \\ 0 & 0 & I \end{bmatrix},$$

we obtain

$$\Pi(s) = D_\pi \left[\begin{array}{c|c} A_2 & B-YF^TD^TJD \\ \hline -F & I \end{array}\right]. \qquad (5.21)$$

Now, we can state the main result of this paper.

THEOREM 5.1. Assume that G(s) given in (5.1) satisfies Assumptions $(A_1)(A_2)(A_3)$. Then, G(s) has a (J, J')-lossless factorization, if and only if there exist a nonsingular matrix $D_\pi \varepsilon R^{(p+q)x(p+q)}$ satisfying (5.4), a solution $X \geq 0$ of the Riccati equation (5.17) which stabilizizes A_1 given in (5.18)(5.16) and a solution $Y \geq 0$ of the Riccati equation (5.19) which stabilizes A_2 given in (5.19). In that case, a (J, J')-lossless factor $\Theta(s)$ is given in (5.20) and a unimodular factor $\Pi(s)$ is given in (5.21).

The case of stable G(s) is much simpler. The result is exactly the same as was obtained in [8].

COROLLARY 5.2. Assume that A is stable in addition to $(A_1)(A_2)$ and (A_3) . Then, G(s) has a (J, J')-lossless factorization (4.1), if and only if there exists a non-singular matrix $D_\pi \varepsilon R^{(p+q)x(p+q)}$ satisfying (5.4) and the Riccati equation (5.17) has the solution $X \geq 0$ which stabilizes A_1 given in (5.18).

If the above condition hold, the factors $\Theta(s)$ and $\Pi(s)$ of (4.1) are given respectively by

$$\Theta(s) = \left[\begin{array}{c|c} A+BF & B \\ \hline C+DF & D \end{array}\right] D_\pi^{-1}, \qquad \Pi(s) = D_\pi \left[\begin{array}{c|c} A & B \\ \hline -F & I \end{array}\right]. \qquad (5.22)$$

(*Proof*) If A is stable, we can take Y=0 in solving (5.19). The assertions of the corollary follow immediately from (5.20) and (5.21) for Y=0.

6. Conclusion

A necessary and sufficient condition for the existence of (J, J')-lossless factorization in RL^∞ has been established in the state space based on the method of conjugation. It includes the result of [8] for RH^∞ as a special case. The derivation is simple, exhibiting the power of conjugation method.

The result obtained in this paper is directly applied to H^∞ control problems without any recourse to Youla parameterization, giving a simpler result for both standard and non-standard plants. This will be published elsewhere.

REFERENCES

[1] J.A. Ball and N. Cohen, "The sensitivity minimization in an H^∞ norm : Parameterization of all optimal solutions," *Int. J. Control*, **46**, pp.785-816 (1987)

[2] J.A. Ball and J.W. Helton, A Beurling-Lax Theorem for the Lie group U(m, n) which contains most classical imterpolation," *J. Operator Th.*, **9**, pp.107-142 (1983)

[3] J.A. Ball, J.W. Helton and M. Verma, "A factorization principle for stabilization of linear control systems," Reprint.

[4] J.A. Ball and A.C.M. Ran, "Optimal Hankel norm model reduction and Wiener-Hopf factorization I : The canonical case,"*SIAM J. Contr.Optimiz.*, **25**, pp.362-382 (1987).

[5] H. Bart, I. Gohberg, M.A. Kaashoek and P. van Dooren, "Factorizations of transfer functions," *SIAM J. Contr. Optimiz.* **18**, pp. 675-696 (1980).

[6] P. Dewilde, "Lossless Chain-scattering matrices and optimum linear prediction : The vector case," *Circuit Th. & Appl.*, **9**, pp.135-175 (1981).

[7] B.A. Francis, *A Course in H∞ Control*, Springer, New York (1987).

[8] M. Green, K. Glover, D. Limebeer and J. Doyle, "A J-spectral factorization approach to H∞ control," *SIAM J. Contr. Optimiz.*, **28**, pp.1350-1371 (1990).

[9] H. Kimura, "Conjugation, interpolation and model-matching in H∞ ," *Int. J. Control*, **49**, pp.269-307 (1989).

[10] H. Kimura, "State space approach to the classical interpolation problem and its applications, " in *Three Decades of Mathematical System Theory*, H. Nijmeijer and J.M. Schumacher (eds.), Springer-Verlag, ,pp. 243-275(1989)

[11] H. Kimura, Y. Lu and R. Kawatani, "On the structure of H∞ control systems and related extensions," *IEEE Trans. Auto. Contr.*, AC-**36**, pp. 653-667 (1991).

[12] H. Kimura, "Chain-scattering formulation of H∞ control problems," Proc. Workshop on Robust Control, San Antonio (1991).

Mixed H_2/H_∞ Filtering by the Theory of Nash Games

[1]D.J.N. Limebeer [2]B.D.O. Anderson [1] B. Hendel

Department of Electrical Engineering, [1]
Imperial College,
Exhibition Rd.,
London.

Department of Systems Engineering, [2]
Australian National University,
Canberra,
Australia.

Abstract

The aim of this paper is to study an H_2/H_∞ terminal state estimation problem using the classical theory of Nash equilibria. The H_2/H_∞ nature of the problem comes from the fact that we seek an estimator which satisfies two Nash inequalities. The first reflects an H_∞ filtering requirement in the sense alluded to in [4], while the second inequality demands that the estimator be optimal in the sense of minimising the variance of the terminal state estimation error. The problem solution exploits a duality with the H_2/H_∞ control problem studied in [2, 3]. By exploiting duality in this way, one may quickly extablish that an estimator exists which staisfies the two Nash inequalities if and only if a certain pair of cross coupled Riccati equations has a solution on some optimisation interval. We conclude the paper by showing that the Kalman filtering, H_∞ filtering and H_2/H_∞ filtering problems may all be captured within a unifying Nash game theoretic framework.

1 Introduction

In this paper we seek to solve a mixed H_2/H_∞ terminal state estimation problem by formulating it as a two player non-zero sum Nash differential game. As is well known [1, 5], two player non-zero sum games have two performance criteria, and the idea is to use one performance index to reflect an H_∞ filtering criterion, while the second reflects the H_2 optimality criterion usually associated with Kalman filtering. Following the precise problem statement given in section 2.1, we reformulate the filtering problem as a deterministic mixed H_2/H_∞ control problem in section 2.2. This approach offers two advantages in that the transformation is relatively simple, and we can then use an adaptation of the mixed H_2/H_∞ theory given in [2] to derive the necessary and sufficient conditions for the existence of the Nash equilibrium solution to the derived control problem, and therefore for the existence of optimal linear estimators which solve the filtering problem. These conditions are presented in section 2.3 where we also derive the dynamics of an on-line estimator. Some properties of the resulting mixed filter are given in section 2.4. The aim of section 2.5 is to provide a reconciliation with Kalman and H_∞ filtering. In particular, we show that Kalman, H_∞ and H_2/H_∞ filters may all be captured as special cases of a another two player Nash game. Brief conclusions appear in section 3.

Our notation and conventions are standard. For $A \in \mathcal{C}^{n\times m}$, A' denotes the complex conjugate transpose. $\mathcal{E}\{\cdot\}$ is the expectation operator. We denote $\|R\|_{2i}$ as the operator norm induced by the usual 2-norm on functions of time. That is :

$$\|R\|_{2i} = \sup_{u\in\mathcal{L}_2[t_0,t_f]} \frac{\|Ru\|_2}{\|u\|_2}.$$

2 The H_2/H_∞ filtering problem

2.1 Problem Statement

We consider a plant of the form

$$\dot{x}(t) = A(t)x(t) + B(t)w(t) \qquad x(t_0) = x_0 \tag{2.1}$$

with noisy observations

$$z_1(t) = C_1(t)x(t) \tag{2.2}$$
$$z_2(t) = C_2(t)x(t) + n_2(t), \tag{2.3}$$

in which the entries of $A(t), B(t), C_1(t)$ and $C_2(t)$ are continuous functions of time. The inputs $w(t)$ and $n_2(t)$ are assumed to be independent zero-mean white noise processes with

$$\mathcal{E}\{w(t)w'(\tau)\} = Q(t)\delta(t-\tau) \tag{2.4}$$
$$\mathcal{E}\{n_2(t)n_2{}'(\tau)\} = R_2(t)\delta(t-\tau) \qquad \forall t,\tau \in [t_0,t_f]. \tag{2.5}$$

The matrices $Q \geq 0$ and $R_2 > 0$ are symmetric and have entries which are continuous functions of time.

Our final assumption concerns the initial state. We let x_0 be a gaussian random variable with

$$\mathcal{E}\{x_0\} = m_0, \qquad \mathcal{E}\{(x_0-m_0)(x_0-m_0)'\} = P_0. \tag{2.6}$$

Finally, we assume that x_0 is independent of the noise processes, i.e.

$$\mathcal{E}\{x_0 w'(t)\} = 0, \qquad \mathcal{E}\{x_0 n_2'(t)\} = 0; \quad \forall t \in [t_0,t_f].$$

The estimate of the terminal state $x(t_f)$ is based on both observations $z_1(t)$ and $z_2(t)$, and the class of estimators under consideration is given by

$$\hat{x}(t_f) = \int_{t_0}^{t_f} [M_1(\tau,t_f)z_1(\tau) + M_2(\tau,t_f)z_2(\tau)]\,d\tau \tag{2.7}$$

where $M_1(\tau,t_f)$ and $M_2(\tau,t_f)$ are linear time-varying impulse responses.

The filtering problem is formulated as a two-player differential Nash game. The first player has access to the observation $z_1(t)$ and tries to maximise the variance of the estimation error by choosing an $M_1(\cdot,t_f)$ which minimises the H_∞ filtering cost function [4]

$$J_1(M_1,M_2)=\gamma^2\|M_1\|_2^2-\mathcal{E}\left\{[x(t_f)-\hat{x}(t_f)]'[x(t_f)-\hat{x}(t_f)]\right\}. \qquad (2.8)$$

$\gamma^2\|M_1\|_2^2$ is a penalty term introduced to prevent player one from assigning an arbitrarily large value to M_1 thus driving J_1 to minus infinity. The second player has access to the observation $z_2(t)$ and attempts to minimise the error variance by selecting an M_2 which minimises the Kalman filtering pay-off function

$$J_2(M_1,M_2)=\mathcal{E}\left\{[x(t_f)-\hat{x}(t_f)]'[x(t_f)-\hat{x}(t_f)]\right\}. \qquad (2.9)$$

We therefore seek two linear estimators M_1^* and M_2^* which satisfy the Nash equilibria

$$\begin{aligned} J_1(M_1^*,M_2^*) &\le J_1(M_1,M_2^*) && (2.10)\\ J_2(M_1^*,M_2^*) &\le J_2(M_1^*,M_2). && (2.11)\end{aligned}$$

2.2 Problem Reformulation

In order to solve the stochastic filtering problem we transform it into a deterministic control problem. This approach has two advantages in that the reformulation is relatively simple, and we can then use an adaptation of the mixed H_2/H_∞ theory [2] to derive the necessary and sufficient conditions for the existence of the Nash equilibria M_1^* and M_2^*.

We begin by introducing the time-varying matrix differential equation [4]

$$\dot{Z}(t,t_f)=-A'(t)Z(t,t_f)+C_1'(t)M_1'(t,t_f)+C_2'(t)M_2'(t,t_f) \qquad (2.12)$$

with $Z(t_f,t_f)=I$. The aim is to rewrite the cost functions (2.8) and (2.9) as quadratic performance indices involving $Z(\cdot)$, $M_1(\cdot)$ and $M_2(\cdot)$. By direct calculation

$$\begin{aligned}\frac{d}{dt}(Z(t,t_f)'x(t)) &= \left[-A'Z+C_1'M_1'+C_2'M_2'\right]'(t)x(t)\\ &+ Z(t,t_f)'\left[A(t)x(t)+B(t)w(t)\right]\\ &= M_1(t,t_f)z_1(t)+M_2(t,t_f)\left[z_2(t)-n_2(t)\right]\\ &+ Z(t,t_f)'B(t)w(t).\end{aligned}$$

Integrating over $[t_0,t_f]$ gives

$$x(t_f)-Z'(t_0,t_f)x_0=\int_{t_0}^{t_f}\left\{M_1z_1+M_2[z_2-n_2]+Z'Bw\right\}(t)dt.$$

Rearranging and using (2.7) yields

$$x(t_f)-\hat{x}(t_f)=Z'(t_0,t_f)x_0-\int_{t_0}^{t_f}\left\{M_2n_2-Z'Bw\right\}(t)dt.$$

In the next step we square both sides and take expectations making use of the assumed statistical properties of the noise processes and the initial state. This gives

$$\begin{aligned}\mathcal{E}\{(x-\hat{x})(x-\hat{x})'(t_f)\} &= \int_{t_0}^{t_f}\{M_2R_2M_2'+Z'BQB'Z\}\,dt \\ &+ Z'(t_0,t_f)P_0Z(t_0,t_f). \qquad (2.13)\end{aligned}$$

We can now use (2.13) to rewrite J_1 and J_2 as

$$\begin{aligned}J_1 &= tr\left\{\int_{t_0}^{t_f}(\gamma^2M_1M_1'-M_2R_2M_2'-Z'BQB'Z)(t)dt\right\} \\ &- tr\{Z'(t_0,t_f)P_0Z(t_0,t_f)\} \qquad (2.14)\\ J_2 &= tr\left\{\int_{t_0}^{t_f}(M_2R_2M_2'+Z'BQB'Z)(t)dt\right\} \\ &+ tr\{Z'(t_0,t_f)P_0Z(t_0,t_f)\}. \qquad (2.15)\end{aligned}$$

Notice that both pay-off functions are now deterministic.

The filtering problem has thus been transformed into the mixed H_2/H_∞ control problem of finding equilibrium solutions $M_1{}^*$ and M_2^* which satisfy

$$J_1(M_1^*,M_2^*) \le J_1(M_1,M_2^*) \qquad (2.16)$$
$$J_2(M_1^*,M_2^*) \le J_2(M_1^*,M_2) \qquad (2.17)$$

where J_1 and J_2 are given by (2.14) and (2.15); the matrix $Z(t)$ is constrained by (2.12).

Remarks

(i) Notice that the "state" $Z(t)$ is matrix valued and the 'dynamics' have a terminal condition $Z(t_f)=I$.

(ii) The optimal linear estimators M_1^* and M_2^* are matrix valued too.

(iii) The initial condition data in the form of $Z'(t_0)P_0Z(t_0)$ appears in the cost functionals.

2.3 Finding an on-line estimate

The necessary and sufficient conditions for the existence of Nash equilibrium solutions to the mixed H_2/H_∞ control problem, and therefore for the existence of linear estimators M_1^* and M_2^* follow directly from [2], and are presented in the following theorem. Scaling arguments have allowed us to assume without loss of generality that $R_2=Q=I$.

Theorem 2.1 *Given the system and observations described by* (2.1) *to* (2.3), *there exist transformations M_1^* and M_2^* which satisfy the Nash equilibria* (2.16) *and* (2.17) *with J_1 and J_2 defined in* (2.14) *and* (2.15) *if and only if the coupled Riccati differential equations*

$$\begin{aligned}\dot{P}_1(t) &= AP_1(t)+P_1(t)A'-BB' \\ &- \begin{bmatrix} P_1(t) & P_2(t)\end{bmatrix}\begin{bmatrix}\gamma^{-2}C_1'C_1 & C_2'C_2 \\ C_2'C_2 & C_2'C_2\end{bmatrix}\begin{bmatrix}P_1(t)\\ P_2(t)\end{bmatrix} \qquad (2.18)\\ \dot{P}_2(t) &= AP_2(t)+P_2(t)A'+BB' \\ &- \begin{bmatrix} P_1(t) & P_2(t)\end{bmatrix}\begin{bmatrix}0 & \gamma^{-2}C_1'C_1 \\ \gamma^{-2}C_1'C_1 & C_2'C_2\end{bmatrix}\begin{bmatrix}P_1(t)\\ P_2(t)\end{bmatrix} \qquad (2.19)\end{aligned}$$

with $P_1(t_0) = -P_0$ *and* $P_2(t_0) = P_0$, *have solutions* $P_1(t) \leq 0, P_2(t) \geq 0$ *on* $[t_0, t_f]$. *If* (2.18) *and* (2.19) *have solutions,*

$$M_1^*(t,t_f) = \gamma^{-2} Z'(t,t_f) P_1(t) C_1' \quad (2.20)$$
$$M_2^*(t,t_f) = Z'(t,t_f) P_2(t) C_2'. \quad (2.21)$$

Moreover,

$$J_1(M_1^*, M_2^*) = P_1(t_f) \quad (2.22)$$
$$J_2(M_1^*, M_2^*) = P_2(t_f). \quad (2.23)$$

With a little manipulation we can now find an on-line estimate for $x(t_f)$ based on the observations $z_1(t_f)$ and $z_2(t_f)$, in the same sense that the Kalman filter provides an on-line state estimate using current observations. Implementing the equilibrium strategies in (2.12) gives

$$\dot{Z}(t,t_f) = (-A' + \gamma^{-2} P_1 C_1' C_1 + P_2 C_2' C_2) Z(t,t_f) \qquad Z(t_f,t_f) = I. \quad (2.24)$$

Since $Z(t,t_f) = Z^{-1}(t_f,t)$, we have

$$\frac{d}{dt_f} Z(t,t_f) = \frac{d}{dt_f} Z^{-1}(t_f,t). \quad (2.25)$$

Direct calculation yields

$$\begin{aligned} \frac{d}{dt_f} Z^{-1}(t_f,t) &= -Z^{-1}(t_f,t) \frac{d}{dt_f} Z(t_f,t) Z^{-1}(t_f,t) \\ &= Z^{-1}(t_f,t)\,(A' - \gamma^{-2} C_1' C_1 P_1 - C_2' C_2 P_2)(t_f), \end{aligned}$$

and then (2.25) $\Rightarrow$

$$\frac{d}{dt_f} Z'(t,t_f) = (A - \gamma^{-2} P_1 C_1' C_1 - P_2 C_2' C_2)(t_f) Z'(t,t_f). \quad (2.26)$$

Now, from (2.7)

$$\hat{x}(t_f) = \int_{t_0}^{t_f} Z'(t,t_f) \{\gamma^{-2} P_1 C_1' z_1 + P_2 C_2' z_2\}(t) dt. \quad (2.27)$$

Differentiating with respect to t_f and using (2.26) we obtain

$$\begin{aligned} \frac{d}{dt_f} \hat{x}(t_f) &= (A - \gamma^{-2} P_1 C_1' C_1 - P_2 C_2' C_2)(t_f) \hat{x}(t_f) \\ &+ (\gamma^{-2} P_1 C_1' z_1 + P_2 C_2' z_2)(t_f) \qquad \hat{x}(t_0) = m_0 \end{aligned}$$

or equivalently

$$\begin{aligned} \frac{d}{dt_f} \hat{x}(t_f) = A(t_f)\hat{x}(t_f) &- K_1(t_f)[C_1(t_f)\hat{x}(t_f) - z_1(t_f)] \\ &- K_2(t_f)[C_2(t_f)\hat{x}(t_f) - z_2(t_f)] \end{aligned} \quad (2.28)$$

with

$$K_1(t_f) = \gamma^{-2} P_1(t_f) C_1'(t_f) \quad (2.29)$$
$$K_2(t_f) = P_2(t_f) C_2'(t_f). \quad (2.30)$$

The matrices $P_1(t_f)$ and $P_2(t_f)$ satisfy the coupled Riccati equations (2.18) and (2.19).

2.4 The mixed H_2/H_∞ filter

In this section we show how the H_2/H_∞ filter is defined in terms of the above analysis, and we can present some of its properties. Suppose M_1^* is replaced by a dormant player as follows : Calculate M_1^* and M_2^* as above, and then in (2.27) replace the term multiplying z_1 by zero, i.e. we take $M_1 = 0 \neq M_1^*$. This means that in (2.28) we replace $K_1(t_f)$ by zero to obtain

$$\frac{d}{dt_f}\hat{x}(t_f) = [A(t_f) - K_2(t_f)C_2(t_f)]\hat{x}(t_f) + K_2(t_f)C_2(t_f)z_2(t_f); \qquad \hat{x}(t_0) = m_0,$$

which is the defining equation of the H_2/H_∞ filter. It is easy to see from this equation that the H_2/H_∞ filter has the usual observer structure associated with Kalman and H_∞ filters. The following result provides an upper bound on the variance of the terminal state estimation error, and it shows that if the white noise assumption on the disturbance inputs w and n_2 is dropped, the energy gain from the disturbances to the generalised estimation error $C_1(\hat{x}(t_f) - x(t_f))$ is bounded above by γ for all $w, n_2 \in \mathcal{L}_2[t_0, t_f]$.

Theorem 2.2 *If* $M_1 \equiv 0$,

(i) the on-line state estimate is generated by

$$\frac{d}{dt_f}\hat{x}(t_f) = (A - P_2C_2'C_2)(t_f)\hat{x}(t_f) + P_2(t_f)C_2'(t_f)z_2(t_f) \qquad \hat{x}(t_0) = m_0, \quad (2.31)$$

(ii) the covariance of the estimation error $e(t_f) = \hat{x}(t_f) - x(t_f)$ *satisfies*

$$\mathcal{E}\{e(t_f)e'(t_f)\} \leq -P_1(t_f) \quad (2.32)$$

and

(iii) when $m_0 = 0$, $\|\mathcal{R}\|_{2i} \leq \gamma$ *for all* $w, n_2 \in \mathcal{L}_2[t_0, t_f]$*, where the operator* $\mathcal{R}$ *is described by*

$$\begin{aligned} \frac{d}{dt_f}e(t_f) &= (A - P_2C_2'C_2)(t_f)e(t_f) + \begin{bmatrix} B & -P_2C_2' \end{bmatrix}\begin{bmatrix} w \\ n_2 \end{bmatrix} && (2.33) \\ z &= C_1e(t_f) && (2.34) \end{aligned}$$

2.5 Reconciliation of Kalman, H_∞ and H_2/H_∞ filtering

Here we establish a link between Kalman, H_∞ and mixed H_2/H_∞ filtering. Each of the three theories may be generated as special cases of the following non-zero sum, two-player Nash differential game:

Given the system described by (2.1) to (2.3) and a state estimate of the form given in (2.7), we seek linear estimators M_1^* and M_2^* which satisfy the Nash equilibria $J_1(M_1^*, M_2^*) \leq J_1(M_1, M_2^*)$ and $J_2(M_1^*, M_2^*) \leq J_2(M_1^*, M_2)$, where

$$\begin{aligned} J_1 &= \gamma^2\|M_1\|_2^2 - \mathcal{E}\left\{[x(t_f) - \hat{x}(t_f)]'[x(t_f) - \hat{x}(t_f)]\right\}, \\ J_2 &= \mathcal{E}\left\{[x(t_f) - \hat{x}(t_f)]'[x(t_f) - \hat{x}(t_f)]\right\} - \rho^2\|M_1\|_2^2. \end{aligned}$$

This game can be reformulated as a deterministic control problem and is solved by $M_1^*(t,t_f) = \gamma^{-2}Z'(t,t_f)S_1(t)C_1'$ and $M_2^*(t,t_f) = Z'(t,t_f)S_2(t)C_2'$, where $Z(t,t_f)$ is generated by (2.12) and $S_1(t)$ and $S_2(t)$ satisfy the coupled Riccati differential equations

$$\begin{aligned}\dot{S}_1(t) &= AS_1(t) + S_1(t)A' - BB' \\ &\quad - \begin{bmatrix} S_1(t) & S_2(t) \end{bmatrix} \begin{bmatrix} \gamma^{-2}C_1'C_1 & C_2'C_2 \\ C_2'C_2 & C_2'C_2 \end{bmatrix} \begin{bmatrix} S_1(t) \\ S_2(t) \end{bmatrix} \\ \dot{S}_2(t) &= AS_2(t) + S_2(t)A' + BB' \\ &\quad - \begin{bmatrix} S_1(t) & S_2(t) \end{bmatrix} \begin{bmatrix} \rho^2\gamma^{-4}C_1'C_1 & \gamma^{-2}C_1'C_1 \\ \gamma^{-2}C_1'C_1 & C_2'C_2 \end{bmatrix} \begin{bmatrix} S_1(t) \\ S_2(t) \end{bmatrix}\end{aligned}$$

with $S_1(t_f) = S_2(t_f) = 0$.

(i) The Kalman filtering problem is recovered by setting $\rho = 0$ and $\gamma = \infty$, in which case $-S_1(t) = S_2(t) = P_e(t)$. The matrix $P_e(t)$ is the solution to the usual Kalman filter Riccati equation.
(ii) The pure H_∞ filtering problem is recovered by setting $\rho = \gamma$. Again, $-S_1(t) = S_2(t) = P_\infty(t)$. The matrix $P_\infty(t)$ is the solution to the H_∞ filtering Riccati equation [4].
(iii) The mixed H_2/H_∞ filtering problem comes from $\rho = 0$.

3 Conclusions

We have shown how a mixed H_2/H_∞ filtering problem may be formulated as a deterministic mixed H_2/H_∞ control problem. The control problem can then solved using the Nash game techniques explained in [2]. The necessary and sufficient conditions for the existence of a solution to the filtering problem are given in terms of the existence of solutions to a pair of cross-coupled Riccati differential equations. We have derived an upper bound on the covariance of the state estimation error in terms of one of the solutions to the coupled equations and have shown that, under certain conditions, the energy gain from the disturbance inputs to the generalised estimation error $C_1(x(t_f) - \hat{x}(t_f))$ is bounded above by γ.

References

[1] T. Basar and G. J. Olsder, "Dynamic noncooperative game theory," Academic Press, New York, 1982

[2] D. J. N. Limebeer, B. D. O. Anderson and B. Hendel, "A Nash game approach to mixed H_2/H_∞ control," submitted for publication

[3] D. J. N. Limebeer, B. D. O. Anderson and B. Hendel, "Nash games and mixed H_2/H_∞ control," preprint

[4] D. J. N. Limebeer and U. Shaked, "Minimax terminal state estimation and H_∞ filtering," submitted to IEEE Trans. Auto. Control

[5] A. W. Starr and Y. C. Ho, "Nonzero-sum differential games," *J. Optimization Theory and Applications*, Vol.3, No. 3, pp 184-206, 1967

The Principle of the Argument and its Application to the Stability and Robust Stability Problems

Mohamed Mansour
Automatic Control Laboratory
Swiss Federal Institute of Technology Zurich, Switzerland

Abstract

It is shown that the principle of the argument is the basis for the different stability criteria for linear continuous and discrete systems. From the principle of argument stability criteria in the frequency domain are derived which lead to Hermite-Bieler theorems for continuous and discrete systems. Routh-Hurwitz criterion and its equivalent for discrete systems, Schur-Cohn criterion and its equivalent for continuous systems are directly obtained. Moreover, the monotony of the argument for different functions can be proved using Hermite-Bieler theorem. Using the principle of the argument and the continuity of the roots of a polynomial w.r.t. its coefficients, all known robust stability results can be obtained in a straightforward and simple way.

1 Principle of the Argument

Consider a function $f(z)$ which is regular inside a closed contour C and is not zero at any point on the contour and let $\Delta_C\{\arg f(z)\}$ denotes the variation of arg $f(z)$ round the contour C, then

Theorem 1 [1]

$$\Delta_C\{\arg f(z)\} = 2\pi \mathrm{N} \tag{1}$$

where N *is the number of zeros of* $f(z)$ *inside* C.

2 Stability Criterion using Principle of the Argument

2.1 Continuous Systems

Consider the characteristic polynomial of a continuous system

$$f(s) = a_0 s^n + a_1 s^{n-1} + \ldots + a_{n-1} s + a_n \tag{2}$$

Theorem 2 *$f(s)$ has all its roots in the left half of the s-plane if and only if $f(j\omega)$ has a change of argument of $n\frac{\pi}{2}$ when ω changes from* 0 *to* ∞.

The proof of this criterion which is known in the literature as Cremer-Leonhard-Michailov Criterion is based on the fact that a negative real root contributes an angle of $\frac{\pi}{2}$ to the change of argument, and a pair of complex roots in the left half plane contributes an angle of π to the change of argument as ω changes from 0 to ∞. It is easy to show geometrically that the argument is monotonically increasing.

$$f(s) \text{ can be written as } f(s) = h(s^2) + sg(s^2) \tag{3}$$

where $h(s^2)$ and $sg(s^2)$ are the even and odd parts of $f(s)$ respectively.

$$f(j\omega) = h(-\omega^2) + j\omega g(-\omega^2) \tag{4}$$

2.2 Discrete Systems

Consider the characteristic polynomial of a discrete system

$$f(z) = a_0 z^n + a_1 z^{n-1} + \dots + a_{n-1} z + a_n \tag{5}$$

Theorem 3 *$f(z)$ has all its roots inside the unit circle of the z-plane if and only if $f(e^{j\theta})$ has a change of argument of $n\pi$ when θ changes from 0 to π.*

The proof of this criterion is based on the fact that a real root inside the unit circle contributes an angle of π to the change of argument, and a pair of complex roots inside the unit circle contributes an angle of 2π to the change of argument as θ changes from 0 to π. It is easy to show geometrically that the argument is monotonically increasing

$$f(z) \text{ can be written as } f(z) = h(z) + g(z) \tag{6}$$

where

$$h(z) = \frac{1}{2}\left[f(z) + z^n f\left(\frac{1}{z}\right)\right] = \alpha_0 z^n + \alpha_1 z^{n-1} + \dots + \alpha_1 z + \alpha_0 \tag{7}$$

and

$$g(z) = \frac{1}{2}\left[f(z) + z^n f\left(\frac{1}{z}\right)\right] = \beta_0 z^n + \beta_1 z^{n-1} + \dots - \beta_1 z - \beta_0 \tag{8}$$

are the symmetric and antisymmetric parts of $f(z)$ respectively. Also

$$f(e^{j\theta}) = 2e^{jn\theta/2}[h^* + jg^*] \tag{9}$$

where for n even, $n = 2\nu$

$$\begin{aligned} h^*(\theta) &= \alpha_0 \cos\nu\theta + \alpha_1 \cos(\nu-1)\theta + \dots + \alpha_{\nu-1}\cos\theta + \frac{\alpha\nu}{2} \\ g^*(\theta) &= \beta_0 \sin\nu\theta + \beta_1 \sin(\nu-1)\theta + \dots + \beta_{\nu-1}\sin\theta \end{aligned} \tag{10}$$

From theorem 3 and equation (9) we get

Theorem 4 *$f(z)$ has all its roots inside the unit circle of the z-plane if and only if $f^* = h^* + jg^*$ has a change of argument of $n\frac{\pi}{2}$ when θ changes from 0 to π.*

f^* is a complex function w.r.t. a rotating axis. It can be shown geometrically that the argument of f^* increases monotonically. Fig. 1 shows the change of coordinates.

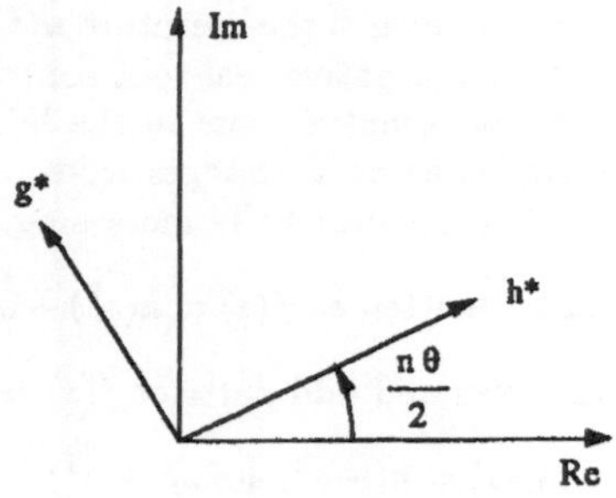

Figure 1: Change of coordinates

3 Hermite-Bieler Theorem

3.1 Continuous Systems

Consider the polynomial in (3).

Theorem 5 *(Hermite-Bieler Theorem) $f(s)$ has all its roots in the left half of the s-plane if and only if $h(\lambda)$ and $g(\lambda)$ have simple real negative alternating roots and* $\frac{a_0}{a_1} > 0$. *The first root next to zero is of $h(\lambda)$.*

The proof of this theorem follows directly from theorem 1; [2]. If theorem 5 is satisfied then $h(\lambda)$ and $g(\lambda)$ are called a positive pair.

3.2 Discrete Systems

Consider polynomial (5).

Theorem 6 *(Discrete Hermite-Bieler Theorem) $f(z)$ has all its roots inside the unit circle of the z-plane if and only if $h(z)$ and $g(z)$ have simple alternating roots on the unit circle and* $\left|\frac{a_n}{a_0}\right| < 1$; [3], [4].

The proof can be sketched as follows:
Due to theorem 4, h^* and g^* and hence h and g will go through 0 alternatively n times as θ changes from 0 to π. The necessary condition $\left|\frac{a_n}{a_0}\right| < 1$ distinguishes between $f(z)$ and its inverse which has all roots outside the unit circle.

4 Monotony of the argument

In [5] it is shown that the argument of a Hurwitz polynomial $f(s) = h(s^2) + sg(s^2)$ in the frequency domain $f(j\omega) = h(-\omega^2) + j\omega g(-\omega^2)$ as well as the argument of the modified functions $f_1^*(\omega) = h(-\omega^2) + jg(-\omega^2)$ and $f_2^*(\omega) = h(-\omega^2) + j\omega^2 g(-\omega^2)$ are monotonically increasing functions of ω, whereby ω varies between 0 and ∞. Similarly it is shown that given a Schur polynomial $f(z) = h(z) + g(z)$ where $h(z)$ and $g(z)$ are the symmetric and the antisymmetric parts of $f(z)$ respectively, the same monotony property can be obtained for an auxiliary function $f^*(\theta)$ defined as $f(e^{j\theta}) = 2e^{jn\theta/2}f^*(\theta)$. The argument of $f^*(\theta) = h^*(\theta) + jg^*(\theta)$ as well as the argument of the modified functions

$$f_3^*(\theta) = \frac{h^*(\theta)}{\cos\frac{\theta}{2}} + j\frac{g^*(\theta)}{\sin\frac{\theta}{2}}$$

and

$$f_4^*(\theta) = \frac{h^*(\theta)}{\sin\frac{\theta}{2}} + j\frac{g^*(\theta)}{\cos\frac{\theta}{2}}$$

are monotonically increasing functions of θ whereby θ varies between 0 and π. These results were proved in [5] using mathematical induction. However, a simpler proof can be obtained using network-theoretical results which is based on Hermite-Bieler theorems, [6]. The results obtained can be directly applied to the robust stability problem [5].

5 Hermite-Bieler Theorem and Criteria for Hurwitz Stability

In this section it is shown how different criteria for Hurwitz stability like Routh-Hurwitz criterion and equivalent of Schur-Cohn criterion for continuous systems are derived from Hermite-Bieler theorem. Also Kharitonov theorem [7], [8] can be similarly derived, [9].

5.1 Routh-Hurwitz Criterion

Without loss of generality consider n even. The first and second rows of the Routh table are given by the coefficients of h and g respectively. The third row is given by the coefficients of $h - \frac{a_0}{a_1}s^2g$. The following theorem shows that Routh table reduces the stability of a polynomial of degree n to another one of degree $n-1$. Repeating this operation we arrive at a polynomial of degree one whose stability is easily determined.

Theorem 7 [2], [10] *$f(s)$ is Hurwitz stable if and only if $h - \frac{a_0}{a_1}s^2g + sg$ is Hurwitz stable.*

Proof: Let $f(s) = h + sg$ be Hurwitz stable i.e. $a_0, a_1, ...a_n > 0$ and according to theorem 5, $h(\lambda)$ and $g(\lambda)$ have simple real negative alternating roots. $h - \frac{a_0}{a_1}s^2g$ and h have the same sign at the roots of g and the number of roots of $h - \frac{a_0}{a_1}s^2g + sg$ is $n-1$. Moreover $a_1a_2 - a_0a_3 > 0$. Therefore, according to theorem 5, $h - \frac{a_0}{a_1}s^2g + sg$ is Hurwitz stable. The proof of the if part is similar. □

5.2 Equivalent of Schur-Cohn-Jury Criterion for Continuous Systems

Theorem 8 [11] *$f(s)$ is Hurwitz stable if and only if $f(s) - \frac{f(-1)}{f(1)}f(-s)$ is Hurwitz stable and $\left|\frac{f(-1)}{f(1)}\right| < 1$*

-1 is a root of $f(s) - \frac{f(-1)}{f(1)} f(-s)$ which can be eliminated, so that the stability of a polynomial of degree n is again reduced to the stability of a polynomial of degree $n-1$. This theorem can be proved using Rouché theorem. Here we give a proof based on Hermite-Bieler theorem and on the following Lemma.

Lemma 1 *$h + sg$ is Hurwitz stable if and only if $\gamma h + \delta sg$ is Hurwitz stable for γ, δ positive numbers.*

Proof: The interlacing property does not change by multiplying h and g by positive numbers. Also $\frac{\delta a_1}{\gamma a_0} > 0$. □

Proof: Proof of theorem 8. Let $f(s)$ be Hurwitz stable.
If $f(s) = h + sg$ then $f(-1) = h - sg$

Therefore

$$f(s) - \frac{f(-1)}{f(1)} f(-s) = \left(1 - \frac{f(-1)}{f(1)}\right) h + \left(1 + \frac{f(-1)}{f(1)}\right) sg.$$

Using Lemma 1 we get $f(s) - \frac{f(-1)}{f(1)} f(-s)$ is Hurwitz stable. The proof of the if part is similar. □

6 Hermite-Bieler Theorem and Criteria for Schur Stability

In this section it is shown how different criteria for Schur stability like the equivalent of Routh-Hurwitz criterion for discrete systems and Schur-Cohn-Jury criterion are derived from Hermite-Bieler theorem. Also analog result of Kharitonov theorem can be obtained; [9], [12].
Consider $f(z)$ given by equation (5). Without loss of generality we assume $a_0 > 0$ and n even. For n odd a similar treatment can be made. A necessary condition for Schur stability of (5) is $\left|\frac{a_n}{a_0}\right| < 1$ which results in $\alpha_0 > 0$ and $\beta_0 > 0$.

6.1 Equivalent of Routh-Hurwitz Criterion for Discrete Systems

If $h(z)$ and $g(z)$ given by (7) and (8) fulfill the discrete Hermite-Bieler theorem then their projections on the real axis $h^*(\chi)$ and $g^*(\chi)$ have distinct interlacing real roots between -1 and $+1$. In this case $h^*(\chi)$ and $g^*(\chi)$ form a discrete positive pair.
From [9] $h^*(\chi)$ and $g^*(\chi)$ are given by

$$\begin{aligned} h^*(\chi) &= \sum_{i=0}^{\nu-1} \alpha_i \ T_{\nu-i} + \frac{\alpha_\nu}{2} \\ g^*(\chi) &= \sum_{i=0}^{\nu-1} \beta_i \ U_{\nu-i-1} \end{aligned} \tag{11}$$

where $n = 2\nu$ and T_k, U_k are Tshebyshev polynomials of the first and second kind respectively.

Let

$$\begin{aligned} h^*(\chi) &= \gamma_0\chi^\nu + \gamma_1\chi^{\nu-1} + \ldots \\ g^*(\chi) &= \delta_0\chi^{\nu-1} + \delta_1\chi^{\nu-2} + \ldots \end{aligned}$$

Theorem 9 [11] *$f(z)$ is Schur stable if and only if $g^*(\chi)$ and $h^*(\chi) - \frac{\gamma_0}{\delta_0}(\chi - 1)g^*(\chi)$ form a discrete positive pair and $h^*(-1) < 0$ for ν odd and $h^*(-1) > 0$ for ν even.*

Proof: Assume $f(z)$ be Schur stable. $h^*(\chi) - \frac{\gamma_0}{\delta_0}(\chi - 1)g^*(\chi)$ has the same value at the roots of $g^*(\chi)$ as $h^*(\chi)$. Therefore, the interlacing property is preserved. The condition $h^*(-1) < 0$ for ν odd and $h^*(-1) > 0$ for ν even guarantees that the reduction of the order is not due to a root of $h^*(\chi)$ between -1 and $-\infty$. Therefore $g^*(\chi)$ and $h^*(\chi) - \frac{\gamma_0}{\delta_0}(\chi - 1)g^*(\chi)$ form a discrete positive pair.
The if part can be proved similarly. □

This theorem can be implemented in table form.

6.2 Schur-Cohn-Jury Criterion

Theorem 10 [4], [11] *The polynomial* (5) *with $\left|\frac{a_n}{a_0}\right| < 1$ is Schur stable if and only if the polynomial $\frac{1}{z}\left[f(z) - \frac{a_n}{a_0}z^n f\left(\frac{1}{z}\right)\right]$ is Schur stable.*

This was proved in [4] using the following Lemma:

Lemma 2 *$h + g$ is Schur stable if and only if $\gamma h + \delta g$ is Schur stable for γ, δ positive numbers.*

Proof: Proof of Lemma 2: The interlacing property does not change by multiplying h and g by positive numbers. Also $\left|\frac{\gamma\alpha_0 - \delta\beta_0}{\gamma\alpha_0 + \delta\beta_0}\right| < 1$ if $\left|\frac{a_n}{a_0}\right| < 1$. The reverse is true. □

Proof: Proof of theorem 10: $f(z) - \frac{a_n}{a_0}z^n f\left(\frac{1}{z}\right) = h(z) + g(z) - \frac{a_n}{a_0}[h(z) - g(z)] = \left(1 - \frac{a_n}{a_0}\right)h(z) + \left(1 + \frac{a_n}{a_0}\right)g(z)$. The only if part of theorem 10 follows directly from $\left|\frac{a_n}{a_0}\right| < 1$ and Lemma 2.
The if part can be proved similarly. □

7 Robust Stability in the Frequency Domain

Let $f(s, \underline{a})$ be given by (2) and $\underline{a} \in A$. A is a closed set in the coefficient space. Let D be an open subset of the complex plane. Robust D-stability means that all roots of the polynomial family lie in D. Let $\Gamma(s^*)$ be the value set of the polynomial family at $s = s^*$ and $\partial\Gamma(s^*)$ its boundary.
Using the principle of the argument and the continuity of the roots w.r.t. the coefficients of a polynomial, we get the following result for robust D-stability.

Theorem 11 [13], [14]. *$f(s, \underline{a})$ with $\underline{a} \in A$ is robust D-stable if and only if the following conditions are satisfied:*

i) $f(s, \underline{a})$ *is D-stable for some* $\underline{a} \in A$

ii) $\Gamma(s^*) \neq 0$ *for some* $s^* \in \partial D$

iii) $\partial\Gamma(s) \neq 0$ *for all* $s \in \partial D$

Using theorem 11 we can solve the following problems in a simple way

i) an edge theorem for the robust stability of polytopes in the coefficient space, [13]

ii) stability of edges

iii) vertex properties of edges, [5], [15]

iv) Kharitonov theorem and its analog and counterpart for discrete systems, [16], [17], [12], [18]

v) robust stability of controlled interval plants continuous and discrete, [14], [19], [20]

A survey of all the above problems is given in [21].

8 Conclusions

In the above it was shown that the stability criteria for linear and discrete systems as well as the solution of the robust stability problem all can be derived from the principle of the argument.

References

[1] E.G. Phillips, Functions of a complex variable with applications, *Oliver and Boyd*, 1963.

[2] H. Chapellat, M. Mansour, and S. P. Bhattacharyya, Elementary proofs of some classical stability criteria, *IEEE Transactions on Education*, **33**(3):232–239, 1990.

[3] H.W. Schüssler, A stability theorem for discrete systems, *IEEE Transactions on Acoustics, Speech and Signal Processing*, **ASSP-24**:87–89, 1976.

[4] M. Mansour, Simple proof of Schur-Cohn-Jury criterium using principle of argument, *Internal Report 90-08, Automatic Control Laboratory, Swiss Federal Institute of Technology, Zurich*, 1990.

[5] M. Mansour and F.J. Kraus, Argument conditions for Hurwitz and Schur stable polynomials and the robust stability problem, *Internal Report 90-05, Automatic Control Laboratory, Swiss Federal Institute of Technology, Zurich*, 1990.

[6] M. Mansour, Monotony of the argument of different functions for Routh and Hurwitz stability, *Internal Report 91-11, Automatic Control Laboratory, Swiss Federal Institute of Technology, Zurich*, 1991.

[7] V. L. Kharitonov, Asymptotic stability of an equilibrium position of a family of systems of linear differential equations, *Differentsial'nye Uravneniya*, **14**:2086–2088, 1978.

[8] B. D. O. Anderson, E. I. Jury, and M. Mansour, On robust Hurwitz polynomials, *IEEE Transactions on Automatic Control*, **32**:909–913, 1987.

[9] M. Mansour and F.J. Kraus, On robust stability of Schur polynomials, *Internal Report 87-05, Automatic Control Laboratory, Swiss Federal Institute of Technology, Zurich*, 1987.

[10] M. Mansour, A simple proof of the Routh-Hurwitz criterion, *Internal Report 88-04, Automatic Control Laboratory, Swiss Federal Institute of Technology, Zurich*, 1988.

[11] M. Mansour, Six stability criteria & Hermite-Bieler theorem, *Internal Report 90-09, Automatic Control Laboratory, Swiss Federal Institute of Technology, Zurich*, 1990.

[12] M. Mansour, F. J. Kraus, and B.D.O. Anderson, Strong Kharitonov theorem for discrete systems, *in Robustness in Identification and Control, Ed. M. Milanese, R. Tempo, A. Vicino, Plenum Publishing Corporation, 1989. Also in Recent Advances in Robust Control, Ed. P. Dorato, R.K. Yedavalli, IEEE Press, 1990.*

[13] S. Dasgupta, P. J. Parker, B. D. O. Anderson, F. J. Kraus, and M. Mansour, Frequency domain conditions for the robust stability of linear and nonlinear systems, *IEEE Transactions on Circuits and Systems*, **38**(4):389–397, 1991.

[14] F. J. Kraus and W. Truöl, Robust stability of control systems with polytopical uncertainty: a Nyquist approach, *International Journal of Control*, **53**(4):967–983, 1991.

[15] F. J. Kraus, M. Mansour, W. Truöl, and B.D.O. Anderson, Robust stability of control systems: extreme point results for the stability of edges, *Internal Report 91-06, Automatic Control Laboratory, Swiss Federal Institute of Technology, Zurich*, 1991.

[16] R. J. Minnichelli, J. J. Anagnost, and C. A. Desoer, An elementary proof of Kharitonov's stability theorem with extensions, *IEEE Transactions on Automatic Control*, **34**(9):995–998, 1989.

[17] F. J. Kraus and M. Mansour, Robuste Stabilität im Frequenzgang (german), *Internal Report 87-06, Automatic Control Laboratory, Swiss Federal Institute of Technology, Zurich*, 1987.

[18] F. J. Kraus and M. Mansour, On robust stability of discrete systems, *Proceedings of the 29th IEEE Conference on Decision and Control*, **2**:421–422, 1990.

[19] F. J. Kraus and M. Mansour, Robust Schur stable control systems, *Proceedings of the American Control Conference, Boston*, 1991.

[20] F. J. Kraus and M. Mansour, Robust discrete control, *Proceedings of the European Control Conference, Grenoble*, 1991.

[21] M. Mansour, Robust stability of systems described by rational functions, *to appear in Advances in Robust Systems Techniques and Applications, Leondes Ed., Academic Press*, 1991.

Robust Control of Interval Systems*

L.H. Keel
Center of Excellence in Information Systems
Tennessee State University
Nashville, TN. U.S.A

J. Shaw and S.P. Bhattacharyya
Department of Electrical Engineering
Texas A&M University
College Station, TX. U.S.A.

ABSTRACT

In this paper we consider a feedback interconnection of two linear time invariant systems, one of which is fixed and the other is uncertain with the uncertainty being a box in the space of coefficients of the numerator and denominator polynomials. This kind of system arises in robust control analysis and design problems. We give several results for the Nyquist plots and stability margins of such families of systems based on the extremal segments of rational functions introduced by Chapellat and Bhattacharyya [1]. These CB segments play a fundamental characterizing role in many control design problems.

1. INTRODUCTION

The problem of designing control systems to account for parameter uncertainty has become very important over the last few years. The main reason for this is the lack of effective methods for dealing with systems containing uncertain or adjustable parameters. Significant progress on this problem occured with the introduction of Kharitonov's Theorem [2] for interval polynomials which provided a new way of dealing with parametric uncertainties. However, despite its mathematical elegance, its applicability to control systems was very limited because the theorem assumes that

*Supported in part by NASA grant NAG-1-863 and NSF Grant ECS-8914357

all coefficients of the given polynomial family perturb independently. Under this assumption the theorem can only provide conservative results for the control problem where the uncertain parameters enter the characteristic polynomoial in a correlated fashion

A breakthrough was reported in the theorem of Chapellat and Bhattacharyya [1] which dealt with the case in which the characteristic polynomial is a linear combination of interval polynomials with polynomial coefficients. The CB Theorem provided Kharitonov-like results for such control problems by constructing certain line segments, the number of which is independent of the polynomial degrees, that completely and nonconservatively characterizethe stability of interval control systems.

Using this novel theorem, it has been discovered that the worst case H^∞ stability margin of an interval control system occurs on one of these CB segments. The details and constructive algorithm to compute H^∞ stability margin are given in [3]. Similarly the worst case parametric stability margin occurs on the CB segments [4].

The main result of this paper is that the Nyquist plot boundaries of all the transfer functions of interest in an interval control system occurs on the CB segments. This key result has farreaching implications in the development of design methods for interval control systems. Throughout this paper, proofs are omitted and may be found in [5]. A similar result has recently been independently discovered by Tesi and Vicino [6].

2. NOTATION AND PRELIMINARY RESULTS

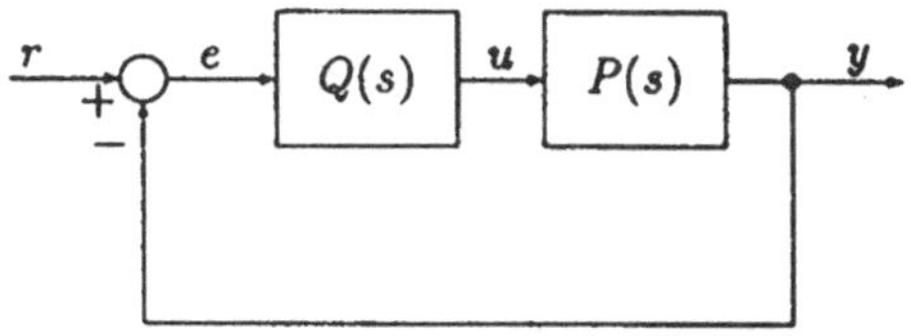

Figure 1. Feedback System

We consider the feedback interconnection (see Figure 1) of the two systems with respective transfer functions

$$Q(s) := \frac{Q_1(s)}{Q_2(s)} \qquad P(s) := \frac{N(s)}{D(s)} \tag{1}$$

where $N(s)$, $D(s)$ and $Q_i(s)$, $i = 1,2$ are polynomials in the complex variable s. In general $Q(s)$ is fixed and $P(s)$ will be assumed to lie in an uncertainty set described as follows.

Write

$$\begin{aligned} N(s) &:= n_p s^p + n_{p-1} s^{p-1} + \cdots + n_1 s + n_0 \\ D(s) &:= d_q s^q + d_{q-1} s^{q-1} + \cdots + d_1 s + d_0 \end{aligned} \tag{2}$$

and let the coefficients lie in prescribed intervals

$$n_i \in [n_i^-, n_i^+], \quad i \in \{0, 1, \cdots, p\} := \underline{p}$$
$$d_i \in [d_i^-, d_i^+], \quad i \in \{0, 1, \cdots, q\} := \underline{q}.$$

Introduce the interval polynomial sets

$$\begin{aligned} \mathcal{N}(s) &:= \{N(s) = n_p s^p + n_{p-1}s^{p-1} + \cdots n_1 s + n_0 \ : \ n_i \in [n_i^-, n_i^+], \ i \in \underline{p}\} \\ \mathcal{D}(s) &:= \{D(s) = d_q s^q + d_{q-1}s^{q-1} + \cdots d_1 s + d_0 \ : \ d_i \in [d_i^-, d_i^+], \ i \in \underline{q}\} \end{aligned} \tag{3}$$

and the corresponding set of interval transfer functions (or interval systems)

$$\mathbf{P}(s) = \{\frac{N(s)}{D(s)} \ : \ (N(s), D(s)) \in \mathcal{N}(s) \mathrm{x} \mathcal{D}(s)\}. \tag{4}$$

The four Kharitonov vertex polynomials associated with $\mathcal{N}(s)$ are

$$\begin{aligned} K_n^1(s) &:= n_0^- + n_1^- s + n_2^+ s^2 + n_3^+ s^3 + n_4^- s^4 + n_5^- s^5 + \cdots \\ K_n^2(s) &:= n_0^- + n_1^+ s + n_2^+ s^2 + n_3^- s^3 + n_4^- s^4 + n_5^+ s^5 + \cdots \\ K_n^3(s) &:= n_0^+ + n_1^- s + n_2^- s^2 + n_3^+ s^3 + n_4^+ s^4 + n_5^- s^5 + \cdots \\ K_n^4(s) &:= n_0^+ + n_1^+ s + n_2^- s^2 + n_3^- s^3 + n_4^+ s^4 + n_5^+ s^5 + \cdots . \end{aligned} \tag{5}$$

and we write

$$\mathcal{K}_{\mathcal{N}}(s) := \{K_n^1(s), K_n^2(s), K_n^3(s), K_n^4(s)\} \tag{6}$$

Similarly the four Kharitonov polynomials associated with $\mathcal{D}(s)$ are denoted $K_d^i(s)$, $i = 1, 2, 3, 4$ and

$$\mathcal{K}_{\mathcal{D}}(s) := \{K_d^1(s), K_d^2(s), K_d^3(s), K_d^4(s)\} \tag{7}$$

The four Kharitonov segment polynomials associated with $\mathcal{N}(s)$ are defined as follows:

$$\mathcal{S}_{\mathcal{N}}(s) :=$$

$$[\lambda K_n^i(s) + (1-\lambda) K_n^j(s) \ : \ \lambda \in [0,1], \ (i,j) \in \{(1,2), (1,3), (2,4), (3,4)\}] \tag{8}$$

and the four Kharitonov segment polynomials associated with $\mathcal{D}(s)$ are

$$\mathcal{S}_{\mathcal{D}}(s) :=$$

$$[\mu K_d^i(s) + (1-\mu) K_d^j(s) \ : \ \mu \in [0,1], \ (i,j) \in \{(1,2), (1,3), (2,4), (3,4)\}] \tag{9}$$

Following Chapellat and Bhattacharyya [1] we introduce the following subset of $\mathcal{N}(s) \mathrm{x} \mathcal{D}(s)$:

$$\begin{aligned} (\mathcal{N}(s) \mathrm{x} \mathcal{D}(s))_{\mathrm{CB}} = \ &\{(N(s), D(s)) \ : \ N(s) \in \mathcal{K}_{\mathcal{N}}(s), \\ &D(s) \in \mathcal{S}_{\mathcal{D}}(s) \text{ or } N(s) \in \mathcal{S}_{\mathcal{N}}(s), \ D(s) \in \mathcal{K}_{\mathcal{D}}(s)\}. \end{aligned} \tag{10}$$

Referring to the control system in Figure 1, we calculate the following transfer functions of interest in analysis and design problems:

$$\frac{y(s)}{u(s)} = P(s) \qquad \frac{u(s)}{e(s)} = Q(s) \tag{11}$$

$$T^o(s) := \frac{y(s)}{e(s)} = P(s)Q(s) \tag{12}$$

$$T^y(s) := \frac{y(s)}{r(s)} = \frac{P(s)Q(s)}{1+P(s)Q(s)} \tag{13}$$

$$T^e(s) := \frac{e(s)}{r(s)} = \frac{1}{1+P(s)Q(s)} \tag{14}$$

$$T^u(s) := \frac{u(s)}{r(s)} = \frac{Q(s)}{1+P(s)Q(s)}. \tag{15}$$

The characteristic polynomial of the system is denoted as

$$\pi(s) = Q_2(s)D(s) + Q_1(s)N(s). \tag{16}$$

As $P(s)$ ranges over the uncertainty set $\mathbf{P}(s)$ (equivalently, $(N(s), D(s))$ ranges over $\mathcal{N}(s)\mathrm{x}\mathcal{D}(s)$) the transfer functions $T^o(s)$, $T^y(s)$, $T^u(s)$, $T^e(s)$ range over corresponding uncertainty sets $\mathbf{T}^o(s)$, $\mathbf{T}^y(s)$, $\mathbf{T}^u(s)$, and $\mathbf{T}^e(s)$, respectively. In other words,

$$\mathbf{T}^o(s) := \left\{ P(s)Q(s) \ : \ P(s) \in \mathcal{P}(s) \right\} \tag{17}$$

$$\mathbf{T}^y(s) := \left\{ \frac{P(s)Q(s)}{1+P(s)Q(s)} \ : \ P(s) \in \mathcal{P}(s) \right\} \tag{18}$$

$$\mathbf{T}^u(s) := \left\{ \frac{Q(s)}{1+P(s)Q(s)} \ : \ P(s) \in \mathcal{P}(s) \right\} \tag{19}$$

$$\mathbf{T}^e(s) := \left\{ \frac{1}{1+P(s)Q(s)} \ : \ P(s) \in \mathcal{P}(s) \right\}. \tag{20}$$

Similarly the characteristic polynomial $\pi(s)$ ranges over the corresponding uncertainty set denoted by

$$\Pi(s) = \{Q_2(s)D(s) + Q_1(s)N(s) \ : \ (N(s), D(s)) \in \mathcal{N}(s)\mathrm{x}\mathcal{D}(s)\}. \tag{21}$$

We now introduce the CB subset of the family of interval systems $\mathbf{P}(s)$:

$$\mathbf{P}_{\mathrm{CB}}(s) := \left\{ \frac{N(s)}{D(s)} \ : \ (N(s), D(s)) \in (\mathcal{N}(s)\mathrm{x}\mathcal{D}(s))_{\mathrm{CB}} \right\}. \tag{22}$$

These subsets will play a central role in all the results to be developed later. We note that each element of $\mathbf{P}_{\mathrm{CB}}(s)$ is a <u>one</u> parameter of transfer functions and there are at most 32 such distinct elements.

The CB subsets of the transfer function sets (17) - (20) and the polynomial set (21) are also introduced:

$$\mathbf{T}^o_{\mathrm{CB}}(s) := \left\{ P(s)Q(s) \ : \ P(s) \in \mathbf{P}_{\mathrm{CB}}(s) \right\} \tag{23}$$

$$\mathbf{T}^y_{\mathrm{CB}}(s) := \left\{ \frac{P(s)Q(s)}{1+P(s)Q(s)} \ : \ P(s) \in \mathbf{P}_{\mathrm{CB}}(s) \right\} \tag{24}$$

$$\mathbf{T}^u_{\mathrm{CB}}(s) := \left\{ \frac{Q(s)}{1+P(s)Q(s)} \ : \ P(s) \in \mathbf{P}_{\mathrm{CB}}(s) \right\} \tag{25}$$

$$\mathbf{T}^e_{\mathrm{CB}}(s) := \left\{ \frac{1}{1+P(s)Q(s)} \ : \ P(s) \in \mathbf{P}_{\mathrm{CB}}(s) \right\} \tag{26}$$

and

$$\Pi_{CB}(s) := \{Q_2(s)D(s) + Q_1(s)N(s) \; : \; (N(s), D(s)) \in (\mathcal{N}(s) \mathrm{x} \mathcal{D}(s))_{CB}\} \qquad (27)$$

In the paper we shall deal with the complex plane image of each of the above sets evaluated at $s = j\omega$. We denote each of these two dimensional sets in the complex plane by replacing s in the corresponding argument by ω. Thus, for example,

$$\Pi(\omega) := \{\Pi(s) \; : \; s = j\omega\} \qquad (28)$$

and

$$\mathbf{T}^{y}_{CB}(\omega) := \{\mathbf{T}^{y}_{CB}(s) \; : \; s = j\omega\} \qquad (29)$$

The Nyquist plot of a set of functions (or polynomials) $\mathbf{T}(s)$ is denoted by $\mathbf{T}$:

$$\mathbf{T} := \cup_{0 \leq \omega < \infty} \mathbf{T}(\omega) \qquad (30)$$

The boundary of a set $\mathcal{S}$ is denoted $\partial \mathcal{S}$.

Stability and Stability Margins

The control system of Figure 1 is stable for fixed $Q(s)$ and $P(s)$ if the characteristic polynomial

$$\pi(s) = Q_2(s)D(s) + Q_1(s)N(s) \qquad (31)$$

is Hurwitz, i.e. has all its $n = q+$ degree $[Q_2(s)]$ roots in the open left half of the complex plane. The system is robustly stable if and only if each polynomial in $\Pi(s)$ is of degree n (degree $D(s)$ remains invariant and equal to q as $D(s)$ ranges over $\mathcal{D}(s)$) and every polynomial in $\Pi(s)$ is Hurwitz.

The following important result was provided in Chapellat and Bhattacharyya [1].

Theorem 1. *(CB Theorem) The control system of Figure 1 is stable for all* $P(s) \in \mathbf{P}(s)$ *if and only if it is stable for all* $P(s) \in \mathbf{P}_{CB}(s)$.

The above Theorem gives a constructive solution to the problem of checking robust stability by reducing it to a problem of checking a set of (at most) 32 root locus problems.

If a fixed system is stable we can determine its gain margin γ and phase margin ϕ as follows:

$$\gamma^+(P(s), Q(s)) := \max\Big\{ \bar{K} \; : \; Q_2(s)D(s) + KQ_1(s)N(s) \text{ is Hurwitz for } K \in [1, \bar{K}] \Big\} \qquad (32)$$

$$\gamma^-(P(s), Q(s)) := \max\Big\{ \underline{K} \; : \; Q_2(s)D(s) + \tfrac{1}{K}Q_1(s)N(s) \text{ is Hurwitz for } K \in [\tfrac{1}{\underline{K}}, 1] \Big\} \qquad (33)$$

Similarly the phase margins are defined as follows:

$$\phi^+(P(s),Q(s)) := \max\left\{ \bar{\phi} : Q_2(s)D(s) + e^{j\phi}Q_1(s)N(s) \text{ is Hurwitz for } \phi \in [0,\bar{\phi}] \right\} \quad (34)$$

$$\phi^-(P(s),Q(s)) := \max\left\{ \underline{\phi} : Q_2(s)D(s) + e^{-j\phi}Q_1(s)N(s) \text{ is Hurwitz for } \phi \in [0,\underline{\phi}] \right\}. \quad (35)$$

Note that γ^+, γ^-, ϕ^+, and ϕ^- are uniquely determined when $Q(s)$ and $P(s)$ are fixed.

The unstructured additive stability margin ν_a (H^∞ stability margin) can also be uniquely determined for a fixed stable closed loop system and is given by

$$\nu_a(P(s)Q(s)) := \frac{1}{\|Q(s)(1+P(s)Q(s))^{-1}\|_\infty}. \quad (36)$$

In the following section we state some fundamental results on the above sets of transfer functions and the calculation of extremal stability margins over the uncertainty set $\mathcal{N}(s)\mathrm{x}\mathcal{D}(s)$.

3. MAIN RESULTS

We shall give the main results here without proof. The proofs are given in Keel, Shaw and Bhattacharyya [5]. Similar results have been recently reported independently by Tesi and Vicino [6].

Theorem 2. *For every $\omega \geq 0$,*

$$\partial \mathbf{P}(\omega) \subset \mathbf{P}_{\mathrm{CB}}(\omega) \quad (37)$$
$$\partial \mathbf{T}^o(\omega) \subset \mathbf{T}^o_{\mathrm{CB}}(\omega) \quad (38)$$
$$\partial \mathbf{T}^y(\omega) \subset \mathbf{T}^y_{\mathrm{CB}}(\omega) \quad (39)$$
$$\partial \mathbf{T}^u(\omega) \subset \mathbf{T}^u_{\mathrm{CB}}(\omega) \quad (40)$$
$$\partial \mathbf{T}^e(\omega) \subset \mathbf{T}^e_{\mathrm{CB}}(\omega) \quad (41)$$

This result shows that at every $\omega \geq 0$ the image set of each transfer function in (17) - (20) is bounded by the corresponding image set of the CB segments.

The next result deals with the Nyquist plots of each of the transfer functions in (17) - (20).

Corollary 1. *The Nyquist plots of each of the transfer function sets $\mathbf{T}^o(s)$, $\mathbf{T}^y(s)$, $\mathbf{T}^u(s)$, and $\mathbf{T}^e(s)$ are bounded by their corresponding CB subsets:*

$$\partial \mathbf{T}^o \subset \mathbf{T}^o_{\mathrm{CB}} \quad (42)$$
$$\partial \mathbf{T}^y \subset \mathbf{T}^y_{\mathrm{CB}} \quad (43)$$
$$\partial \mathbf{T}^u \subset \mathbf{T}^u_{\mathrm{CB}} \quad (44)$$
$$\partial \mathbf{T}^e \subset \mathbf{T}^e_{\mathrm{CB}} \quad (45)$$

This result has many important implications in control system design and will be explored in forthcoming papers. One important consequence is given below:

Theorem 3. *Suppose that the closed loop system is robustly stable, i.e. stable for all $P(s) \in \mathbf{P}(s)$. Then*

$$\max_{P(s)\in\mathbf{P}(s)} \gamma^{\pm} = \max_{P(s)\in\mathbf{P}_{CB}(s)} \gamma^{\pm} \tag{46}$$

$$\max_{P(s)\in\mathbf{P}(s)} \phi^{\pm} = \max_{P(s)\in\mathbf{P}_{CB}(s)} \phi^{\pm} \tag{47}$$

$$\max_{P(s)\in\mathbf{P}(s)} \nu_a = \max_{P(s)\in\mathbf{P}_{CB}(s)} \nu_a \tag{48}$$

and

$$\min_{P(s)\in\mathbf{P}(s)} \gamma^{\pm} = \min_{P(s)\in\mathbf{P}_{CB}(s)} \gamma^{\pm} \tag{49}$$

$$\min_{P(s)\in\mathbf{P}(s)} \phi^{\pm} = \min_{P(s)\in\mathbf{P}_{CB}(s)} \phi^{\pm} \tag{50}$$

$$\min_{P(s)\in\mathbf{P}(s)} \nu_a = \min_{P(s)\in\mathbf{P}_{CB}(s)} \nu_a. \tag{51}$$

In control systems the maximum margins are useful for synthesis and design (with $P(s)$ being a compensator) and the minimum margins are useful in calculating worst case stability margins ($P(s)$ is the plant). In each case the crucial point that makes the result constructive and useful is the fact that $\mathbf{P}_{CB}(s)$ consists of a finite and "small" number of line segments.

4. CONCLUDING REMARKS

We have showe that the boundaries of various transfer function sets and various extremal gain and phase margins occur on the CB segments first introduced in [1]. Using this idea, construction of Nyquist and Bode bands can be given. These important tools allow us to reexamine all of classical control theory with the added ingredient of robustness in both stability and performance using the framework of interval systems. More interestingly, these new results provide a way of developing design and synthesis techniques in the field of parametric robust control. These techniques also connect with H^{∞} control theory and promise many new results in the analysis and design of control systems containing uncertain or adjustable real parameters.

REFERENCES

[1] H. Chapellat and S. P. Bhattacharyya, "A generalization of Kharitonov's theorem: robust stability of interval plants," *IEEE Transactions on Automatic Control*, vol. AC - 34, pp. 306 – 311, March 1989.

[2] V. L. Kharitonov, "Asymptotic stability of an equilibrium position of a family of systems of linear differential equations," *Differential Uravnen*, vol. 14, pp. 2086 – 2088, 1978.

[3] H. Chapellat, M. Dahleh, and S. P. Bhattacharyya, "Robust stability under structured and unstructured perturbations," *IEEE Transactions on Automatic Control*, vol. AC - 35, pp. 1100 – 1108, October 1990.

[4] H. Chapellat, M. Dahleh, and S. Bhattacharyya, "Extremal manifolds in robust stability," TCSP Report, Texas A&M University, July 1990.

[5] L. Keel, J. Shaw and S. P. Bhattacharyya, "Frequency domain design of interval control systems," Tech. Rep., Tennessee State University, July 1991. Also in TCSP Tech. Rep., Texas A&M University, July 1991.

[6] A. Tesi and A. Vicino, "Kharitonov segments suffice for frequency response analysis of plant-controller families". To appear in *Control of Uncertain Dynamic Systems*, September 1991, CRC Press.

Rejection of Persistent, Bounded disturbances for Sampled-Data Systems*

Bassam Bamieh[†], Munther A. Dahleh[‡], and J. Boyd Pearson[†]

June 4, 1991

Abstract

In this paper, a complete solution for the ℓ^1 sampled-data problem is furnished for arbitrary plants. Then ℓ^1 sampled-data problem is described as follows: Given a continuous-time plant, with continuous-time performance objectives, design a digital controller that delivers this performance. This problem differs from the standard discrete-time methods in that it takes into consideration the inter-sampling behaviour of the closed loop system. The resulting closed loop system dynamics consists of both continuous-time and discrete-time dynamics and thus such systems are known as "Hybrid" systems. It is shown that given any degree of accuracy, there exists a standard discrete-time ℓ^1 problem, which can be determined apriori, such that for any controller that achieves a level of performance for the discrete-time problem, the same controller achieves the same performance within the prescribed level of accuracy if implemented as a sampled-data controller.

*The first and last authors' research is supported by NSF ECS-8914467 and AFOSR-91-0036. The second author is supported by Wright-Patterson A.F.B. F33615-90-C-3608, , C.S. Draper Laboratory DL-H-418511 and by the ARO DAAL03-86-K-0171.

[†]Dept. of Electrical and Computer Engineering, Rice University, Houston, TX 77030

[‡]Laboratory of Information and Decision Systems, Massachusetts Institute of Technology, Cambridge, MA

1 Introduction

This paper is concerned with designing digital controllers for continuous-time systems to optimaly achieve certain performance specifications in the presence of uncertainty. Contrary to discrete time designs, such controllers are designed taking into consideration the inter-sample behavior of the system. Such hybrid systems are generally known as sampled-data systems, and have recently received renewed interest by the control community.

The difficulty in considering the continuous time behavior of sampled-data systems, is that it is time varying, even when the plant and the controller are both continuous-time and discrete-time time-invariant respectively. We consider in this paper the *standard problem with sampled-data controllers* (or the sampled-data problem, for short) shown in figure 1. The continuous time controller is constrained to be sampled-data controller, that is, it is of the form $\mathcal{H}_\tau C \mathcal{S}_\tau$. The generalized plant is continuous-time time-invariant and C is discrete-time time-invariant,$\mathcal{H}_\tau$ is a zero order hold (with period τ), and $\mathcal{S}_\tau$ is an ideal sampler (with period τ). $\mathcal{H}_\tau$ and $\mathcal{S}_\tau$ are assumed synchronized. Let $\mathcal{F}(G, \mathcal{H}_\tau C \mathcal{S}_\tau)$ denote the mapping between the exogenous input and the regulated output. $\mathcal{F}(G, \mathcal{H}_\tau C \mathcal{S}_\tau)$ is in general time varying, in fact it is τ-periodic where τ is the period of the sample and hold devices.

Sampled-data systems have been studied by many researchers in the past in the context of LQG controllers (e.g. [12]). Recently, [4] studied this problem in the context of $\mathcal{H}^\infty$ control, and were able to provide a solution in the case where the regulated output is in discrete time and the exogenous input is in continuous time. The exact problem was solved in [2],[3], and independently by [9] and [13]. The L^∞-induced norm problem (the one we are concerned with in this paper) was considered in [7] for the case of stable plants.

In this paper we will use the framework developed in [2],[3], to study the ℓ^1 sampled-data problem. Precisely, the controller is designed to minimize the induced norm of the periodic system over the space of bounded inputs (i.e. L^∞). This minimization results from posing time domain specifications and design constraints, which is quite natural for control system design. The solution provided in this paper is to solve the sampled-data problem by solving an (almost) equivalent discrete time ℓ^1 problem. While this was the approach followed in [7], the main contribution of this paper is that it provides bounds that can be computed apriori to determine the equivalent discrete-time problem, given any desired degree of accuracy and thus provides a solution for the synthesis problem. The solution in this paper is presented in the context of the lifting framework of [2], [3], as an approximation procedure for certain infinite dimensional problems. This approach has the advantage of handling both the stable and the unstable case in the same framework, and results in techniques that are more transparent than those in [7].

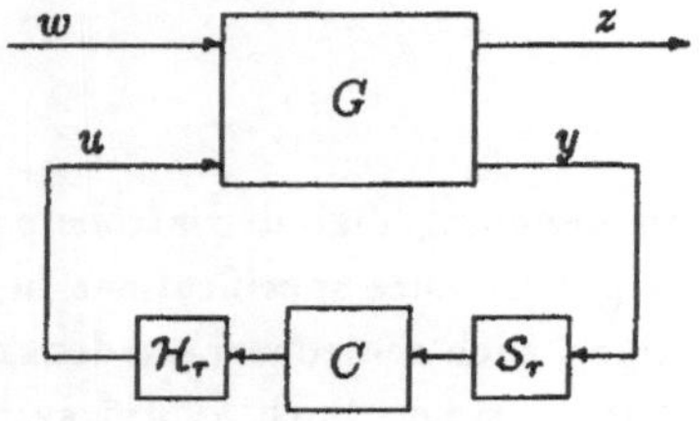

Figure 1: Hybrid discrete/continuous time system

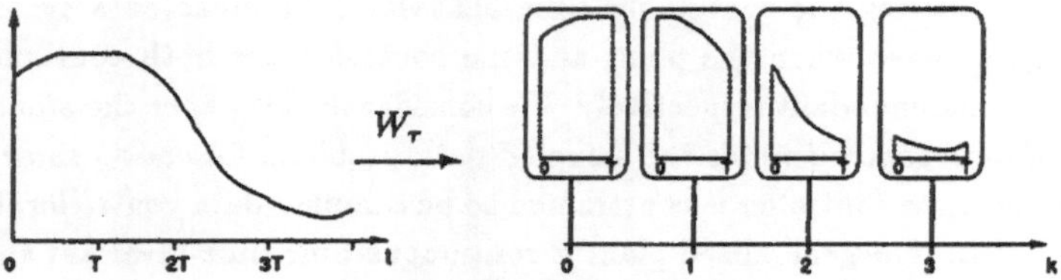

Figure 2: $W_\tau : L^p_e[0,\infty) \longrightarrow \ell_{L^p[0,\tau]}$

2 The Lifting Technique

In this section we briefly summarize the lifting technique for continuous-time periodic systems developed in [2], [3], and apply it to the sampled-data problem.

For continuous time signals, we consider the usual $L^\infty[0,\infty)$ space of essentially bounded functions [6], and it's extended version $L^\infty_e[0,\infty)$. We will also need to consider discrete time signals that take values in a function space, for this, we define ℓ_X to be the space of all X-valued sequences, where X is some Banach space. We define ℓ^∞_X as the subspace of ℓ_X with bounded norm sequences, i.e. where for $\{f_i\} \in \ell_X$, the norm $\|\{f_i\}\|_{\ell^\infty_X} := \sup_i \|f_i\|_X < \infty$. Given any $f \in L^\infty_e[0,\infty)$, we define it's *lifting* $\hat{f} \in \ell_{L^\infty[0,\tau]}$, as follows: $\hat{f}$ is an $L^\infty[0,\tau]$-valued sequence, we denote it by $\{\hat{f}_i\}$, and for each i,

$$\hat{f}_i(t) := f(t+\tau i) \quad 0 \le t \le \tau.$$

The lifting can be visualized as taking a continuous time signal and breaking it up into a sequence of 'pieces' each corresponding to the function over an interval of length τ (see figure 2). Let us denote this lifting by $W_\tau : L^\infty_e[0,\infty) \longrightarrow \ell_{L^\infty[0,\tau]}$. W_τ is a linear isomorphism, furthermore, if restricted to $L^\infty[0,\infty)$, then $W_\tau : L^\infty[0,\infty) \longrightarrow \ell^\infty_{L^\infty[0,\tau]}$ is an isometry, i.e. it preserves norms.

Using the lifting of signals, one can define a lifting on systems. Let G be a linear continuous time system on $L^\infty_e[0,\infty)$, then it's *lifting* $\hat{G}$ is the discrete time system $\hat{G} :=$

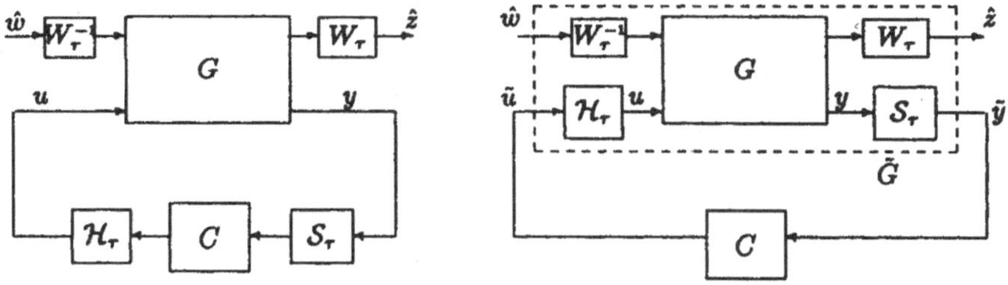

Figure 3: Equivalent Problem

$W_\tau G W_\tau^{-1}$, this is illustrated in the commutative diagram below:

$$\begin{array}{ccc} \ell_{L^\infty[0,\tau]} & \xrightarrow{\hat{G}} & \ell_{L^\infty[0,\tau]} \\ W_\tau^{-1}\downarrow & & \uparrow W_\tau \\ L_e^\infty[0,\infty) & \xrightarrow{G} & L_e^\infty[0,\infty) \end{array}$$

Thus $\hat{G}$ is a system that operates on Banach space ($L^\infty[0,\tau]$) valued signals, we will call such systems infinite dimensional. Note that since W_τ is an isometry, if G is stable, i.e. a bounded linear map on L^∞ then $\hat{G}$ is also stable, and furthermore, their respective induced norms are equal, $\|\hat{G}\| = \|G\|$. The correspondence between a system and it's lifting also preserves algebraic system properties such as addition, cascade decomposition and feedback (see [2] for the details).

To apply the lifting to the sampled-data problem, consider again the standard problem of figure 1, and denote the closed loop operator by $\mathcal{F}(G, \mathcal{H}_\tau C S_\tau)$. Since the lifting is an isometry, we have that $\|\mathcal{F}(G, \mathcal{H}_\tau C S_\tau)\| = \|W_\tau \mathcal{F}(G, \mathcal{H}_\tau C S_\tau) W_\tau^{-1}\|$, this is shown in figure 3(a).

We now present (from [2]) a state space realization for the new generalized plant $\tilde{G}$. The original plant G has the realization:

$$G = \left[\begin{array}{c|cc} A & B_1 & B_2 \\ \hline C_1 & D_{11} & D_{12} \\ C_2 & 0 & 0 \end{array}\right].$$

It is assumed that the sampler is proceeded with a presampling filter. It can be shown ([2]) that a realization for the generalized plant $\tilde{G}$ (figure 3) is given by

$$\tilde{G} = \begin{bmatrix} \tilde{G}_{11} & \tilde{G}_{12} \\ \tilde{G}_{21} & \tilde{G}_{22} \end{bmatrix} = \left[\begin{array}{c|cc} \hat{A} & \hat{B}_1 & \tilde{B}_2 \\ \hline \hat{C}_1 & \hat{D}_{11} & \hat{D}_{12} \\ \tilde{C}_2 & 0 & 0 \end{array}\right]$$

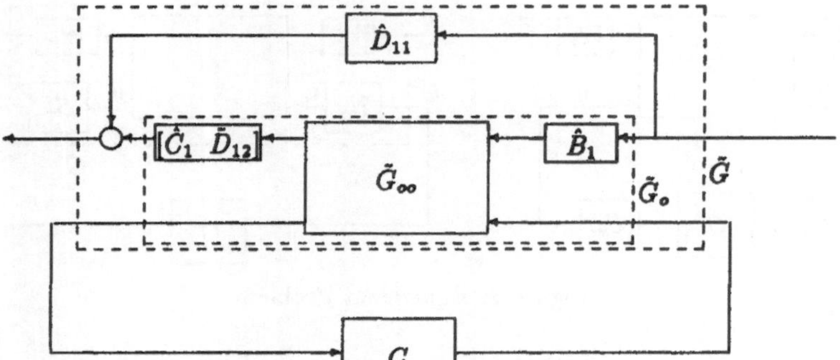

Figure 4: Decomposition of $\tilde{G}$

$$= \left[\begin{array}{c|cc} e^{A\tau} & e^{A(\tau-s)}B_1 & \Psi(\tau)B_2 \\ \hline C_1e^{At} & C_1e^{A(t-s)}1_{(t-s)}B_1 + D_{11}\delta(t-s) & C_1\Psi(t)B_2 + D_{12} \\ C_2 & 0 & 0 \end{array}\right]$$

where $\Psi(t) := \int_0^t e^{As}ds$. The system $\tilde{G}$ has the following input and output spaces

$$\begin{aligned} \tilde{G}_{11} &: \ell_{L^\infty[0,\tau]} \longrightarrow \ell_{L^\infty[0,\tau]} \\ \tilde{G}_{12} &: \ell_{\mathbb{R}^u} \longrightarrow \ell_{L^\infty[0,\tau]} \\ \tilde{G}_{21} &: \ell_{L^\infty[0,\tau]} \longrightarrow \ell_{\mathbb{R}^y} \\ \tilde{G}_{22} &: \ell_{\mathbb{R}^u} \longrightarrow \ell_{\mathbb{R}^y} \end{aligned}$$

Using this lifting, $\tilde{G}$ can be decomposed as

$$\tilde{G} = \begin{bmatrix} \hat{D}_{11} & 0 \\ 0 & 0 \end{bmatrix} + \begin{bmatrix} [\ \hat{C}_1 \ \ \hat{D}_{12}\ ] & 0 \\ 0 & I \end{bmatrix} \left[\begin{array}{c|cc} \hat{A} & I & \hat{B}_2 \\ \hline \begin{bmatrix} I \\ 0 \end{bmatrix} & \begin{bmatrix} 0 \\ 0 \end{bmatrix} & \begin{bmatrix} 0 \\ I \end{bmatrix} \\ \hat{C}_2 & 0 & 0 \end{array}\right] \begin{bmatrix} \hat{B}_1 & 0 \\ 0 & I \end{bmatrix}$$

This decomposition is illustrated in figure 4. The closed loop mapping $\mathcal{F}(\tilde{G}, C)$ is correspondingly decomposed as

$$\mathcal{F}(\tilde{G},C) \;=\; \hat{D}_{11} \;+\; \mathcal{F}(\tilde{G}_o,C) \;=\; \hat{D}_{11} \;+\; [\ \hat{C}_1 \ \ \hat{D}_{12}\ ]\ \mathcal{F}(\tilde{G}_{\infty},C)\ \hat{B}_1. \qquad (1)$$

We will use the notation $\hat{\mathcal{O}} := [\ \hat{C}_1 \ \ \hat{D}_{12}\ ]$, and call $\hat{\mathcal{O}}$ the output operator and $\hat{B}_1$ the input operator. With this decomposition, $\tilde{G}_{\infty}$ is finite dimensional, and $\hat{\mathcal{O}}$, $\hat{B}_1$ are finite rank operators

$$\hat{\mathcal{O}} : \mathbb{R}^{x+u} \longrightarrow L^\infty[0,\tau], \qquad \hat{B}_1 : L^\infty[0,\tau] \longrightarrow \mathbb{R}^x.$$

It is evident from this decomposition that only the "discrete" part of the plant depends on the controller, while the infinite dimensional hybrid part depends only on the system's dynamics.

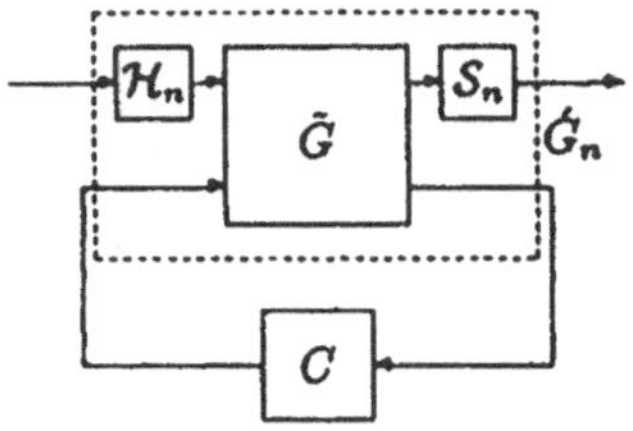

Figure 5: The system $\acute{G}_n$

3 Solution Procedure

Using the lifting we are able to convert the problem of finding a controller to minimize the L^∞ induced norm of the hybrid system (figure 1) into the following standard problem with an infinite dimensional generalized plant $\tilde{G}$:

$$\mu := \inf_{C \text{ stablizing}} \|\mathcal{F}(G, \mathcal{H}_\tau C S_\tau)\| =: \inf_{C \text{ stablizing}} \|\mathcal{F}(\tilde{G}, C)\| \tag{2}$$

The above infinite dimensional problem is solved by the following approximation procedure through solving a standard MIMO ℓ^1 problem [5, 10]. Let $\mathcal{H}_n$ and S_n be the following operators defined between $L^\infty[0,\tau]$ and $\ell^\infty(n)$ ($\ell^\infty(n)$ is $\mathbb{R}^n$ with the maximum norm),

$$S_n : L^\infty[0,\tau] \longrightarrow \ell^\infty(n) \qquad (S_n u)(i) = u(\frac{\tau}{n} i) \; ; \; u \in L^\infty[0,\tau]$$

$$\mathcal{H}_n : \ell^\infty(n) \longrightarrow L^\infty[0,\tau] \qquad (\mathcal{H}_n u)(t) = u(\lfloor \frac{tn}{\tau} \rfloor) \; ; \; \{u(i)\} \in \ell^\infty(n),$$

One can in fact show that the operators are well defined. Now to approximate the infinite dimensional problem, we use the approximate closed loop system $S_n \mathcal{F}(\tilde{G}, C)\mathcal{H}_n$ (see figure 5), and for each n we define

$$\mu_n := \inf_{C \text{ stabilizing}} \|S_n \mathcal{F}(\tilde{G}, C)\mathcal{H}_n\|, \tag{3}$$

This new problem now involves the induced norm over $\ell^\infty_{\ell^\infty(n)}$, i.e. it is a standard MIMO ℓ^1 problem. Let us denote the generalized plant associated with $\mathcal{H}_n \mathcal{F}(\tilde{G}, C) S_n$ by $\acute{G}_n$, such that

$$S_n \mathcal{F}(\tilde{G}, C)\mathcal{H}_n = \mathcal{F}(\acute{G}_n, C),$$

where $\acute{G}_n$ and a realization for it is given by,

$$\acute{G}_n := \begin{bmatrix} S_n & 0 \\ 0 & I \end{bmatrix} \tilde{G} \begin{bmatrix} \mathcal{H}_n & 0 \\ 0 & I \end{bmatrix} = \left[\begin{array}{c|cc} \hat{A} & \hat{B}_1 \mathcal{H}_n & \tilde{B}_2 \\ \hline S_n \tilde{C}_1 & S_n \tilde{D}_{11} \mathcal{H}_n & S_n \tilde{D}_{12} \\ \tilde{C}_2 & 0 & 0 \end{array}\right] =: \left[\begin{array}{c|cc} \hat{A} & \acute{B}_1 & \tilde{B}_2 \\ \hline \acute{C}_1 & \acute{D}_{11} & \acute{D}_{12} \\ \tilde{C}_2 & 0 & 0 \end{array}\right].$$

The solution in now described below.

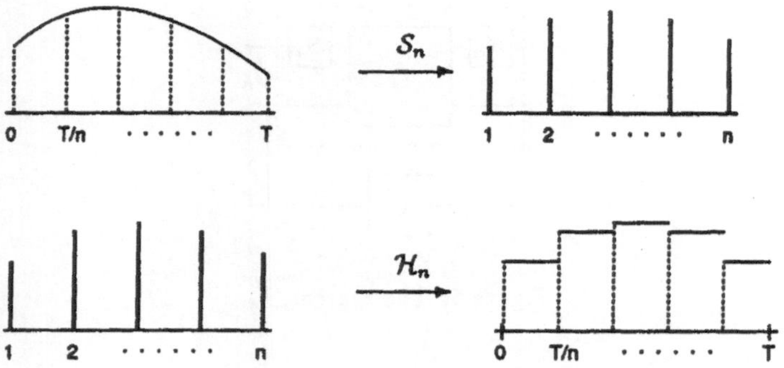

Figure 6: The operators $\mathcal{S}_n$ and $\mathcal{H}_n$

4 Design Bounds

Let us begin with analysis. Note that since $\|\mathcal{F}(\tilde{G},C)\|$ is a periodically time varying system, its L^∞-induced norm is not immediately computable. An alternative method of computing $\|\mathcal{F}(\tilde{G},C)\|$ comes from the limit

$$\|\mathcal{F}(\tilde{G},C)\| \;=\; \lim_{n\to\infty} \|\mathcal{S}_n\mathcal{F}(\tilde{G},C)\mathcal{H}_n\| \;=:\; \lim_{n\to\infty} \|\mathcal{F}(\acute{G}_n,C)\|, \tag{4}$$

for a fixed C. This equation (4) however is by far not sufficient to show the convergence of the synthesis procedure, since given only (4), the rate of convergence may depend on the choice of C. Our objective is to obtain explicit bounds on $\|\mathcal{F}(\tilde{G},C)\|$ in the following form

Main Inequality : *There are constants K_o and K_1 which depend only on G, such that*

$$\|\mathcal{F}(\acute{G}_n,C)\| \;\le\; \|\mathcal{F}(\tilde{G},C)\| \;\le\; \frac{K_1}{n} + \left(1+\frac{K_o}{n}\right)\|\mathcal{F}(\acute{G}_n,C)\|, \tag{5}$$

The significance of the bound (5) is that it is exactly what is needed for synthesis. When one performs an ℓ^1 design on $\acute{G}_n$, the result is a controller that keeps $\|\mathcal{F}(\acute{G}_n,C)\|$ small, but the objective is to keep the L^∞-induced norm of the hybrid system (or equivalently $\|\mathcal{F}(\tilde{G},C)\|$) small, and the inequality (5) guarantees this. The proof of this bound utilizes heavily the decomposition discussed earlier, and the fact that the infinite dimensional operators $\hat{\mathcal{O}}$ and $\hat{B}_1$ are finite dimensional, and are dependent only on the original plant. The details of these proofs can be found in [1].

References

[1] B. Bamieh, M.A. Dahleh and J.B. Pearson, 'Minimization of the L^∞ induced norm for sampled-data systems,' Rice University Technical Report, Houston, TX, June 1991.

[2] B. Bamieh and J.B. Pearson,'A General Framework for Linear Periodic Systems with Application to $\mathcal{H}^\infty$ Sampled-Data Control', to appear in *IEEE Trans. on Aut. Control.*

[3] B. Bamieh, J.B. Pearson, B.A. Francis, A. Tannenbaum, 'A Lifting Technique for Linear Periodic Systems with Applications to Sampled-Data Control', to appear in *Systems and Controls Letters.*

[4] T. Chen and B. Francis, 'On the L^2 induced 26 26 norm of a sampled-data system', *Systems and Control Letters* 1990, v.15, 211-219.

[5] M.A. Dahleh and J.B. Pearson, 'ℓ^1-optimal feedback control for MIMO discrete-time systems', in *Trans. Aut. Control,* AC-32, 1987.

[6] C.A. Desoer and M. Vidyasagar, *Feedback Systems: Input-Output Properties*, Academic Press, New York, 1975.

[7] G. Dullerud and B. Francis, '$\mathcal{L}^1$ Performance in Sampled-Data Systems', preprint, Submitted to *IEEE Trans. on Aut. Control.*

[8] B.A. Francis and T.T. Georgiou, 'Stability theory for linear time-invariant plants with periodic digital controllers', *IEEE Trans. Aut. Control,* AC-33, pp. 820-832, 1988.

[9] P.T. Kabamba and S. Hara, 'Worst case analysis and design of sampled data control systems', Preprint.

[10] J.S. McDonald and J.B. Pearson, 'ℓ^1-Optimal Control of Multivariable Systems with Output Norm Constraints', *Automatica,* vol. 27, no. 3, 1991.

[11] N. Sivashankar and Pramod P. Khargonekar, '$\mathcal{L}_\infty$-Induced Norm of Sampled-Data Systems', *Proc. of the CDC,* 1991.

[12] P.M. Thompson, G. Stein and M. Athans, 'Conic sectors for sampled-data feedback systems,' *System and Control Letters,* Vol 3, pp. 77-82, 1983.

[13] H.T. Toivonen, 'Sampled-data control of continuous-time systems with an $\mathcal{H}_\infty$ optimality criterion', Chemical Engineering, Abo Akademi, Finland, Report 90-1, January 1990.

Rational Approximation of L_1-optimal controller

Yoshito Ohta†, Hajime Maeda‡ and Shinzo Kodama‡

† Department of Mechanical Engineering for Computer-Controlled Machinery
Osaka University
2-1 Yamada-Oka, Suita, Osaka 565, Japan
‡ Department of Electronic Engineering
Osaka University
2-1 Yamada-Oka, Suita, Osaka 565, Japan

Abstract

This paper studies the L_1-optimal control problem for SISO systems with rational controllers. It is shown that the infimal achievable norm with rational controllers is as small as that with irrational controllers. Also a way to construct a rational suboptimal controller is studied.

1. Introduction

The problem of L_1-optimal control was first formulated by Vidyasagar [7] to solve the disturbance rejection problem with persistent input. A complete solution for general problems was presented by Dahleh and Pearson for discrete-time 1-block MIMO systems [2] and for continuous-time SISO systems [3]. In [3] they also pointed out the usefulness of L_1-optimal control in the following areas: (i) disturbance rejection of bounded input, (ii) tracking of bounded input, and (iii) robustness.

For continuous-time systems, it turns out that an L_1-optimal solution has a certain disadvantage: delay terms appear both in the numerator and denominator of an optimal controller [3]. This raises a question as to what limitation we impose if controllers are confined to be rational. One obvious way to look at this is to approximate the irrational controller by rational one. However, as it will be shown in a later section, an irrational transfer function cannot be approximated by the set of stable rational functions in the topology with which the L_1-control problems are formulated. This means the question is not trivial.

In this paper, it is shown that rational controllers are as effective as irrational controllers in continuous-time L_1-control problems. Specifically, it is established that the infimal achievable norm with rational controllers is as small as that with irrational controllers. In the course of proving the above, a way to construct a sequence of rational controllers that achieves the optimal performance index is presented.

2. Mathematical preliminaries

In this section, some underlying linear spaces are reviewed to define the L_1-control problems. The following notations are used throughout the paper. R_+ is the set of nonnegative real numbers.

$L_1(R_+)$	the set of Lebesgue integrable functions on R_+.
$L_\infty(R_+)$	the set of Lebesgue measurable and bounded functions on R_+.
$C_0(R_+)$	the set of continuous functions on R_+ that converge to 0.
$C^{(n)}(R_+)$	the set of n-time continuously differentiable functions on R_+.
$BV(R_+)$	the set of bounded variations on R_+.
$\mathfrak{A}$	the Banach algebra with convolution whose elements have the form $F = F_a + \sum_{k=0}^{\infty} f_i d(\bullet - h_i)$, where F_a Œ $L_1(R_+)$ and (f_i) is absolutely summable. $\|F\|_A = \|F_a\|_1 + \sum_{k=0}^{\infty} \lvert f_i \rvert$
$\hat{\mathfrak{A}}$	the class of the Laplace transformations of $\mathfrak{A}$. $\|\hat{F}\|_A = \|F\|_A$

For the algebra $\mathfrak{A}$, see [4] and [1]. $\mathfrak{A}$ is a subalgebra of $BV(R_+)$. Indeed, for $F \in \mathfrak{A}$, define

$$d\tilde{F} = F dt. \tag{1}$$

Then $\tilde{F} \in BV(R_+)$, and the correspondence gives an isometric injection that preserves convolution. The importance of the algebra $\mathfrak{A}$ is that it is the set of stable linear time-invariant operators from $L_\infty(R_+)$ to $L_\infty(R_+)$. The operation of $F \in \mathfrak{A}$ is defined by the convolution

$$\begin{aligned} y(t) &= F*x(t) \\ &= \int_0^t F(t-t)\, x(\tau)\, d\tau. \end{aligned} \tag{2}$$

The induced norm as an operator is

$$\begin{aligned} \|F\|_{ind} &= \sup\{\|y\|_\infty, y = F*x, x \in L_\infty(R_+), \|x\|_\infty \le 1\} \\ &= \|F\|_A. \end{aligned} \tag{3}$$

The set of stable rational functions $\hat{\mathfrak{S}}$ is a subset of $\hat{\mathfrak{A}}$ that is not closed nor dense in $\hat{\mathfrak{A}}$. To see that it is not dense, an example shows that

$$\inf\{\|\exp(-s) - \hat{F}\|_A, \hat{F} \in \hat{\mathfrak{S}}\} = 1 > 0. \tag{4}$$

This implies that irrational transfer functions cannot be in general approximated by rational transfer functions. Systems are assumed to admit coprime factorization in $\hat{\mathfrak{A}}$ (or in $\hat{\mathfrak{S}}$, if it is finite-dimensional).

3. L_1-rational approximation problems

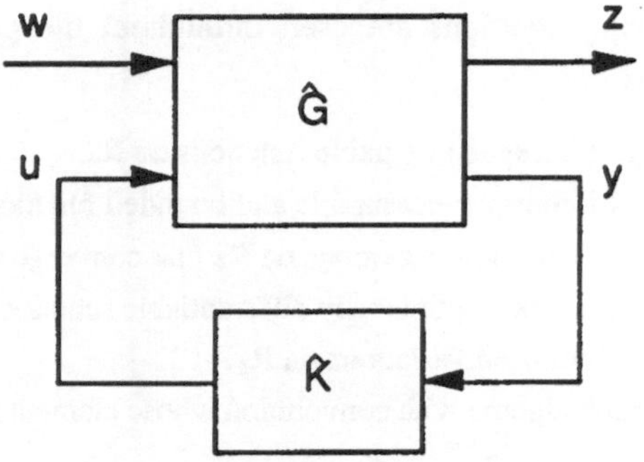

Fig.1 The standard problem

Consider the standard control problem shown in Fig.1, where $\hat{G}$ is the generalized plant and $\hat{K}$ is the controller. The signal w is exogenous input such as disturbances and command input; z is the output to be controlled; u is the control signal; and y is the measured output. The signals are scalar valued. We assume that the plant $\hat{G}$ is rational. The problem is to stabilize the feedback loop, and make the norm of the transfer function $\hat{T}_{zw}$ from w to z as small as possible. The norm should reflect the goal that the maximum magnitude of the output z is small. An appropriate choice of the norm is the norm of $\hat{\mathfrak{A}}$, and this leads to the L_1-control problem,

$$\inf \{\|\hat{T}_{zw}\|_A,\ \hat{K} \text{ stabilizes the feedback loop}\}. \tag{5}$$

With some mild assumptions, the parametrization of stabilizing controllers for SISO systems brings the transfer function $\hat{T}_{zw}$ into the affine form

$$\hat{T}_{zw} = \hat{T}_1 - \hat{T}_2\hat{Q}, \tag{6}$$

where $\hat{T}_1, \hat{T}_2 \in \hat{\mathfrak{S}}$ are fixed transfer functions, and $\hat{Q} \in \hat{\mathfrak{A}}$ (if irrational controllers are admissible) or $\hat{\mathfrak{S}}$ (if controllers are confined to be rational) is a free parameter. Then the L_1-control problem (5) is transformed into the following model-matching problems:

$$\gamma_{MM} = \inf \{\|\hat{T}_1 - \hat{T}_2\hat{Q}\|_A,\ \hat{Q} \in \hat{\mathfrak{A}}\}, \tag{7}$$
$$\gamma_{RMM} = \inf \{\|\hat{T}_1 - \hat{T}_2\hat{Q}\|_A,\ \hat{Q} \in \hat{\mathfrak{S}} \subset \hat{\mathfrak{A}}\}. \tag{8}$$

Obviously the optimal value of the rational model-matching problem (7) is not less that that of the irrational one (8), i.e., $\gamma_{MM} \le \gamma_{RMM}$. Because of (4), it is not clear if the equality holds. If it does, then it turns out that rational controllers are as effective as irrational controllers in the L_1-control problem. Thus we are interested in the following problems: (i)

$\gamma_{MM} = \gamma_{RMM}$? (ii) How can we construct a minimizing sequence of (8)? The focus of this paper is to show that (i) is affirmative. In the course of proving that, the problems (ii) will be addressed.

4. Representation of dual problem

It was shown in [3] that the L_1-model matching problem (7) achieves the infimum, and that an optimal solution can be constructed via the duality theorem [5]. The dual problem turned out to be a finite-dimensional linear programming with infinite number of constraints.

The purpose of this section is to present the dual problem in a form suited for subsequent discussions. Some part of the following argument is in line with [8].

First, another model matching problem is introduced to incorporate with the duality theorem. Let $\hat{T}_1$ and $\hat{T}_2 \in \mathfrak{B} \subset \mathfrak{A}$ be as in (7). Since $\mathfrak{A}$ is a subalgebra of $BV(R_+)$, let $d\check{T}_1 = T_1 dt$ and $d\check{T}_2 = T_2 dt$. Then $\check{T}_1, \check{T}_2 \in BV(R_+)$. Define the subspace

$$W = \left\{ \check{T}_2 * \check{Q},\ \check{Q} \in BV(R_+) \right\} \subset BV(R_+). \tag{9}$$

The embedded L_1-model matching problem is defined as follows:

$$\gamma_{EMM} = \inf \{ \|\check{T}_1 - \check{R}\|_{TV},\ \check{R} \in W \subset BV(R_+) \}. \tag{10}$$

Since $C_0(R_+)^* = BV(R_+)$, the duality theorem says that the problem (10) takes the same optimal value as the (pre-)dual problem:

$$\gamma_{DP} = \sup \left\{ \int_{R+} F(t)\, d\check{T}_1,\ F \in {}^{\perp}W \subset C_0(R_+),\ \|F\|_\infty \le 1 \right\}, \tag{11}$$

where

$${}^{\perp}W = \left\{ F \in C_0(R_+) : \int_{R+} F(t)\, d\check{R} = 0,\ \ \check{R} \in W \right\} \tag{12}$$

is the (left) annihilator of W. The duality theorem asserts that there is a minimizer of (10), and that

$$\gamma_{DP} = \gamma_{EMM}. \tag{13}$$

The following is a characterization of the annihilator.

Proposition 1. (annihilator) Let $\hat{T}_2(s) = [A_2, B_2, C_2, D_2]$ be an nth order minimal realization. Assume that the zeros of $\hat{T}_2$ lie in the open right half plane. Then the set ${}^{\perp}W \cap C^{(n)}(R_+)$ is characterized as follows:

$${}^{\perp}W \cap C^{(n)}(R_+) = \{ F : F(t) = L^T \exp(-\hat{A}^T t) z,\ z \in R^n \}, \tag{14}$$

where

$$L = B_2 D_2^{-1} \text{ and } \hat{A} = A_2 - LC_2. \tag{15}$$

If one can show ${}^{\perp}W \subset C^{(n)}(R_+)$, then Proposition 1 is a complete characterization. In [5], differentiability was first assumed, and then a limit is taken to give the annihilator. However, we could not follow the argument there. The result of Proposition 1 lacks mathematical beauty, and the following discussion becomes a little bit awkward, but it will be shown that it is enough for our purpose. See the argument at the end of this section.

The dual problem (11) will be slightly modified to coordinate Proposition 1, and will be represented by a linear programming. The modified dual problem is:

$$\gamma_{MDP} = \sup \left\{ \int_{R+} F(t)\, d\hat{T}_1,\ F \in {}^{\perp}W \cap C^{(n)}(R_+),\ \|F\|_\infty \le 1 \right\}. \qquad (16)$$

Substituting the characterization (14) into (16), we obtain the following linear programming.

Proposition 2 (Linear programming) Let $\hat{T}_1(s)=[A_1,B_1,C_1,D_1]$ be a minimal realization. Let $\hat{T}_2(s)=[A_2,B_2,C_2,D_2]$ and W be as in Proposition 1. Define

$$M = LD_1 + RB_1, \qquad (17)$$

where R is a unique solution of the Lyapunov equation

$$\hat{A}R - RA_1 = LC_1. \qquad (18)$$

Then the modified dual problem (17) is represented as the linear programming

$$\gamma_{LP} = \max M^T z \qquad (19)$$

subject to

$$|L^T \exp(-\hat{A}^T t) z| \le 1,\ t \ge 0. \qquad (20)$$

The three problems (11), (16), and (19) apparently satisfy

$$\gamma_{DP} \ge \gamma_{MDP} = \gamma_{LP}. \qquad (21)$$

The equality holds in (21) because the linear programming (19) is just another representation of Theorem 5 in [3]. Hence

$$\gamma_{EMM} = \gamma_{DP} = \gamma_{MDP} = \gamma_{LP}. \qquad (22)$$

Here is a brief review how an optimal solution of (7) is obtained. Suppose z_{opt} is optimal for (19). Then writing

$$M = \sum_{i=1}^{q} \mu_i \left(\exp(-\hat{A}\, t_i)\, L\right), \qquad (23)$$

where

$$\sum_{i=1}^{q} |\mu_i| = \gamma_{LP}, \qquad (24)$$

for some $\{t_1, t_2, \ldots, t_q\}$, z_{opt} satisfies

$$|L^T \exp(-\hat{A}^T t_i) z_{opt}| = 1,\ i = 1,2,\ldots,q. \tag{25}$$

Let $F_{opt} \in C_0(R_+)$, $F_{opt}(t) = L^T \exp(-\hat{A}^T t) z_{opt}$. Then it is optimal for (16). The result in [3] indicates that this is also an optimal solution of (11). By the alignment condition [5], an optimal solution $\hat{R}_{opt}$ of (10) is such that $\hat{\Phi}_{opt} = \hat{T}_1 - \hat{R}_{opt}$ is aligned with F_{opt}. From (25), $\hat{\Phi}_{opt}$ is supported by the finite set $\{t_1, t_2, \ldots, t_q\}$, and hence Φ_{opt} is a sum of delta functions:

$$\Phi_{opt}(t) = \sum_{i=1}^{q} \mu_i\ \delta(t - t_i), \tag{26}$$

where μ_i are defined by (23). Since $\Phi_{opt} \in \mathfrak{A}$, let $\hat{\Phi}_{opt} = \hat{T}_1 - \hat{T}_2 \hat{Q}_{opt}$ for some $\hat{Q}_{opt} \in \hat{\mathfrak{A}}$. Actually $\hat{Q}_{opt}$. is optimal for (7).

5. Main results

In this section, we shall show that rational controllers are effective in continuous-time L_1-optimal control. For this, it will be shown that $\gamma_{MM} = \gamma_{RMM}$. The following notions of convex analysis are used (see for example [6]). The convex hull of a set S is the smallest convex set that contains S, and is denoted by conv S. A polytope is a set that is the convex hull of finitely many points. For a set S, the polar set S° is defined as

$$S^\circ = \{x: y^T x \le 1,\ y \in S\}. \tag{27}$$

If S is a closed convex set containing the origin, then $S^{\circ\circ} = S$.

Lemma 1. Let $\hat{T}_1(s) = [A_1, B_1, C_1, D_1]$ and $\hat{T}_2(s) = [A_2, B_2, C_2, D_2]$ be minimally realized stable SISO transfer functions. Assume that the zeros of $\hat{T}_2$ lie in the open right half plane. Define L, $\hat{A}$, and M as (15) and (17). Suppose there is a symmetric polyhedron $\mathbf{X}$ and a positive number λ such that

$$(I + \lambda\hat{A})\mathbf{X} \subset \mathrm{conv}(\pm L, \mathbf{X}) \tag{28}$$

and

$$\sup\{M^T z,\ z \in (\mathrm{conv}(\pm L, \mathbf{X}))^\circ\} = \gamma < 1, \tag{29}$$

Then there is a $\hat{Q} \in \mathfrak{S}$ such that

$$\|\hat{T}_1 - \hat{T}_2 \hat{Q}\|_A \le \gamma. \tag{30}$$

Proof: Define a matrix X in such a way that the columns of X and $-X$ generate the polyhedron $\mathbf{X}$. We may assume that X does not contain zero columns. From (28) there are matrices $\overline{A}$ and $\overline{C}_3$ satisfying

$$(I + \lambda\hat{A})X = X\overline{A} - L\overline{C}_3, \text{ and} \tag{31}$$

$$\left\|\left[\begin{matrix}\overline{A}\\ \overline{C}_3\end{matrix}\right]\right\|_1 \leq 1, \tag{32}$$

where $\|\bullet\|_1$ is the matrix 1-norm defined as the maximal absolute column sum. Define matrices

$$A = (\overline{A} - I)/\lambda, \text{ and} \tag{33}$$

$$C_3 = \overline{C}_3/\lambda. \tag{34}$$

From (29) it follows that $M \in (\mathrm{conv}(\pm L, \boldsymbol{X}))^{\circ\circ} = (\mathrm{conv}(\pm L, \boldsymbol{X}))$. Hence there are B_4 and D_{34} such that

$$M = -XB_4 + LD_{34}, \text{ and} \tag{35}$$

$$\left\|\left[\begin{matrix}B_4\\ D_{34}\end{matrix}\right]\right\|_1 \leq \gamma. \tag{36}$$

Then it will be shown that $\hat{\Phi}(s) = [A, B_4, C_3, D_{34}]$ satisfies the interpolation condition, i.e., $\hat{\Phi} = \hat{T}_1 - \hat{T}_2\hat{Q} \in \mathcal{S}$ for some $\hat{Q} \in \mathcal{S}$, and $\|\hat{\Phi}\|_A \leq \gamma$. It is shown in Lemma 2 below that A is a stable matrix and that $\|\hat{\Phi}\|_A \leq g$. For the interpolation condition, the (A, B) part of a realization of $\hat{T}_2^{-1}(\hat{\Phi} - \hat{T}_1)$ is:

$$\left[\left[\begin{matrix}\hat{A} & LC_3 & LC_1\\ 0 & A & 0\\ 0 & 0 & A_1\end{matrix}\right]\left[\begin{matrix}L(D_{34} - D_1)\\ B_4\\ -B_1\end{matrix}\right]\right]. \tag{37}$$

Let T be a nonsingular matrix

$$T = \left[\begin{matrix}I & -X & R\\ 0 & I & 0\\ 0 & 0 & I\end{matrix}\right]. \tag{38}$$

Using (17), (18), (31), (33), (34), and (35), (37) is transformed into

$$\left[T\left[\begin{matrix}\hat{A} & LC_3 & LC_1\\ 0 & A & 0\\ 0 & 0 & A_1\end{matrix}\right]T^{-1}\; T\left[\begin{matrix}L(D_{34} - D_1)\\ B_4\\ -B_1\end{matrix}\right]\right]$$
$$= \left[\left[\begin{matrix}\hat{A} & 0 & 0\\ 0 & A & 0\\ 0 & 0 & A_1\end{matrix}\right]\left[\begin{matrix}0\\ B_4\\ -B_1\end{matrix}\right]\right]. \tag{39}$$

Since A and A_1 are stable, it follows that $\hat{Q} = \hat{T}_2^{-1}(\hat{\Phi} - \hat{T}_1)$ is stable.

Lemma 2. The matrix A of (33) is stable. The transfer function $\hat{\Phi}(s) = [A, B_4, C_3, D_{34}]$ defined in the proof of Lemma 1 satisfies

$$\|\hat{\Phi}\|_A \leq \gamma. \tag{40}$$

Theorem 1. Suppose that the zeros of $\hat{T}_2$ lie in the open right half plane. Then the two model matching problems (7) and (8) take the same optimal values, i.e.

$$\gamma_{MM} = \gamma_{RMM}. \tag{41}$$

Since the set of stable rational function $\mathfrak{S}$ is not dense in $\mathfrak{A}$, a minimizing sequence of the rational model matching problem (8) may not converge to a minimizer of the model matching problem (7) in the norm topology. This is what happens usually, because the optimal objective transfer function (e.g. weighted sensitivity) is a finite sum of pure delays (see (4)). On the other hand, Theorem 1 says that the two problems attain the same optimal value.

6. Conclusions

It was shown that the continuous-time L_1-optimal control problems admit a sequence of rational controllers that attains the optimal performance index. Since the set of rational stable transfer functions $\mathfrak{S}$ is not dense in the algebra $\mathfrak{A}$, the approximation is not in the norm topology. Also a way to construct such an approximating sequence was presented.

References

[1] F.M. Callier and C.A. Desoer, "An algebra of transfer functions of distributed linear time-invariant systems." IEEE Trans. Circuits Syst., vol.CAS-25, pp.651-662, 1978.

[2] M.A. Dahleh and J.B. Pearson, "l^1 optimal feedback controllers for MIMO discrete-time systems," IEEE Trans. Automat. Contr., vol.AC-32, pp.314-322, 1987.

[3] M.A. Dahleh and J.B. Pearson, "L1-optimal compensators for continuous-time systems," IEEE Trans. Automat. Contr., vol.AC-32, pp.889-895, 1987.

[4] C.A. Desoer and M. Vidyasagar, Feedback Systems: Input-Output Properties, Orlando, Florida: Academic Press, 1975.

[5] D.G. Luenberger, Optimization by Vector Space Methods, New York: John Wiley & Sons, 1969.

[6] R.T. Rockafellar, Convex Analysis, Princeton, New Jersey: Princeton, 1970.

[7] M. Vidyasagar, "Optimal rejection of persistent bounded disturbances," IEEE Trans. Automat. Contr., vol.AC-31, pp.527-534, 1986.

[8] M. Vidyasagar, "Further results on the optimal rejection of persistent bounded disturbances, Part II: Continuous-time case," preprint.

Robust Stability of Sampled Data Systems

Mituhiko Araki and Tomomichi Hagiwara

Department of Electrical Engineering, Kyoto University,
Yoshida, Sakyo-ku, Kyoto 606-01, Japan

Abstract: In this paper, we study robust stability and stabilizability of sampled-data control systems composed of a continuous-time plant and a discrete-time compensator together with sample/hold devices, using the frequency domain method. Our study is not confined only to the case of standard (i.e., single-rate) sampled-data controllers but also extended to the case of multirate sampled-data controllers. Specifically, we compare, by numerical examples, the tolerable amounts of the uncertainty of a given continuous-time plant for which the robust stabilization can be attained in the three cases: the cases of a continuous-time controller, a single-rate sampled-data controller, and a multirate sampled-data controller. The result will be useful for estimating the deterioration caused by use of sampled-data controllers in place of continuous ones, or evaluating the advantages of the multirate sampled-data scheme over the standard one.

1 Introduction

To ensure robust stability of closed-loop systems is one of the most important requirements in the control system design. The robust stability condition of continuous-time closed-loop systems and the robust stabilizability condition of continuous-time plants with bounded uncertainty are clarified in [1] and [2], respectively. However, to the knowledge of the authors, much has not been obtained about robust stability and stabilizability of sampled-data control systems composed of a continuous-time plant and a discrete-time compensator together with sample/hold devices. This paper aims at presenting a frequency domain method to analyze robust stability and stabilizability of sampled-data control systems on the same basis with continuous-time control systems. Our study is not confined only to the case of standard (i.e., single-rate) sampled-data controllers but also extended to the case of multirate sampled-data controllers [3],[4],[5]. Specifically, we compare, by numerical examples, the tolerable amounts of the uncertainty of a given continuous-time plant for which the robust stabilization can be attained in the three cases: the cases of a continuous-time controller, a single-rate sampled-data controller, and a multirate sampled-data controller. The result will be useful for estimating the deterioration caused by use of sampled-data controllers in place of continuous ones, or evaluating the advantages of the multirate sampled-data scheme over the standard one.

2 Multirate Sampled-Data Controllers

Many types of multirate sampled-data controllers have been studied in the literature [3]–[8], but in this paper we only deal with the multirate input sampled-data controllers which will be described in the following. Note that this controller reduces to the standard one if all the input multiplicities are set to 1.

We suppose that the given plant is an n-th order controllable, observable system with m inputs and p outputs described by

$$\frac{dx}{dt} = Ax + Bu, \quad y = Cx. \tag{1}$$

We sample all the plant outputs with sampling period T_0, and we place a zero-order hold circuit $h_i(s) = (1 - e^{-T_i s})/s$ preceded by a sampler with sampling period T_i at the i-th plant input, where

$$T_i = T_0/N_i \quad (N_i\text{: integer}). \tag{2}$$

Thus, the i-th plant input is changed N_i times during the interval T_0 with which the output is sampled. We refer to T_0 as frame period and N_i as input multiplicity. Let $u^{\mathrm{D}}(k)$ be the vector composed of the all inputs applied during the frame interval $[kT_0, \overline{k+1}T_0)$:

$$u^{\mathrm{D}}(k) := \left[u_1^{\mathrm{D}}(k)^T, \ \cdots, \ u_m^{\mathrm{D}}(k)^T\right]^T, \tag{3}$$

$$u_i^{\mathrm{D}}(k) := \left[u_i(kT_0), \ \cdots, \ u_i(kT_0 + \overline{N_i - 1}T_i)\right]^T$$
$$(i = 1, \cdots, m), \tag{4}$$

and let $y^{\mathrm{D}}(k) := y(kT_0)$. Denoting the z-transforms, with respect to the frame period, of $u^{\mathrm{D}}(k)$ and $y^{\mathrm{D}}(k)$ by $U^{\mathrm{D}}(z)$ and $Y^{\mathrm{D}}(z)$, respectively, we have

$$Y^{\mathrm{D}}(z) = G^{\mathrm{D}}(z)U^{\mathrm{D}}(z) \tag{5}$$

where $G^{\mathrm{D}}(z)$ is the pulse transfer function matrix associated with the expanded discrete-time realization [9] of the plant (1). We call $G^{\mathrm{D}}(z)$ the "pulse transfer function of the multirate plant" or, simply, "multirate plant."

With the above sampled-data scheme, we use the control law given by by

$$\begin{aligned} w(k+1) &= A_w w(k) + B_w y^{\mathrm{D}}(k), \\ u^{\mathrm{D}}(k) &= -C_w w(k) - D_w y^{\mathrm{D}}(k), \end{aligned} \tag{6}$$

where w is the state of the controller with an appropriate dimension. The controller with the above structure is called multirate input sampled-data controller. We denote the pulse transfer function of the controller by $G_w^{\mathrm{D}}(z) = C_w(zI - A_w)^{-1}B_w + D_w$. Our control system is equivalent to an ordinary digital closed-loop system given in Fig. 1. The above controller reduces to the usual single-rate sampled-data controller if $N_i = 1$ for $i = 1, \cdots, m$.

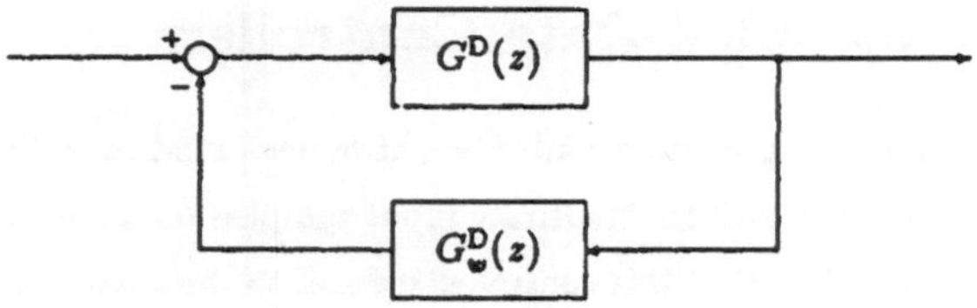

Fig. 1: Multirate sampled-data control system

3 Multirate Plant Uncertainty

In this section we study how the upper bound of the uncertainty of the multirate plant which originates from the uncertainty of the continuous-time plant is. Let $G_0(s)$ be the nominal model of the continuous-time plant; i.e., $G_0(s) = C(sI - A)^{-1}B$, and assume that the transfer function matrix $G(s)$ of the actual plant is given by

$$G(s) = G_0(s) + \Delta G(s), \quad \pi_G = \pi_{G_0} \tag{7}$$

where $\Delta G(s)$ is bounded on the imaginary axis:

$$\|\Delta G(j\omega)\| < |\ell(j\omega)| \quad (^\forall\omega), \tag{8}$$

where $\ell(s)$ is a strictly proper rational function, and π_G and π_{G_0} denote the number of the unstable poles of $G(s)$ and that of $G_0(s)$, respectively. In correspondence to the above uncertainty of $G(s)$, the multirate plant $G^D(z)$ is expected to have the uncertainty $\Delta G^D(z)$ of the form

$$G^D(z) = G_0^D(z) + \Delta G^D(z). \tag{9}$$

We want to evaluate the upper bound of $\|\Delta G^D(e^{j\omega T_0})\|$ for $0 \leq \omega \leq 2\pi/T_0$, where the upper bound should be taken over all possible $G(s)$, or equivalently $\Delta G(s)$, given by (8). Now, using the notion of symmetric-coordinate multirate impulse modulation [9] as demonstrated in [10], and defining

$$\omega_i = 2\pi/T_i \quad (i = 0, 1, \cdots, m), \tag{10}$$

we can evaluate this quantity as

$$\|\Delta G^D(e^{j\omega T_0})\| < \ell^D(\omega) \tag{11}$$

where $\ell^D(\omega)$ is given as follows.

(i) The case where $N_i = N_0$ $(i = 1, \cdots, m)$

$$\ell^D(\omega) = \sqrt{\sum_{k=0}^{N_0-1} \frac{1}{N_0} \{\ell_0^D(\omega - k\omega_0)\}^2} \tag{12}$$

where

$$\ell_0^{\rm D}(\omega) = \sum_{k=-\infty}^{\infty} \left| \frac{2}{(\omega T_0/N_0) + 2k\pi} \sin \frac{\omega T_0/N_0}{2} \right| \left| \ell(j\overline{\omega + kN_0\omega_0}) \right| \tag{13}$$

(ii) The case where $N_i \neq N_j$ for some i and j

$$\ell^{\rm D}(\omega) = \sqrt{\sum_{i=1}^{m} \sum_{k=0}^{N_i-1} \frac{1}{N_i} \{\ell_i^{\rm D}(\omega - k\omega_0)\}^2} \tag{14}$$

where

$$\ell_i^{\rm D}(\omega) = \sum_{k=-\infty}^{\infty} \left| \frac{2}{\omega T_i + 2k\pi} \sin \frac{\omega T_i}{2} \right| \left| \ell(j\overline{\omega + k\omega_i}) \right| \tag{15}$$

Note that (12) and (14) coincide with each other if $m = 1$ but that the latter does not reduce to the former (but to the former multiplied by $m^{1/2}$) when $N_i = N_0$ for $i = 1, \cdots, m$. This is because our evaluation of the above $\ell^{\rm D}(\omega)$ is, in general, conservative and derivations of (i) and (ii) use different approaches to taking multirate structure into account in order to make the result as less conservative as possible. Obtaining least conservative evaluation would be a future problem, while it seems quite difficult. For the derivation of (12) and (14) in the general case, the readers are referred to [10], but the derivation of (12) for the case of $N_0 = 1$ (this is the case of the ordinary single-rate sampled-data plant) shall be given in the following to show the base idea. First, note that

$$Z[G(s)H(s)] = Z[G_0(s)H(s)] + Z[\Delta G(s)H(s)] \tag{16}$$

follows from the linearity of z-transformation as in (9), where $Z[\cdot]$ denotes z-transformation and $H(s)$ denotes the zero-order hold with sampling period T_0, where $Z[\Delta G(s)H(s)] =: \Delta G^*(z)$ is the uncertainty for the single-rate plant $Z[G_0(s)H(s)]$. Applying the impulse modulation formula, we obtain

$$\Delta G^*(e^{j\omega T_0}) = \frac{1}{T_0} \sum_{k=-\infty}^{\infty} H(j\overline{\omega + k\omega_0}) \Delta G(j\overline{\omega + k\omega_0}) \tag{17}$$

where $\omega_0 := 2\pi/T_0$ denotes the sampling frequency. Evaluating the norm of the series on the right side, we obtain

$$\begin{aligned} \|\Delta G^*(e^{j\omega T_0})\| &\leq \sum_{k=-\infty}^{\infty} \left| \frac{1 - e^{-j\omega T_0}}{j\omega + k\omega_0 T_0} \right| \|\Delta G(j\overline{\omega + k\omega_0})\| \\ &< \sum_{k=-\infty}^{\infty} \left| \frac{2}{\omega T_0 + 2k\pi} \sin \frac{\omega T_0}{2} \right| \left| \ell(j\overline{\omega + k\omega_0}) \right| \end{aligned} \tag{18}$$

where we used the inequality (8). The right hand side of (18) is nothing but (12) for $N_0 = 1$. In [11], a similar method was used to derive an upper bound for the uncertainty of a discretized plant, where multiplicative uncertainty was assumed.

4 Robust Stabilizability—Numerical Examples

The upper bound of the uncertainty of the multirate plants (including the single-rate plants as a special case) given in the previous section enables us to compare the robustly-stabilizing ability of continuous-time controllers, single-rate sampled-data controllers, and multirate sampled-data controllers on the same basis (i.e., on the basis of the uncertainty assumption about the continuous-time plants). In this section, we make such comparison by numerical examples. We first consider the following three scalar plants $G_0(s)$ with uncertainty bound $\ell(s)$.

$$\left.\begin{array}{ll} \text{P1.} & 1/(s-1) \\ \text{P2.} & 1/(s-5) \\ \text{P3.} & 1/(s-10) \end{array}\right\}, \quad \ell(s) = \frac{k}{1+s} \; (k > 0) \tag{19}$$

For each plant, the upper bound of k for which the plant is robustly stabilizable by a continuous-time linear time-invariant controller is given as follows, where the upper bounded is denoted by k_c.

Plant	P1	P2	P3
k_c	1.0	0.6	0.55

(20)

Next, we examine the upper bound of k for which the plant is robustly stabilizable by a multirate input sampling controller with a given input multiplicity (including a single-rate sampled-data controller as a special case). As described in the previous section, we can derive an upper bound $\ell^{\mathrm{D}}(\omega)$ for the multirate plant corresponding to the given input multiplicity (Note that (12) and (14) coincide for scalar plants). In general, however, the obtained upper bound $\ell^{\mathrm{D}}(\omega)$ can not be expressed in the form of $|f(e^{j\omega T_0})|$ with any rational function $f(z)$. Since this is not convenient for studying robust stabilizability, we seek a rational function $f(z)$ such that $|f(e^{j\omega T_0})| \geq \ell^{\mathrm{D}}(\omega)$, yet $|f(e^{j\omega T_0})| \approx \ell^{\mathrm{D}}(\omega)$. To get such $f(z)$ graphically, we introduce the bilinear transformation

$$\chi = \frac{z-1}{z+1}, \quad z = \frac{1+\chi}{1-\chi}. \tag{21}$$

Under this transformation, $z = e^{j\omega T_0}$ is mapped to $\chi = j\xi$ and vice versa, where $\xi = \tan\frac{\omega T_0}{2}$, or equivalently,

$$\omega = \frac{2}{T_0}\arctan\xi. \tag{22}$$

Therefore, we obtain

$$\ell^{\mathrm{D}}(\omega) = \ell^{\mathrm{D}}\left(\frac{2}{T_0}\arctan\xi\right) =: \ell^{\mathrm{D}}_{\xi}(\xi). \tag{23}$$

By plotting this $\ell^{\mathrm{D}}_{\xi}(\xi)$ $(0 < \xi < \infty)$ on Bode diagram, we can see that it can be very accurately approximated (and also bounded from above) by $|\ell^{\mathrm{D}}_{\chi}(j\xi)|$ with

$$\ell^{\mathrm{D}}_{\chi}(\chi) = K_{\ell}\frac{1+T_{\ell}\chi}{1+\beta T_{\ell}\chi}, \tag{24}$$

where β, T_l and K_l are constant numbers depending on the frame period T_0 and the input multiplicity N_0. Since $\ell_\chi^{\mathrm{D}}(\chi)$ is a rational function of χ, the desired $f(z)$ is obtained by the inverse bilinear transformation of $\ell_\chi^{\mathrm{D}}(\chi)$. Thus, we can calculate an upper bound of k for which the robust stabilizability of the plant by a multirate (or single-rate if $N_0 = 1$) sampled-data controller is assured. However, it should be noted that we are dealing with only sufficient condition for robust stabilizability, since the upper bounds (12) and (14) are, in general, conservative as mention in the previous section (This is also the case in all examples to follow, except the case of continuous-time controllers). The calculated results are as follows, where the upper bound is denoted by k_m.

k_m (P1)			k_m (P2)			k_m (P3)		
N_0	$T_0 = 0.2$	$T_0 = 0.8$	N_0	$T_0 = 0.2$	$T_0 = 0.8$	N_0	$T_0 = 0.2$	$T_0 = 0.8$
1	0.8177	0.4383	1	0.2072	6.290×10^{-3}	1	0.05846	5.911×10^{-5}
2	0.8185	0.4466	2	0.2132	7.935×10^{-3}	2	0.06436	8.241×10^{-5}
3	0.8244	0.4483	3	0.2144	8.506×10^{-3}	3	0.06573	9.593×10^{-5}
4	0.8186	0.4488	4	0.2148	8.746×10^{-3}	4	0.06622	1.036×10^{-4}

(25)

From the above result, we see that the robustly-stabilizing ability of multirate input sampling controller is higher than that of single-rate sampled-data controller (which is a natural consequence from theoretical point of view), and that the relative difference of the ability between single-rate sampled-data controllers and multirate sampled-data controllers with higher input multiplicity becomes larger as the unstable pole of the plant and/or the frame period become larger.

The effect of the frame period on the robustly-stabilizing ability of multirate input sampling controller is also studied for the plant P3, and the results are given as follows.

T_0	0.1	0.01	0.001
$N_0 = 1$	0.1887	0.4974	0.5445
$N_0 = 4$	0.1955	0.4972	0.5442

(26)

The results indicate that the upper bound of k approaches the upper bound for continuous-time linear time-invariant controllers, k_c, regardless of the value of N_0 (i.e., including the case of single-rate sampled-data controllers). For single-rate sampled-data controllers, this result has been shown by Hara et al. [12] in a different approach.

Next, we consider the case of the following multivariable plants.

$$\left.\begin{array}{l} \text{P4.}\quad G_{4o}(s) = \begin{bmatrix} \frac{1}{s-10} & 0 \\ 0 & \frac{1}{s+1} \end{bmatrix} \\ \text{P5.}\quad G_{5o}(s) = \begin{bmatrix} \frac{1}{s-10} & \frac{1}{s-10} \\ \frac{1}{s+1} & \frac{1}{s+1} \end{bmatrix} \end{array}\right\} , l_o(s) = \frac{k}{s+1} \tag{27}$$

For each plant, the upper bound of k for which the plant is robustly stabilizable by a continuous-time linear time-invariant controller is as follows.

Plant	P4	P5
k_c	0.55	0.7778

(28)

As in the case of scalar plants, we first calculate $\ell^D(\omega)$ and then find graphically a rational function $f(z)$ such that $|f(e^{j\omega T_0})| \geq \ell^D(\omega)$, yet $|f(e^{j\omega T_0})| \approx \ell^D(\omega)$ using bilinear transformation and Bode diagram. Finally, we can calculate the upper bound of k such that the existence of a robustly stabilizing multirate sampled-data controller with given input multiplicities is ensured, and the results are as follows.

k_m (P4)			
N_1	N_2	$T_0 = 0.2$	$T_0 = 0.8$
1	1	0.5846×10^{-1}	0.5911×10^{-4}
2	2	0.6436×10^{-1}	0.8240×10^{-4}
2	1	0.3717×10^{-1}	0.4745×10^{-4}
1	2	0.4772×10^{-1}	0.4831×10^{-4}

k_m (P5)			
N_1	N_2	$T_0 = 0.2$	$T_0 = 0.8$
1	1	0.8267×10^{-1}	0.8360×10^{-4}
2	2	0.9102×10^{-1}	0.1165×10^{-3}
2	1	0.6049×10^{-1}	0.6772×10^{-4}
1	2	0.6049×10^{-1}	0.6772×10^{-4}

(29)

As in the case of scalar plants, the above results show that the relative difference of the robustly-stabilizing ability between single-rate sampled-data controllers and multirate sampled-data controllers is large when the frame period is large, provided that input multiplicities N_1 and N_2 are the same. However, when N_1 and N_2 are different, the range of k for which the existence of a robustly stabilizing controller can be ensured does not always increase as the input multiplicities become large. Theoretically speaking, the range never decreases as long as an input multiplicity becomes larger by multiplied by an integer. Therefore, the strange phenomena in the above results are due to the fact that the evaluation of $\ell^D(\omega)$ in (12) and (14) is conservative. Less conservative results could be obtained if we knew the upper bound for the continuous-time uncertainty $\Delta G(s)$ columnwise. However, details are omitted because of space limitation.

References

[1] M. J. Chen and C. A. Desoer, "Necessary and Sufficient Condition for Robust Stability of Linear Distributed Feedback Systems," *Int. J. Control*, Vol. 35, No. 2, pp. 255–267 (1982)

[2] M. Vidyasagar and H. Kimura, "Robust Controllers for Uncertain Linear Multivariable Systems," *Automatica*, Vol. 22, No. 1, pp. 85–94 (1986)

[3] A. B. Chammas and C. T. Leondes, "Pole Assignment by Piecewise Constant Output Feedback," *Int. J. Control*, Vol. 29, No. 1, pp. 31–38 (1979)

[4] P. T. Kabamba, "Control of Linear Systems Using Generalized Sampled-Data Hold Functions," *IEEE Trans. Automatic Control*, Vol. 32, No. 9, pp. 772–783 (1987)

[5] M. Araki and T. Hagiwara, "Pole Assignment by Multirate Sampled-Data Output Feedback," *Int. J. Control*, Vol. 44, No. 6, pp. 1661–1673 (1986)

[6] T. Mita, B. C. Pang and K. Z. Liu, "Design of Optimal Strongly Stable Digital Control Systems and Application to Output Feedback Control of Mechanical Systems," *Int. J. Control*, Vol. 45, pp. 2071–2082 (1987)

[7] T. Hagiwara and M. Araki, "Design of a Stable State Feedback Controller Based on the Multirate Sampling of the Plant Output," *IEEE Trans. Automatic Control*, Vol. 33, No. 9, pp. 812–819 (1988)

[8] M. Araki and T. Hagiwara, "Periodically Time-Varying Controllers," China-Japan Joint Symposium on Systems Control Theory and Its Application, Hangzhou, China (1989)

[9] M. Araki and K. Yamamoto, "Multivariable Multirate Sampled-Data Systems: State-Space Description, Transfer Characteristics, and Nyquist Criterion," *IEEE Trans. Automatic Control*, Vol. AC-31, No. 2, pp. 145–154 (1986)

[10] M. Araki, T. Hagiwara, T. Fujimura and Y. Goto, "On Robust Stability of Multirate Digital Control Systems," Proc. 29th Conference on Decision and Control, pp. 1923–1924, Honolulu, Hawaii (1990)

[11] C. E. Rohrs, G Stein and K. J. Astrom, "Uncertainty in Sampled Systems," Proc. American Control Conference, Boston, pp. 95–97 (1985)

[12] S. Hara, M. Nakajima and P. T. Kabamba, "Robust Stabilization in Sampled-Data Control Systems," Proc. 13th Dynamical System Theory Symposium (in Japanese), pp. 115–120 (1991)

Robust Control System Design for Sampled-Data Feedback Systems

Shinji HARA* and Pierre T. KABAMBA**
* Dept. of Control Engineering, Tokyo Institute of Technology,
Oh-okayama, Meguro-ku, Tokyo 152, Japan
** Aerospace Engineering Dept., The University of Michigan,
Ann Arbor, MI 48109–2140, USA

In this paper, we consider an optimization problem for sampled data control systems in the sense of the L_2 induced norm of the linear operator with continuous-time inputs and outputs. The problem is a worst case design and a counterpart of the H_∞-optimization problem for purely continuous-time or discrete-time systems. Hence, it can be applied to the robust controller design taking account of the intersample behavior for sampled-data feedback systems. We show that the optimization problem for a 4-block generalized continuous-time plant with a digital controller can be solved with a γ-iteration on a certain discrete-time 4-block plant which depends on γ. The computation algorithm with three exponentiations is also derived.

1 Introduction

The first motivation of this paper is the intersampling behavior of sampled data systems, by which we mean systems with continuous time inputs and outputs, but some state variables evolve in continuous time and other evolve in discrete time. Such systems are important in practice because of the widespread use of digital computers for active control (see e.g.[1]). Typically, sampled data systems are analyzed and designed through their discrete time behavior, i.e. their behavior at the sampling times. The classical tool for assessing intersampling performance is the modified z-transform, but it implicitly requires discretization of the inputs. While a few performance criteria (such as stability or deadbeat response) can be judged based on discrete time behavior, many others (such as disturbance and noise attenuation, overshoot, settling time, etc.) require a closer look at the intersampling behavior. Moreover, it has been shown that a sampled data system may very well have innocuous sample time dynamics, but unacceptable intersampling ripples (see e.g. [2], [3], [4], [5]). Recently, there has been several publications specifically accounting for the intersampling behavior of sampled data systems [2]–[14].

The second motivation of the paper is the recently developed theory of H_∞ analysis and design in the state space (see e.g. [15], [16] and the references therein). This approach considers that the signals of interest are square integrable, and aims at minimizing the worst possible amplification factor between selected signals. Mathematically, this corresponds to minimizing the L_2 induced norm of a linear operator. Although most of the results were first derived for linear time invariant continuous time systems, they have been extended to linear time invariant discrete-time systems (see e.g. [17], [18]). Such an extension, however, does not cover the important class of sampled data systems. The technical difficulty is that sampled data systems are inherently time varying. Note that optimal H_∞ design of time varying systems has been considered, however these results are not directly applicable to sampled data systems. Also, induced norms of sampled data systems are given in [9], [10], a treatment which is more operator-theoretic than the state space treatment of this paper.

In this paper, we show how to optimize the L_2 induced norm of a sampled data system obtained by using a digital controller on a standard 4-block analog plant. This problem is a worst case design and it is a counterpart of the H_∞-norm optimization problem for purely continuous or discrete-time systems [19], [15], [16]. It will be shown that the optimal attenuation problem for a 4-block continuous time plant with a digital controller can be solved with a γ-iteration on a certain discrete time 4-block plant which depends on γ. The computation algorithm with three exponentiations is also derived.

Throughout the paper, superscript T denotes matrix transpose, parentheses $(\ \cdot\ )$ around an independent variable indicate an analog function of continuous time or the Laplace transform of such a function, whereas square brackets $[\ \cdot\ ]$ indicate a discrete sequence or the z-transform of a sequence. $\sigma_{\max}(\ \cdot\ )$ indicates the maximum singular value of the matrix argument.

2 Optimization of the L_2 Induced Norm

Consider a sampled-data feedback control system which consists of a generalized continuous-time plant (including frequency shaped weighting matrices and antialiasing filters),

$$G(s) = \begin{bmatrix} G_{11}(s) & G_{12}(s) \\ G_{21}(s) & G_{22}(s) \end{bmatrix} \tag{2.1}$$

a discrete-time controller to be designed $K[z]$, a given hold function $H(t)$ and a sampler with sampling period $\tau > 0$. The state-space representations of $G(s)$ and $K[z]$ are given by

$$\begin{cases} \dot{x}(t) = Ax(t) + B_1r(t) + B_2u(t) \\ e(t) = C_1x(t) + D_{11}r(t) + D_{12}u(t) \\ y(t) = C_2x(t) + D_{21}r(t) + D_{22}u(t) \end{cases} \tag{2.2}$$

$$\begin{cases} x_d[k+1] = A_dx_d[k] + B_dy(k\tau) \\ v[k] = C_dx_d[k] + D_dy(k\tau) \end{cases} \tag{2.3}$$

respectively, where $w(t)$, $u(t)$, $z(t)$ and $y(t)$ denote the exogenous input, control input, controlled output and measured output (with appropriate dimensions), respectively, all of which are analog signals. The control input $u(t)$ is determined by

$$u(t) = H(t)v[k], \quad t \in [k\tau,\ (k+1)\tau) \tag{2.4}$$

where $H(t)$ is a τ-periodic matrix.

We make the following assumptions:

Assumptions A:

1C) (A, B_2) is stabilizable and (A, C_2) is detectable

1S) The sampling period τ is chosen so that $(\hat{A}, \hat{B}_2)$ is stabilizable and $(\hat{A}, \hat{C}_2)$ is detectable, where

$$\hat{A} = e^{A\tau}, \ \hat{B}_2 = \int_0^\tau e^{A(\tau-\xi)} B_2 H(\xi)\, d\xi, \ \hat{C}_2 = C_2 \tag{2.5}$$

2) $D_{21} = 0$ and $D_{22} = 0$

3) $D_d = 0$

4) The pair (A_d, C_d) is observable, and the matrix C_d is specified.

Let us discuss these assumptions in some detail. The first assumptions 1C) $\sim$ 1S) is required for stabilization. If 1C) is satisfied, then 1S) holds for almost all τ. The second assumption 2) is needed to help guarantee boundedness of the closed loop operator. In practice, the second and third assumptions are both quite reasonable because the sampler is always preceded by a low pass filter and the digital controller is often strictly proper for practical reasons. The fourth assumption is consistent with using a minimal realization of the compensator, for which there is no loss of generality in assuming we use an observable canonical form. It should be noticed that when the digital controller has a computational delay, the third and fourth assumptions are automatically satisfied. Indeed in that case, the digital controller has an observable realization of the form

$$\begin{cases} x_d[k+1] = \bar{A}_d x_d[k] + \bar{B}_d y(k\tau) \\ v[k+1] = \bar{C}_d x_d[k] + \bar{D}_d y(k\tau) \end{cases} \tag{2.6}$$

or equivalently, an observable realization

$$\begin{bmatrix} x_d[k+1] \\ v[k+1] \end{bmatrix} = \begin{bmatrix} \bar{A}_d & 0 \\ \bar{C}_d & 0 \end{bmatrix} \begin{bmatrix} x_d[k] \\ v[k] \end{bmatrix} + \begin{bmatrix} \bar{B}_d \\ \bar{D}_d \end{bmatrix} y(k\tau)$$

$$w[k] = \begin{bmatrix} 0 & I \end{bmatrix} \begin{bmatrix} x_d[k] \\ v[k] \end{bmatrix}$$

where $C_d = \begin{bmatrix} 0 & I \end{bmatrix}$ is specified and $D_d = 0$.

With Assumptions A, the feedback system is represented by

$$\begin{aligned} \dot{x}(t) &= Ax(t) + B_2 H(t) C_d x_d[k] + B_1 w(t) \\ x_d[k] &= A_d x_d[k] + B_d C_2 x(k\tau) \\ z(t) &= C_1 x(t) + D_{12} H(t) C_d x_d[k] + D_{11} w(t) \end{aligned} \tag{2.7}$$

For every compensator $K[z]$ which internally stabilizes (2.7), we define a performance index which is the $\mathcal{L}_2$ induced norm of the closed loop operator, that is

$$J(K) = \sup_{r \in \mathcal{L}_2} \left\{ \frac{\|z\|_2}{\|w\|_2} \right\}. \tag{2.8}$$

Our objective is to find a digital stabilizing controller $K[z]$ which achieves a specified attenuation level $\gamma \geq 0$, i.e. such that $J(K) < \gamma$. This is accomplished by applying the result in [13] to the sampled data feedback system (2.7) and yields the following development.

For every $\gamma > \sigma_{\max}(D_{11})$ define $\hat{G}_\gamma[z]$ as a fictitious 4-block discrete time plant, which *is independent of the compensator, but depends on* γ, and with state space realization

$$\begin{aligned} \hat{x}[k+1] &= \hat{A}\hat{x}[k] + \hat{B}_1\hat{w}[k] + \hat{B}_2\hat{u}[k] \\ \hat{z}[k] &= \hat{C}_1\hat{x}[k] + \hat{D}_{11}\hat{w}[k] + \hat{D}_{12}\hat{u}[k] \\ \hat{y}[k] &= \hat{C}_2\hat{x}[k] \end{aligned} \tag{2.9}$$

where $\hat{A}$, $\hat{B}_2$, $\hat{C}_2$ are defined by (2.5) and other matrices in (2.9), $\hat{B}_1$, $\hat{C}_1$, $\hat{D}_{11}$ and $\hat{D}_{12}$, are obtaind as follows.

Let

$$\hat{B}_1 := \int_0^{\tau} e^{A(\tau-\xi)} B_1 \hat{F}_\gamma(\xi)\, d\xi \tag{2.10}$$

where $\hat{F}_\gamma(t)$ is given in Appendix B. Define

$$\begin{cases} \hat{C}_1(t) := C_1 e^{At} \\ \hat{C}_2(t) := \int_0^t C_1 e^{A(t-\xi)} B_1 \hat{F}_\gamma(\xi)\, d\xi + D_{11}\hat{F}_\gamma(t) \\ \hat{C}_3(t) := C_1 \int_0^t e^{A(t-\xi)} B_2 H(\xi)\, d\xi + D_{12} H(t) \end{cases} \tag{2.11}$$

and

$$\hat{M}_\gamma := \int_0^{\tau} \begin{bmatrix} \hat{C}_1^T(t) \\ \hat{C}_2^T(t) \\ \hat{C}_3^T(t) \end{bmatrix} \begin{bmatrix} \hat{C}_1(t) & \hat{C}_2(t) & \hat{C}_3(t) \end{bmatrix} dt$$

Since $\hat{M}_\gamma$ is positive semidefinite, it can be factored into conformably partitioned matrices

$$\hat{M}_\gamma = \begin{bmatrix} \hat{C}_1^T \\ \hat{D}_{11}^T \\ \hat{D}_{12}^T \end{bmatrix} \begin{bmatrix} \hat{C}_1 & \hat{D}_{11} & \hat{D}_{12} \end{bmatrix} \tag{2.12}$$

which completes the characterization of the fictitious 4-block plant $\hat{G}_\gamma[z]$.

We can now state the following result for the optimal 4-block sampled data synthesis problem.

Theorem 2.1 Suppose Assumpions A hold, and $\gamma > \sigma_{\max}(D_{11})$. Then the digital compensator $K[z]$ stabilizes (2.7) and achieves $J(K) < \gamma$ if and only if $K[z]$ stabilizes the fictitious digital plant $\hat{G}_\gamma[z]$ defined in (2.9)–(2.12) with attenuation level γ. By this we mean that (2.9)–(2.12) together with the feedback law

$$\begin{aligned} \hat{x}_d[k+1] &= A_d\hat{x}_d[k] + B_d\hat{y}[k] \\ \hat{u}[k] &= C_d\hat{x}_d[k] \end{aligned} \tag{2.13}$$

yields an intenally stable closed loop satisfying

$$\sup_{\hat{w}\in l_2^{m}} \left\{ \frac{\sum_{k=0}^{\infty} \hat{z}^T[k]\hat{z}[k]}{\sum_{k=0}^{\infty} \hat{w}^T[k]\hat{w}[k]} \right\} < \gamma^2 \tag{2.14}$$

or equivalently

$$\|G_{\hat{z}\hat{w}}\|_\infty < \gamma \tag{2.15}$$

Remark 2.1 All the matrices in (2.9)–(2.12) are independent of the parameters of the controller $K[z]$ to be designed, except for a dependence on C_d which has been assumed specified in Assumption A. However, they depend on $H(t)$ and γ as well as the parameters of $G(s)$. The practical implication of Theorem 2.1 is therefore that an optimal digital controller $K[z]$ can be designed using a discrete time version of the γ-iteration proposed in [15], *but with a plant which depends on* γ. The computation is initialized by finding an upper bound for the achievable attenuation; and this can be done very simply by finding a digital stabilizer for $G(s)$ and computing its attenuation level γ_0. At each step, one decrements γ, computes the realization (2.9)–(2.12) of the fictitious plant $\hat{G}_\gamma[z]$, and tests whether there exists a digital stabilizer for $\hat{G}_\gamma[z]$ which achieves attenuation level γ. This test involves solving two discrete Ricatti equations, and is constructive [17], [18]. When the test is successful, the digital controller is also guaranteed to stabilize and achieve attenuation level γ when applied to $G(s)$; otherwise, attenuation level γ cannot be achieved by any LTI digital compensator in Fig.4.1.

Remark 2.2 An optimal digital controller can also be designed using a bisection algorithm. Start with a lower bound and an upper bound of the optimal attenuation, e.g. $\sigma_{\max}(D_{11})$ and the same upper bound as in Remark 2.1, respectively. At every step, define γ as the center of the current bounding interval; test whether there exists a digital compensator which achieves attenuation level γ using Proposition 2.1; and update the lower or upper bound accordingly.

3 Computational Algorithm

We now show the computational algorithm with three exponentiations for deriving the equivalent discrete-time H_∞ problem, where we assume that $H(t) = I$, i.e., zero-order hold case. The algorithm can be derived from the main theorem in Section 2 and the following relations:

$$\int_0^\tau e^{A\xi} B d\xi = \begin{bmatrix} 0 & I \end{bmatrix} \exp\left\{ \begin{bmatrix} A & B \\ 0 & 0 \end{bmatrix} \tau \right\} \begin{bmatrix} I \\ 0 \end{bmatrix} \tag{3.1}$$

$$\int_0^\tau e^{A^T\xi} Q e^{A\xi} d\xi = \Phi^T \Gamma^T \tag{3.2}$$

where

$$\exp\left\{ \begin{bmatrix} -A^T & Q \\ 0 & A \end{bmatrix} \tau \right\} =: \begin{bmatrix} * & \Gamma \\ 0 & \Phi \end{bmatrix} \tag{3.3}$$

< Computational Algorithm >

• Step 1: Computation of W_γ: $W_\gamma = \Phi_\gamma^T \Gamma_\gamma$

$$\exp\left\{\begin{bmatrix} -A_V^T & C_V^T C_V \\ 0 & A_V \end{bmatrix} \tau\right\} = \begin{bmatrix} * & \Gamma_\gamma \\ 0 & \Phi_\gamma \end{bmatrix}$$

where

$$A_V = \begin{bmatrix} A + B_1 R_\gamma^{-1} D_{11}^T C_1 & -B_1 R_\gamma^{-1} B_1^T & B_2 + B_1 R_\gamma^{-1} D_{12} \\ C_1^T C_1 + C_1^T D_{11} R_\gamma^{-1} D_{11}^T C_1 & -A^T - C_1^T D_{11} R_\gamma^{-1} B_1^T & C_1^T (I + D_{11} R_\gamma^{-1} D_{11}^T) D_{12} \\ 0 & 0 & 0 \end{bmatrix}$$

$$C_V = \begin{bmatrix} R_\gamma^{-1} D_{11}^T C_1 & -R_\gamma^{-1} B_1^T & R_\gamma^{-1} D_{11}^T D_{12} \end{bmatrix}$$

$$R_\gamma = \gamma^2 I - D_{11}^T D_{11}$$

• Step 2: Computation of B_{s1}:

$$B_{s1} = \begin{bmatrix} C_B & 0 \end{bmatrix} \exp\{A_B \tau\} B_B$$

where

$$A_B = \begin{bmatrix} A & B_1 C_V \\ 0 & A_V \end{bmatrix}, \quad B_B = \begin{bmatrix} 0 \\ W_\gamma^{-1/2} \end{bmatrix}, \quad C_B = \begin{bmatrix} I_n & 0 \end{bmatrix}$$

• Step 3: Computation of $\begin{bmatrix} C_{s1} & D_{s11} & D_{s12} \end{bmatrix} := M_\gamma^{1/2}$:
Let

$$\exp\left\{\begin{bmatrix} -A_N^T & C_N^T C_N \\ 0 & A_N \end{bmatrix} \tau\right\} = \begin{bmatrix} * & \Gamma_N \\ 0 & \Phi_N \end{bmatrix}$$

and define $\hat{M}_\gamma = \Phi_N^T \Gamma_N$. Then, we have

$$M_\gamma = B_N^T \hat{M}_\gamma B_N$$

where

$$A_N = \begin{bmatrix} A & B_1 C_V & B_2 \\ 0 & A_V & 0 \\ 0 & 0 & 0 \end{bmatrix}, \quad B_N = \begin{bmatrix} I & 0 & 0 \\ 0 & W_\gamma^{-1/2} & 0 \\ 0 & 0 & I \end{bmatrix}$$

$$C_N = \begin{bmatrix} C_1 & D_{11} C_V & D_{12} \end{bmatrix}$$

Remark 1: The sizes of the exponentiations in steps 1, 2 and 3 are $(2n + m_2) \times 2$, $(3n + m_2)$, and $(3n + 2m_2) \times 2$, respectively.
Remark 2: Though $\hat{G}_\gamma[z]$ depends on γ, we can apply the γ iteration for the optimization.

4 Conclusions

The work reported herein represents an attempt to extend the H_∞ methodology (worst case design) to sampled data systems, so as to make it applicable to the common situation where a digital controller regulates an analog plant.

This work obviously leads to original research problems and possible extensions. Among these, we will mention the development of an extensive theory of sampled data systems, where each signal is assumed to have an analog and a discrete part. Other obvious extensions are the use of periodic or multirate digital controllers and the optimization of the induced norm with respect to the hold function, in the spirit of [20], [3]. These questions are under study and will be reported on in the future.

References

[1] G.F. Franklin and J.D. Powell: Digital Control of Dynamic Systems, Addison-Wesley (1980)

[2] S. Urikura, and A. Nagata: Ripple-Free Deadbeat Control for Sampled Data Systems, IEEE Trans. Auto. Contr., vol.32, 474/482 (1987)

[3] Y.C. Juan, and P.T. Kabamba: Optimal Hold Functions for Sampled Data Regulation, to appear in Automatica

[4] Y. Yamamoto: New Approach to Sampled-Data Control Systems — A Function Space Method; Proc. 29th CDC, 1882/1887 (1990)

[5] H.K. Sung and S. Hara: Ripple-free Condition in Sampled-Data Control Systems; Proc. 13th Dynamical System Theory Symp., 269/272 (1991)

[6] B.A. Francis, and T.T. Georgiou: Stability Theory for Linear Time Invariant Plants with Periodic Digital Controllers, IEEE Trans. Auto. Contr., vol. 33, 820/832 (1988)

[7] T. Chen, and B.A. Francis: Stability of Sampled Data Feedback Systems, IEEE Trans. Auto. Contr., vol.36, 50/58 (1991)

[8] T. Chen, and B.A. Francis: H_2 Optimal Sampled Data Contol, Preprint (1990)

[9] T. Chen, and B.A. Francis: On the L_2 Induced Norm of Sampled Data System, Systems & Control Letters, vol.15, 211/219 (1990)

[10] T. Chen, A. Feintuch and B.A. Francis: On the Existence of H_∞-Optimal Sampled-Data Controllers, Proc. 29th CDC, 1794/1795, Honolulu (1990)

[11] S. Hara and P.T. Kabamba: Worst Case Analysis and Design of Sampled Data Control Systems, Proc. 12th Dynamical System Theory Symp., 167/172, (1989)

[12] P.T. Kabamba and S.Hara: On computing the induced norm of sampled data systems, Proc. 1990 ACC, San Diego, CA, May, 1990

[13] S. Hara and P.T. Kabamba: Worst Case Analysis and Design of Sampled Data Control Systems, 29th IEEE CDC, Honolulu, 202/203 (1990)

[14] B. Bamieh and J.B. Pearson: A General Framework for Linear Periodic Systems with Application to H_∞ Sampled-Data Control, Technical report No.9021, Rice University, November, 1990

[15] K. Glover and J.C. Doyle: State-Space Formulae for Stabilizing Controllers that Satisfy a H_∞ Norm Bound and Relations to Risk Sensitivity, Systems and Control Problems, IEEE Trans. AC-34-8, 831/897, (1989)

[16] J.C. Doyle, K. Glover, P.P. Khargonekar, and B.A. Francis: State Space Solutions to Standard H_2 and $H\infty$ Control Problems, IEEE Trans. Auto. Contr., vol. 34, 831/847 (1989)

[17] R. Kondo, S. Hara, and T. Itou: Characterization of Discrete Time $H\infty$ Controllers via Bilinear Transformation, Proc. 29th Conf. Dec. Contr., Honolulu, Hawaii, (1990)

[18] R. Kondo and S. Hara: A Unified Approach to Continuous/Discrete Time H_∞ Problems via J-Spectral Factorizations, to be presented at MTNS '91

[19] S.P. Boyd, V. Balakrishnan, and P.T. Kabamba: A Bisection Method for Computing the H_∞ Norm of a Transfer Matrix and Related Problems, Math. Contr. Sig. Syst., vol.2, 207/219 (1989)

[20] P.T. Kabamba: Control of Linear Systems Using Generalized Sampled Data Hold Functions, IEEE Trans. Auto. Contr., vol.32, 772/783 (1987)

[21] W. Sun et al: H_∞ Control and Filtering with Sampled Measurements, Proc. ACC '91 (1991)

Appendix E

Define the following matrices of function γ.

$$\hat{R}_\gamma := \gamma^2 I - D_{11}^T D_{11}$$

$$\hat{E} := \begin{bmatrix} A + B_1 \hat{R}_\gamma^{-1} D_{11}^T C_1 & -B_1 \hat{R}_\gamma^{-1} B_1^T \\ C_1^T C_1 + C_1^T D_{11} \hat{R}_\gamma^{-1} D_{11}^T C_1 & -A^T - C_1^T D_{11} \hat{R}_\gamma^{-1} B_1^T \end{bmatrix}$$

$$\begin{bmatrix} \hat{G}_1(t) \\ \hat{G}_2(t) \end{bmatrix} := \begin{bmatrix} (B_2 + B_1 \hat{R}_\gamma^{-1} D_{11}^T D_{12}) H(t) C_d \\ C_1^T (I + D_{11} \hat{R}_\gamma^{-1} D_{11}^T) D_{12} H(t) C_d \end{bmatrix} \tag{E.1}$$

and define

$$\begin{bmatrix} \hat{\Phi}_{11}(t) & \hat{\Phi}_{12}(t) \\ \hat{\Phi}_{21}(t) & \hat{\Phi}_{22}(t) \end{bmatrix} = e^{\hat{E}t}$$

$$\begin{bmatrix} \hat{\Phi}_{13}(t) \\ \hat{\Phi}_{23}(t) \end{bmatrix} = \int_0^t e^{\hat{E}(t-\xi)} \begin{bmatrix} \hat{G}_1(\xi) \\ \hat{G}_2(\xi) \end{bmatrix} d\xi \tag{E.2}$$

$$\begin{aligned} \hat{V}_{\gamma 1}(t) &:= D_{11}^T C_1 \hat{\Phi}_{11}(t) - B_1^T \hat{\Phi}_{21}(t) \\ \hat{V}_{\gamma 2}(t) &:= D_{11}^T C_1 \hat{\Phi}_{12}(t) - B_1^T \hat{\Phi}_{22}(t) \\ \hat{V}_{\gamma 3}(t) &:= D_{11}^T C_1 \hat{\Phi}_{13}(t) + D_{11}^T D_{12} H(t) C_d - B_1^T \hat{\Phi}_{23}(t) \end{aligned} \tag{E.3}$$

Without loss of generality, assume that the columns of $\hat{V}_\gamma(t)$ are linearly independent on $[0, \tau]$. (If this is not the case, just select a maximal set of linearly independent columns, and redefine $V_\gamma(t)$ accordingly). Then define

$$\begin{aligned} \hat{W}_\gamma &:= \int_0^\tau \hat{V}_\gamma^T(\xi) \hat{R}_\gamma^{-2} \hat{V}_\gamma(\xi)\, d\xi \\ \hat{F}_\gamma(t) &:= \hat{R}_\gamma^{-1} \hat{V}_\gamma(t) \hat{W}_\gamma^{-1/2} \end{aligned} \tag{E.4}$$

A Function State Space Approach to Robust Tracking for Sampled-Data Systems

Yutaka Yamamoto

Abstract

It is well known that tracking to continuous-time signals by sampled-data systems presents various difficulties. For example, the usual discrete-time model is not suitable for describing intersample ripples. In order to adequately handle this problem we need a framework that explicitly contains the intersample behavior in the model. This paper presents an *infinite-dimensional yet time-invariant* discrete-time model which contains the full intersample behavior as information in the state. This makes it possible to clearly understand the intersample as a result of a *mismatch* between the intersample tracking signal and the system zero-directional vector. This leads to an internal model principle for sampled-data systems, and some nonclassical feature arising from the interplay of digital and continuous-time behavior.

1 Introduction

It is well known that tracking to continuous-time signals by sampled-data systems presents various difficulties. For example, the usual discrete-time model is not suitable for describing intersample ripples. In order to adequately handle this problem we need a framework that explicitly contains the intersample behavior in the model. There are now several approaches in this direction: [1], [2], [4], [12].

This paper employs the approach of [12], which gives an *infinite-dimensional yet time-invariant* discrete-time model, and this model contains the full intersample behavior as information in the state (a similar approach is also employed by [7] for computing H^∞-norms). This makes it possible to clearly understand intersample ripples as a result of a *mismatch* between the intersample tracking signal and the system zero-directional vector. This leads to an internal model principle for sampled-data systems, and some nonclassical feature arising from the interplay of digital and continuous-time behavior.

2 Preliminaries

Let

$$\dot{x}(t) = Ax(t) + Bu(t), \quad y(t) = Cx(t) \tag{1}$$

be a given continuous-time system, and h a fixed sampling period. Given a function $\psi(t)$ on $[0,\infty)$, we define $\tilde{S}(\psi)$ as a sequence of functions on $[0,h)$ as follows:

$$\tilde{S}(\psi) := \{\psi_k(\theta)\}_{k=0}^{\infty}, \quad \psi_k(\theta) := \psi((k-1)h+\theta). \tag{2}$$

For such a sequence $\{\psi_k(\theta)\}_{k=1}^{\infty}$, its z-transform is defined as

$$\mathcal{Z}(\psi) = \sum_{k=0}^{\infty} \psi_k(\theta) z^{-k}. \tag{3}$$

We take $\tilde{S}(u) = u_k(\theta)$, $\tilde{S}(x) = x_k(\theta)$, $\tilde{S}(y) = y_k(\theta)$ as input, state, and output vectors (although infinite-dimensional), and give a discrete-time transition rule. Indeed, at time $t = kh$, let the state of the system (A,B,C) be $x_k(h)$ and let input $u_{k+1}(\theta)$ $(0 \le \theta < h)$ be applied to the system. Then the state and output trajectories $x_{k+1}(\theta)$ and $y_{k+1}(\theta)$ are given by the following equations:

$$\begin{aligned} x_{k+1}(\theta) &= e^{A\theta}x_k(h) + \int_0^{\theta} e^{A(\theta-\tau)}Bu_{k+1}(\tau)d\tau, \\ y_k(\theta) &= Cx_k(\theta), \quad 0 < \theta \le h. \end{aligned} \tag{4}$$

Let U be the space of piecewise continuous functions on $(0,h]$ that are right-continuous. Introducing the operators

$$\begin{aligned} F:\ & U^n \to U^n & &: x(\theta) \mapsto e^{A\theta}x(h), \\ G:\ & U^m \to U^n & &: u(\theta) \mapsto \int_0^{\theta} e^{A(\theta-\tau)}Bu(\tau)d\tau, \\ H:\ & U^n \to U^p & &: x(\theta) \mapsto Cx(\theta), \end{aligned} \tag{5}$$

equation (4) can be written simply as

$$x_{k+1} = Fx_k + Gu_{k+1}, \quad y_k = Hx_k. \tag{6}$$

Consider the hybrid control system depicted in Fig. 1. Here $C(z)$ and $P(s)$ denote discrete-time and continuous time systems (A_d, B_d, C_d) and (A_c, B_c, C_c), respectively. It is easy to

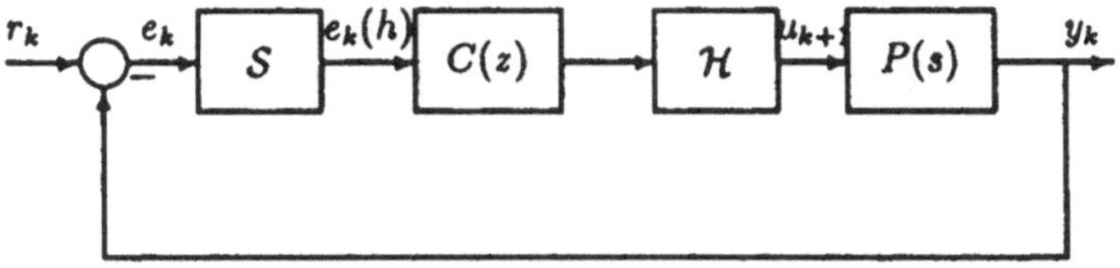

Figure 1: Hybrid Closed-Loop System Σ_d

see that the closed-loop system Σ_d obeys the following equations:

$$\begin{bmatrix} w_{k+1} \\ x_{k+1}(\theta) \end{bmatrix} = \begin{bmatrix} A_d & -B_d C_c \delta_h \\ B(\theta) C_d & e^{A_c\theta}\delta_h \end{bmatrix} \begin{bmatrix} w_k \\ x_k(\theta) \end{bmatrix} + \begin{bmatrix} B_d \delta_h \\ 0 \end{bmatrix} r_k(\theta) \tag{7}$$

$$e_k(\theta) = [0 \ -C_c] \begin{bmatrix} w_k \\ x_k(\theta) \end{bmatrix} + r_k(\theta) \tag{8}$$

where $B(\theta)$ is given by

$$B(\theta) := \int_0^\theta e^{A_c(\theta-\tau)} B_c H(\tau) d\tau, \tag{9}$$

where $H(t)$ is the holder function given on $[0, h]$ and δ_h denote the sampling at $\theta = h$:

$$\delta_h \cdot x(\theta) = x(h). \tag{10}$$

With this model, we obtain the following (operator) transfer matrix

$$\hat{e}(\theta) = \hat{r}(\theta) - C_c \left[B(\theta)C_d \ e^{A_c\theta}\right] \begin{bmatrix} zI - A_d & B_d C_c \\ -B(h)C_d & zI - e^{A_c h} \end{bmatrix}^{-1} \begin{bmatrix} B_d \hat{r}(h) \\ 0 \end{bmatrix}. \tag{11}$$

3 Tracking and Internal Model Principle

In what follows, we assume that the two systems (A_d, B_d, C_d) and (A_c, B_c, C_c) are stabilizable and detectable, and that the latter satisfies the spectrum nondegeneracy condition, i.e., there are no pairs of eigenvalues of A_c that are different by an integer multiple of $2\pi j/h$. This assures the stabilizability and detectability of the sampled system. We further assume that there is no pole-zero cancellation between (A_d, B_d, C_d) and (A_c, B_c, C_c).

Let $W_{er}(z)$ be the closed-loop transfer operator from r to e given by (11). We first consider tracking to $e^{\mu t}v$ (Re $\mu \geq 0$), which is expressed as $\{\lambda^k v(\theta)\}_{k=0}^{\infty}$, $\lambda = e^{\mu h}$, $v(\theta) = e^{\mu\theta}v$. The z-transform of this function is, according to (3), $zv(\theta)/(z-\lambda)$.

We say that $\lambda \in \mathbb{C}$ is a *transmission zero* of the closed-loop transfer operator $W_{er}(z)$ if there exists $v(\theta) \in U$ such that

$$W_{er}(\lambda)v(\theta) = 0. \tag{12}$$

If this holds, $v(\theta)$ is called a *zero directional vector* associated with λ.

The following theorem was proved by [12]:

Theorem 3.1 *Suppose that the closed-loop system transfer operator $W_{er}(z)$ is stable, i.e., it has no poles outside the open unit disk $\{|z| < 1\}$. Assume also that the holder is the zero-order hold: $H(t) \equiv 1$. Then*

1. *If $\lambda \neq 1$, tracking without stationary ripples is possible only by incorporating μ into $P(s)$ as a pole.*

2. *If $\lambda = 1$, tracking without ripples is possible by incorporating λ into $C(z)$ as a pole.*

3. When $\lambda = 1$, the stationary ripple is given by $W_{er}(1)v(\theta)$.

This theorem tells us the following: Assuming the zero-order holder, i) tracking to $e^{\mu t}$ ($\mu \neq 0$) is possible only by incorporating a continuous-time compensator which contains the internal model of this exogenous signal; ii) in the case $\lambda = 1$ mostly encountered, tracking is possible by incorporating a digital internal model $z/(z-1)$; iii) even if complete tracking is not possible, its ripple can be explicitly computable.

It is also possible to characterize transmission zeros in terms of the state space representation and invariant zeros. Hence in the case of tracking to signals with simple poles is quite similar to the standard case. However, tracking to signals with repeated poles is quite different, and the situation cannot be described with simple pole-zero arguments. To see this, consider the ramp signal t. The z-transform of this signal is

$$\frac{h}{(z-1)^2} + \frac{\theta}{(z-1)},$$

and the double and simple poles here do not appear separately. Furthermore, in this case, tracking can be achieved by incorporating one pole $z = 1$ into the digital compensator, and the other $s = 0$ into the analog part. Therefore, a more elaborate study is necessary in the general case.

Using a mixed continuous-time/discrete-time model, Sung and Hara [8] derived a state space necessary and sufficient condition for tracking. A geometric approach was taken by [6] to get a necessary and sufficient condition. Somewhat earlier than these, Urikura and Nagata [10] gave a geometric condition for the case of deadbeat tracking. We here look for a condition in the frequency domain. For simplicity of arguments, we assume that the underlying systems are single input/single output.

To see the difficulty involved, let us write the loop transfer function of Fig. 1 as $p(z,\theta)/q(z)$. Note that the denominator does not depend on the parameter θ. This can also be seen from formula (11) where the characteristic polynomial depends only on z. This means that if we consider intersample tracking, its continuous-time behavior is reflected upon the *numerator* $p(z,\theta)$, and not the denominator $q(z)$. This is not suitable for stating the internal model principle as in the usual way, where the system characteristics are reflected upon the denominator, and the internal model principle is stated in terms of the relationships of denominators of the loop transfer function and the exogenous signal generator. This type of statement is clearly impossible for the sampled-data systems if we consider the intersample tracking.

It is therefore desirable to recover the continuous-time transfer function to prove the internal model principle. How can it be done? To this end, let us first introduce the notion of finite Laplace transforms.

Definition 3.2 Let φ be any function or distribution on the interval $[0,h]$. The *finite Laplace transform*, denoted $\mathcal{L}_h[\varphi]$ is defined by

$$\mathcal{L}_h[\varphi](s) := \int_0^h e^{-st}\varphi dt. \qquad (13)$$

The integral must be understood in the sense of distributions if φ is a distribution.

Note that, in view of the well known Paley-Wiener theorem [9], $\mathcal{L}_h[\varphi](s)$ is always an entire function of exponential type.

The z-transform of a function ϕ on $[0,\infty)$ and its Laplace transform can be related by the following lemma:

Lemma 3.3 *Suppose that ϕ satisfies the estimate*

$$\|\phi\|_{L^2,[kh,(k+1)h]} \leq Ce^{\beta k} \tag{14}$$

for some $C,\beta > 0$, where $\|\phi\|_{L^2,[kh,(k+1)h]}$ is the L^2-norm on $[kh,(k+1)h]$. Then the Laplace transform $L[\phi]$ exists, and

$$\mathcal{L}_h[\mathcal{Z}[\phi][z]]|_{z=e^{hs}} = L[\phi](s). \tag{15}$$

Proof That ϕ is Laplace transformable is obvious from (14). To show (15), observe

$$\begin{aligned} L[\phi](s) &= \sum_{k=0}^{\infty}\int_0^h \phi_k e^{-(kh+\theta)s}d\theta = \sum_{k=0}^{\infty} e^{-khs}\mathcal{L}_h[\phi_k](s) \\ &= \mathcal{L}_h[(\sum_{k=0}^{\infty} z^{-k}\phi_k)]|_{z=e^{hs}} = \mathcal{L}_h[\mathcal{Z}[\phi][z]]|_{z=e^{hs}}. \end{aligned} \tag{16}$$

The interchange of integral and summation in the second line is justified by the absolute convergence of the Laplace transform. □

Note that replacing z by e^{hs} actually corresponds to the left shift operator: $\phi(t) \mapsto \phi(t+h)$. Therefore, the digital compensator $C(z)$ has the (continuous-time) transfer function $C(e^{hs})$. The holder element has bounded support, so that it has some Laplace transform representation that is entire function. For example, the zero-order hold element has the transfer function $(1-e^{-hs})/s$. This is clearly entire due to pole-zero cancellation at $s=0$. Therefore, the loop transfer function of Fig. 1 falls into the category of the pseudorational class as defined in [11]. Roughly speaking, a transfer function $g(s)$ is *pseudorational* if it admits a factorization $G(s) = p(s)/q(s)$ where $p(s)$ and $q(s)$ are Laplace transforms of distributions with compact support in $(-\infty,0]$. Such a function $d(s)$ *is said to divide* $q(s)$ if $q(s)/d(s)$ is again the same type of function, i.e., the Laplace transform of a distribution with compact support in $(-\infty,0]$. (In particular, it is an entire function of exponential type; see [11] for details.) Using the results obtained there, we can now state and prove the necessity part of the internal model principle. A difficulty here is that although the loop transfer function can be well described in the s-domain, the closed-loop transfer function is not, because sampling is *not* a time-invariant operator in the continuous time sense. Therefore, we must bypass this problem without explicitly using the closed-loop transfer function $W_{er}(z)$.

Theorem 3.4 *Consider the unity feedback system given in Fig. 1. Suppose that the exogenous signal is generated by $d^{-1}(s)$ with poles only in the closed right-half plane. Suppose also that the closed-loop system is internally stable, and let $G(s) = q^{-1}(s)p(s)$ be a coprime factorization of the loop transfer function. If the closed-loop system asymptotically tracks any signal generated by $d^{-1}(s)$, then $d(s)$ divides $q(s)$.*

Proof Let $r(t)$ be any signal generated by $d^{-1}(s)$, and let $y(t)$ the corresponding output tracking $r(t)$. We first consider the system in the discrete-time mode. If the asymptotic tracking occurs, then it follows that the error $e(t)$ must also converge to zero at the sampling instants, i. e., $e(kh) \to 0$, $k \to \infty$. Therefore, we have

$$y(t) = y_1(t) + y_2(t),$$

where $y_2(t)$ is the response (of the closed-loop system) corresponding to the input $e(kh)$, and $y_1(t)$ is the initial-state response. Since the closed-loop system is stable and $e(kh) \to 0$ as $k \to \infty$, we must have $y_2(t) \to 0$ as $t \to \infty$. This means that $y_1(t)$ must be asymptotic to $r(t)$. Also, since the sampled input $e(kh)$ to the closed-loop system approaches zero, $y_1(t)$ should be asymptotic to the initial-state response of the open-loop system $G(s)$ (see Fig. 2; i.e., the system is almost in the stationary mode). This means that for sufficiently large sampling step k, $y_1(kh+\theta) \sim r(\theta)$. Furthermore, this output must be generated by an initial state of $G(s)$. Therefore, it should follow that any initial state response generated by $d^{-1}(s)$ must be contained in the set of initial state responses generated by $q^{-1}(s)$. In the terminology of [11],

$$X^d \subset X^q.$$

According to [11, Theorem 4.6], this implies $d(s)|q(s)$. □

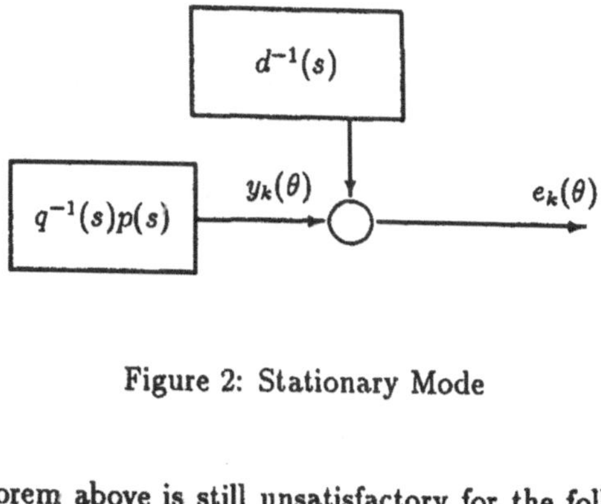

Figure 2: Stationary Mode

The necessity theorem above is still unsatisfactory for the following reasons: i) We have replaced the digital compensator $C(z)$ by the delay element $C(e^{hs})$. But those poles of $C(e^{hs})$ are not necessarily poles of $C(z)$ since the latter is just a digital compensator and only by combination with the analog holder element, does it generate a continuous-time response. For example, consider the zero-order hold: $(1-e^{-hs})/s$. If the preceding digital element has $z-1$ in the denominator, it produces the signal $1/s$ via cancellation $(1-e^{-hs})/(z-1)|_{z=e^{hs}} = e^{-hs}$. We thus make the following hypothesis:

Hypothesis: *Let $C(z)$ and $\psi(s)$ be the transfer functions of digital compensator and the holder. Then the product $C(e^{hs})\psi(s)$ becomes a product of $C'(e^{hs})$ and $\phi(s)$ which are rational functions of e^{hs} and s, respectively.*

This hypothesis is satisfied, for example, for

$$C(z) = \frac{z}{z-1}; \quad \psi(s) = \frac{1-e^{-hs}}{s^2+\omega^2}, \omega = 2\pi/h.$$

Remark 3.5 A similar hypothesis is made in [6].

Let us now state and prove the sufficiency of our internal model principle.

Theorem 3.6 *Let $C(z)$, $C'(z)$, $\psi(s)$ and $\phi(s)$ be as above, satisfy the hypothesis above. Write the loop transfer function $P(s)C(e^{hs})$ as $q^{-1}(s)v(s)C'(e^{hs})$, where $q(s)$ contains the denominator of $P(s)$ and $\phi(s)$. Suppose that the closed-loop system Fig. 1 is stable, the exogenous signal is generated by $d^{-1}(s)$, and that $d(s)|q(s)$. Then the closed-loop system asymptotically tracks any signal generated by $d^{-1}(s)$.*

Proof Since $d(s)|q(s)$, if we look at only the sampled instants, the closed-loop system contains the discrete-time internal model. Therefore, by the internal stability, at least tracking at sampled instants occurs. Therefore, the error input to the loop transfer function tends to zero. Hence for sufficiently large $t_0 = kh$, the output $y(t), t \geq t_0$ is close to the tracking signal $d^{-1}r_0$ at sampled instants. Reset this t_0 to zero, and then we have that the output $y(t)$ becomes

$$y(t) = q^{-1}x_0 + \epsilon(t)$$

where $\epsilon(t)$ is the output corresponding to the input $e_k(h)$ so that $\epsilon(t) \to 0$. (The output of in $C'(z)$ contributes to the output only through the continuous-time part $q^{-1}(s)$, so it need not be counted in the expression above.) Therefore, $q^{-1}x_0$ must asymptote to $d^{-1}r_0$ at least on the sampled instants $t = kh$. Since $d^{-1}r_0$ is a linear combination of exponential polynomials, we may confine our consideration to the tracking to signals of type $t^i e^{\mu t}$. However, since we have the spectrum nondegeneracy condition, this occurs only when the output $q^{-1}x_0$ asymptotes also on the intersample intervals. □

Remark 3.7 If we do not have the spectrum nondegeneracy condition, then we cannot conclude tracking in the intersample periods. Indeed, if we have $s = 0, 2\pi j/h$ both as poles in the continuous-time part, then it may well happen that the closed-loop system may track to the sine wave even if the exogenous signal is a step, because the feedback loop error signal does not contain any intersample information. However, in such a case, the detectability property obviously fails.

Acknowledgments. I wish to thank Professors M. Ikeda and S. Hara for helpful discussions.

References

[1] T. C. Chen and B. A. Francis, "Stability of sampled-data systems," *Technical Report No. 8905, University of Toronto*, 1989.

[2] B. A. Francis and T. T. Georgiou, "Stability theory for linear time-invariant plants with periodic digital controllers," *IEEE Trans. Autom. Control*, **AC-33**: 820-832, 1988.

[3] T. Hagiwara and M. Araki, "Design of a stable feedback controller based on the multirate sampling of the plant output," *IEEE Trans. Autom. Control*, **AC-33**: 812-819, 1988.

[4] S. Hara and P. T. Kabamba, "Worst case analysis and design of sampled data control systems," *Proc. 29th CDC*: 202-203, 1990.

[5] P. T. Kabamba, "Control of linear systems using generalized sampled-data hold functions," *IEEE Trans. Autom. Control*, **AC-32**: 772-783, 1987.

[6] A. Kawano, T. Hagiwara and M. Araki, "Robust servo condition for sampled-data systems," (in Japanese) *SICE Control Theory Symp.*: 35-40, 1990.

[7] B. Bamieh and J. B. Pearson, "A general framework for linear periodic systems with applications to H^∞ sampled-data control," *Tec. Rep. 9021, Rice Univ.*, 1990.

[8] H.-K. Sung and S. Hara, "Ripple-free condition in sampled-data control systems," *SICE 13th DST Symp.*: 269-272, 1991.

[9] F. Treves, *Topological Vector Spaces, Distributions and Kernels*, Academic Press, New York, 1971.

[10] S. Urikura and A. Nagata, "Ripple-free deadbeat control for sampled-data systems," *IEEE Trans. Autom. Control*, **AC-32**: 474-482, 1987.

[11] Y. Yamamoto, "Pseudo-rational input/output maps and their realizations: a fractional representation approach to infinite-dimensional systems," *SIAM J. Control & Optimiz.*, **26**: 1415-1430, 1988.

[12] Y. Yamamoto, "New approach to sampled-data systems: a function space method" *Proc. 29th CDC*: 1882-1887, 1990.

SUPER-OPTIMAL HANKEL-NORM APPROXIMATIONS

Fang-Bo YEH and Lin-Fang WEI
Department of Mathematics, Tunghai University
Taichung, Taiwan, Republic of China

Abstract: It is well-known that optimal Hankel-norm approximations are seldom unique for multivariable systems. This comes from the Hankel-norm being somewhat of a crude criterion for the reduction of multivariable systems. In this paper, the strengthened condition originated with N. J. Young is employed to restore the uniquess. A state-space algorithm for the computation of super-optimal solution is presented.

I. Introduction

Recent developments in optimal Hankel-norm approximations [1]–[3] have held great attention in the control society. As pointed out in [1], based on the Kronecker theorem and the singular value analysis, Hankel-norm criterion appears to be very natural and useful. Roughly speaking, the Hankel-norm or Hankel singular values can be thought as a measure of the controllability and observability of a LTI system, which has strong relations to the McMillan degree of a system. In addition to this sound physical meaning, another merit of using Hankel-norm criterion is that the calculation of a lower degree approximation with minimum error or a minimum degree approximation within a given tolerance can be easily computed. These features bail the design out of endless iterations in the face of large-scale multivariable systems. The celebrated paper [2] gives detailed state-space solutions and their L^{∞} error bounds to these problems. For single input or single output systems, it is well-known that the optimal Hankel-norm approximation is unique and can be easily determined by the Schmidt pair of the associated Hankel operator. However, the optimal solutions are seldom unique for multivariable systems. The problem how to choose the best solution then naturally arises. A simple example reported in [2] is used to clarify the situation. Consider the system

$$G(s) = \begin{bmatrix} \frac{2s+0.45}{s^2+1.25s+0.09} & 0 \\ 0 & \frac{1}{s+0.5} \end{bmatrix}$$

A bit of calculation shows that all optimal Hankel-norm approximations with McMillan degree two are given by

$$\hat{G}(s) = \begin{bmatrix} \frac{0.5s+0.675}{s+0.15} & 0 \\ 0 & \hat{g}(s) \end{bmatrix}$$

where $\hat{g}$ is any stable function of McMillan degree one and $\| \frac{1}{s+0.5} - \hat{g} \|_H \leq \frac{1}{2}$. Here $\| \cdot \|_H$ stands for the Hankel-norm of a system. It is trivial to see that $\hat{g}(s) = \frac{1}{s+0.5}$ would be the best choice in a natural sense. However, it is not at all clear how this could be generalized in [2]. The non-uniqueness comes from that the Hankel-norm is somewhat of a crude criterion for the reduction of multivariable systems. To make the solution unique, some finer measure should be imposed. In this paper, the strengthened condition in [9] is employed to restore the uniquess.

To formulate the problem more precisely, we begin with the nomenclature. The symbol $\mathbf{RL}^{\infty,p\times q}$ denotes the space of $p \times q$ matrix-valued, proper and real-rational functions in s with no poles on the $j\omega$ axis. Its subspace with no poles in the right-half plane is denoted as $\mathbf{RH}_{+}^{\infty,p\times q}$. The superscript $p \times q$ will be omitted if the size is irrelevant. The problem of optimal Hankel-norm approximation can be defined as follows: given $G \in \mathbf{RH}_{+}^{\infty,p\times q}$ and an integer $k \geq 0$, find a $\hat{G} \in \mathbf{RH}_{+}^{\infty,p\times q}$ with McMillan degree k such that the error's Hankel-norm is minimized, i.e.,

$$\min_{\hat{G}\in\mathbf{RH}_{+}^{\infty,p\times q} \text{ and } \deg(\hat{G})=k} \|G-\hat{G}\|_{H} \tag{1.1}$$

It is known that Hankel-norm is only related to the stable part of a system, and the addition of an antistable function does not affect the norm. Thus, the problem of optimal Hankel-norm approximations is equivalent to

$$\min_{\hat{G}\in\mathbf{RH}_{-,k}^{\infty,p\times q}} s_{1}^{\infty}(G-\hat{G}) \tag{1.2}$$

where $\mathbf{RH}_{-,k}^{\infty,p\times q}$ denotes the subset of $\mathbf{RL}^{\infty,p\times q}$ with no more than k poles in the left half-plane, $\mathbf{RH}_{-,0}^{\infty,p\times q}$ is abbreviated to $\mathbf{RH}_{-}^{\infty,p\times q}$, and

$$s_{j}^{\infty}(E) := \max_{\omega\in\mathbf{R}} s_{j}\left(E(j\omega)\right)$$

The symbol $s_j(\cdot)$ denotes the jth largest singular value of a constant matrix. Then, deleting the antistable part of $\hat{G}$ obtained from (1.2) will give the optimal solution to (1.1). In (1.2), it is seen that only the first frequency dependent singular value of the error system $G - \hat{G}$ is minimized. A generalization is that we seek to minimize all the singular values whenever possible. Therefore, the problem of super-optimal Hankel-norm approximation is defined as follows: given $G \in \mathbf{RL}^{\infty,p\times q}$ and an integer $k \geq 0$ find a $\hat{G} \in \mathbf{RH}_{-,k}^{\infty,p\times q}$ such that the sequence

$$s_{1}^{\infty}(G-\hat{G}),\ s_{2}^{\infty}(G-\hat{G}),\ \cdots$$

is minimized lexicographically. To our knowledge, this problem was first studied in [10], wherein the existence and uniqueness of the super-optimal solution had been proved using the conceptual operator-theoretic constructions. In this paper, a different approach which requires only simple state-space calculations will be studied. The approach is much more straightforward and comprehensible. Besides, the pole-zero cancellations in the algorithm will be analyzed in detail.

II. Mathematical Preliminaries

In the development of this work, the state-space approach is adopted. For a proper and real-rational matrix function $G(s)$, $G^{\sim}(s)$ is synonymous with $G^{T}(-s)$, and the data structure $[A, B, C, D]$ denotes a realization of $G(s)$, i.e.,

$$G(s) = C(sI - A)^{-1}B + D = [A, B, C, D] = \left[\begin{array}{c|c} A & B \\ \hline C & D \end{array}\right]$$

where A, B, C and D are constant matrices with compatible dimensions. A collection of state-space operations using this data structure can be found in [6]. For a stable system $G(s)$ with the realization $[A, B, C, D]$, the corresponding controllability and observability Gramian are defind as the uniquely non-negative definite solutions to the following Lyapunov equations, respectively,

$$AP + PA^T + BB^T = 0$$

$$A^TQ + QA + C^TC = 0$$

If the realization is minimal, then P and Q must be positive definite. When both P and Q are equal and diagonal, we say that the realization $[A, B, C, D]$ is balanced and the jth largest diagonal element, denoted $\sigma_j(G)$, is defined as the jth Hankel singular values of $G(s)$. It is always possible to get a balanced realization by use of the state similarity transformation.

The problem of approximating a given Hankel matrix by a lower rank one had been studied in [4] and [5]. Their striking results stated that the restriction of the solution to be a Hankel matrix does not affect the achievable error. These remarkable results will be briefly reviewed here. We denote $\mathbf{RL}^{2,p}$ the space of p vector-valued, strictly proper and real-rational functions with no poles on the $j\omega$ axis, and is a Hilbert space under the inner product

$$\langle u, v\rangle := \frac{1}{2\pi}\int_{-\infty}^{\infty} u^T(-j\omega)v(j\omega)\, d\omega$$

The subspace of $\mathbf{RL}^{2,p}$ with no poles in the right-half plane is denoted $\mathbf{RH}^{2,p}$, and its orthogonal complement is denoted $\mathbf{RH}_{\perp}^{2,p}$. For a system G in $\mathbf{RH}_{+}^{\infty,p\times q}$, the Hankel operator with symbol G, denoted Γ_G, maps $\mathbf{RH}_{\perp}^{2,q}$ to $\mathbf{RH}^{2,p}$. For $x \in \mathbf{RH}_{\perp}^{2,q}$, $\Gamma_G\, x$ is defined as

$$\Gamma_G\, x := \Pi(G\, x)$$

where Π is the orthogonal projection operator which maps $\mathbf{RL}^{2,p}$ onto $\mathbf{RH}^{2,p}$. An important result is that the Hankel operator Γ_G is of finite rank and its singular values are just the Hankel singular values of $G(s)$. The pair (v_j, w_j) satisfies

$$\Gamma_G\, v_j = \sigma_j(G)\, w_j$$

$$\Gamma_G^*\, w_j = \sigma_j(G)\, v_j$$

is called the jth Schmidt pair of Γ_G corresponding to $\sigma_j(G)$. The following Lemma relates the Schmidt pair of Γ_G to any optimal solution, and is central to this study.

Lemma 2.1: Given $G \in \mathbf{RH}_{+}^{\infty,p\times q}$ and an integer $k \geq 0$ then

(1) $\min_{\hat{G}\in\mathbf{RH}_{-,k}^{\infty,p\times q}} s_1^{\infty}(G - \hat{G}) = \sigma_{k+1}(G) := \sigma,$

(2) if $\sigma_k(G) > \sigma$ then for any optimal $\hat{G}$ we have $(G-\hat{G})v = \sigma w$ and $(G-\hat{G})^{\sim}w = \sigma v$. Here (v, w) denotes the $(k+1)$st Schmidt pair of the Hankel operator Γ_G. In case of $k = 0$, $\sigma_0(G)$ is interpreted as $+\infty$. □

A matrix G in $\mathbf{RL}^\infty$ is all-pass if $G^{\sim}(s)G(s) = I$, and is inner if G is restricted to $\mathbf{RH}_+^\infty$. An inner matrix that has a left-inverse in $\mathbf{RH}_+^\infty$ is called a minimum-phase inner matrix [8]. The following lemmas give some important properties of minimum-phase inner matrices.

Lemma 2.2: [8] Let $G(s) = [A, B, C, D]$ be inner together with the controllability Gramian P and observability Gramian Q. Then G is minimum-phase inner if and only if $\|PQ\| < 1$, and

$$G^{-L}(s) = [A, BD^T + PC^T, D^T C(PQ - I)^{-1}, D^T]$$

is one of stable left-inverses of G. □

Lemma 2.3: For a strictly tall inner matrix $G(s) = [A, B, C, D]$ having nonsingular controllability Gramian P, the right-coprime factorization of $G(-s)$ over $\mathbf{RH}_+^\infty$, i.e. $G(-s) = N_G(s)M_G^{-1}(s)$, can be written as

$$N_G(s) = [A^T, -P^{-1}B, CP + DB^T, D]$$

$$M_G(s) = [A^T, -P^{-1}B, B^T, I]$$

where M_G is inner and N_G is minimum-phase inner. □

III. Diagonalizing Matrices

In this section, we shall study how to construct all-pass matrices which will diagonalize each possible error system. It is seen that in order to decide the second singular value, namely $s_2^\infty(G - \hat{G})$, and keep $s_1^\infty(G - \hat{G})$ unchanged at the same time, a natural way is to diagonalize the error system $G - \hat{G}$ by pre-multiplying and post-multiplying two suitable all-pass matrices. Not to sacrifice any generality, it is important to have this diagonalizing process hold for all optimal solutions. Recalling that any optimal solution $\hat{G}$ should satisfy Lemma 2.1 part (2), it is clear that the Schmidt pair (v, w) serve as a starting point in this diagonlization process.

We begin by assuming that a minimal balanced realization $[A, B, C, D]$ of a stable system G with McMillan degree n is given. Then the associated controllability and observability Gramians are both equal and can be arranged as $\mathrm{diag}(\sigma, \Sigma)$, where $\sigma = \sigma_{k+1}(G)$ and Σ is also diagonal. To have a clear presentation, hereinafter, we shall assume that σ is distinct, i.e.,

(A1) $\sigma_k(G) > \sigma > \sigma_{k+2}(G)$.

Relaxing this assumption is possible but only leads to a more messy indexing notation. Partition matrices A, B and C as

$$A = \begin{bmatrix} a_{11} & A_{12} \\ A_{21} & A_{22} \end{bmatrix}, \qquad B = \begin{bmatrix} B_1 \\ B_2 \end{bmatrix}, \qquad C = [C_1 \quad C_2]$$

where a_{11} is a scalar, B_1 is a row vector and C_1 is a column vector. Obviously, the following Lyapunov equations are hold

$$\begin{bmatrix} a_{11} & A_{12} \\ A_{21} & A_{22} \end{bmatrix} \begin{bmatrix} \sigma & 0 \\ 0 & \Sigma \end{bmatrix} + \begin{bmatrix} \sigma & 0 \\ 0 & \Sigma \end{bmatrix} \begin{bmatrix} a_{11} & A_{21}^T \\ A_{12}^T & A_{22}^T \end{bmatrix} + \begin{bmatrix} B_1B_1^T & B_1B_2^T \\ B_2B_1^T & B_2B_2^T \end{bmatrix} = 0$$

$$\begin{bmatrix} a_{11} & A_{21}^T \\ A_{12}^T & A_{22}^T \end{bmatrix} \begin{bmatrix} \sigma & 0 \\ 0 & \Sigma \end{bmatrix} + \begin{bmatrix} \sigma & 0 \\ 0 & \Sigma \end{bmatrix} \begin{bmatrix} a_{11} & A_{12} \\ A_{21} & A_{22} \end{bmatrix} + \begin{bmatrix} C_1^TC_1 & C_1^TC_2 \\ C_2^TC_1 & C_2^TC_2 \end{bmatrix} = 0$$

To simplify the notation, we will assume that G has been scaled such that $B_1B_1^T = C_1^TC_1 = 1$, which is without loss of any generality. The $(k+1)st$ Schmidt pair v and w for the above balanced realization can be written as

$$v(-s) = \left[A^T, e_1/\sqrt{\sigma}, B^T, 0\right], \qquad w(s) = \left[A, e_1/\sqrt{\sigma}, C, 0\right]$$

where e_1 is the first column of an $n \times n$ identity matrix. By direct state-space calculation, it is easy to verify that $v^\sim v = w^\sim w$. Thus, it is possible to factorize $v(-s)$ and $w(s)$ such that

$$v(-s) = \hat{v}(s)\ \alpha(s), \qquad w(s) = \hat{w}(s)\ \alpha(s)$$

where $\hat{v}$ and $\hat{w}$ are all-pass vectors, and α is a scalar function. The state-space realizations of $\hat{v}$ and $\hat{w}$ can be written as

$$\hat{v}(s) = \left[A_v, A_{12}^T, C_v, B_1^T\right], \qquad \hat{w}(s) = \left[A_w, A_{21}, C_w, C_1\right]$$

in which

$$A_v = A_{22}^T + \sigma A_{12}^T(A_{21}^T - A_{12}X_v), \qquad C_v = B_2^T + \sigma B_1^T(A_{21}^T - A_{12}X_v)$$

$$A_w = A_{22} + \sigma A_{21}(A_{12} - A_{21}^TX_w), \qquad C_w = C_2 + \sigma C_1(A_{12} - A_{21}^TX_w)$$

where X_v and X_w satisfy the following algebraic Riccati equations,

$$(A_{22} + \sigma A_{21}A_{12})X_v + X_v(A_{22} + \sigma A_{21}A_{12})^T - \sigma X_vA_{12}^TA_{12}X_v - \sigma A_{21}A_{21}^T = 0 \quad (3.1)$$

$$(A_{22} + \sigma A_{21}A_{12})^TX_w + X_w(A_{22} + \sigma A_{21}A_{12}) - \sigma X_wA_{21}A_{21}^TX_w - \sigma A_{12}^TA_{12} = 0 \quad (3.2)$$

A direct computation will show that any solutions to (3.1) and (3.2) will make $\hat{v}$ and $\hat{w}$ all-pass. Two special kinds of solutions are of interest, namely, the stabilizing and antistabilizing solutions. For our purpose, we choose X_v and X_w to be the antitabilizing solutions, which is inspired from [13]. It will be shown that the use of antistabilizing solutions can be of great help in analyzing the pole-zero cancellations. Hence to ensure the existence of antistabilizing solutions, it is natural to assume that

(A2) (A_{22}^T, A_{12}^T) and (A_{22}, A_{21}) are controllable.

Then, it can be verified that $Q_v = \sigma X_v + \Sigma$ and $Q_w = \sigma X_w + \Sigma$ satisfy the following Lyapunov equations, respectively

$$A_v^TQ_v + Q_vA_v + C_v^TC_v = 0$$

$$A_w^TQ_w + Q_wA_w + C_w^TC_w = 0$$

In case of $k = 0$, it can be proved that Q_v and Q_w are nonsingular provided that σ is distinct. However, for general k, the situation is not clear and we will assume that

(A3) Q_v and Q_w are non-singular.

The all-pass completions [7] of $\hat{v}$ and $\hat{w}$ are given by

$$\hat{V}_\perp(s) = [A_v, -Q_v^{-1}B_2B_\perp^T, C_v, B_\perp^T]$$

$$\hat{W}_\perp(s) = [A_w, -Q_w^{-1}C_2^TC_\perp, C_w, C_\perp]$$

where $[B_1^T \quad B_\perp^T]$ and $[C_1 \quad C_\perp]$ are orthogonal. Denote $\hat{V} = [\hat{v} \quad \hat{V}_\perp]$ and $\hat{W} = [\hat{w} \quad \hat{W}_\perp]$. Then, the error $G - \hat{G}$ can be diagonalized as shown in the following lemma.

Lemma 3.1: Given that G satisfies Assumptions (A1), (A2) and (A3), then for any optimal solution $\hat{G}$ we have

$$\hat{W}^\sim(G-\hat{G})\hat{V}(-s) = \begin{bmatrix} \sigma g_1 & 0 \\ 0 & \hat{W}_\perp^\sim(G-\hat{G})\hat{V}_\perp(-s) \end{bmatrix}$$

where g_1 is all-pass and independent of $\hat{G}$. □

IV. Super-Optimal Solutions

The super-optimal model matching problem, i.e. $k = 0$, was first studied in [9] wherein a high-level algorithm is released. The implementations of Young's algorithm have been reported in [11] using the polynomial approach, and in [12] and [13] using the state-space approach. The basic idea of Young's algorithm can be summarized as follows. First, the minimum achievable L^∞ norm of the error is calculated. Then, two all-pass matrices are constructed to diagonalize the error system, which results in the same model-matching problem but with one less dimension. Hence by induction, this dimension peeling process can be continued layer by layer until a unique solution is found. Finally, the super-optimal solution of the original problem is constructed from the solution of each layer. However, difficulty arises when apply this idea to the general case $k > 0$. The reason is that the addition and multiplication of two $\mathbf{RH}_-^\infty$ functions is still an $\mathbf{RH}_-^\infty$ function, but is generally not true for $\mathbf{RH}_{-,k}^\infty$ functions. This causes the minimum achievable norm of the subsequent layer hard to determine. In order to ensure that the final super-optimal solution of the original problem is in $\mathbf{RH}_{-,k}^{\infty,p\times q}$, the solution set of each subsequent layer should be precisely characterized, and will be studied in this section.

We now continue our work in the previous section. By [2], an optimal solution is given by

$$\hat{G}_1(s) = \left[\begin{array}{c|c} \Gamma^{-1}(\sigma^2A_{22}^T + \Sigma A_{22}\Sigma + \sigma C_2^TC_1B_1B_2^T) & \Gamma^{-1}(\Sigma B_2 - \sigma C_2^TC_1B_1) \\ \hline C_2\Sigma - \sigma C_1B_1B_2^T & D + \sigma C_1B_1 \end{array}\right]$$

where $\Gamma = \Sigma^2 - \sigma^2 I$ is nonsingular by (A1). Recalling Lemma 3.1, we see that any optimal solution $\hat{G}$ should satisfy

$$\hat{W}^\sim(\hat{G}-\hat{G}_1)\hat{V}(-s) = \begin{bmatrix} 0 & 0 \\ 0 & \hat{W}_\perp^\sim(\hat{G}-\hat{G}_1)\hat{V}_\perp(-s) \end{bmatrix}$$

Thus, $\hat{G}\hat{V}(-s)$ and $\hat{G}_1\hat{V}(-s)$ have the same first column, i.e.,

$$\hat{G}\hat{V}(-s) = \begin{bmatrix}\hat{G}\hat{v}(-s) & \hat{G}\hat{V}_\perp(-s)\end{bmatrix} := \begin{bmatrix}h & H\end{bmatrix}$$

$$\hat{G}_1\hat{V}(-s) = \begin{bmatrix}\hat{G}_1\hat{v}(-s) & \hat{G}_1\hat{V}_\perp(-s)\end{bmatrix} := \begin{bmatrix}h & H_1\end{bmatrix}$$

Besides, it is required that $\hat{w}^\sim(H - H_1) = 0$. Now observe that

$$\begin{aligned}\hat{W}^\sim(G-\hat{G})\hat{V}(-s) &= \hat{W}^\sim(G-\hat{G}_1)\hat{V}(-s) - \hat{W}^\sim(\hat{G}-\hat{G}_1)\hat{V}(-s) \\ &= \begin{bmatrix}\sigma g_1 & 0 \\ 0 & F_1\end{bmatrix} - \begin{bmatrix}0 & 0 \\ 0 & \hat{W}_\perp(H-H_1)\end{bmatrix}\end{aligned}$$

where

$$F_1(s) = \hat{W}_\perp^\sim(G-\hat{G}_1)\hat{V}_\perp(-s)$$

Hence to compute the super-optimal solution, we require finding H such that

(C1) $\begin{bmatrix}h & H\end{bmatrix} \in \mathbf{RH}_{-,k}^{\infty,p\times q}\hat{V}(-s)$,
(C2) $\hat{w}^\sim(H - H_1) = 0$,
(C3) $s_j^\infty(F_1 - \hat{W}_\perp^\sim(H - H_1))$, for $j = 1, 2, \ldots$ is minimized lexicographically.

To solve the above problem, the first step is to parametrize all the functions H that satisfy both Conditions (C1) and (C2) in terms of some free function. And then the minimizing process (C3) can be carried out. These parametrizations are studied in Theorems 4.1 and 4.2.

First of all, according to Lemma 2.3, we introduce the following factorizations which play an important role in the riddance of (C1) and (C2)

$$\hat{v}(s) = n_v(-s)\ m_v^{-1}(-s)$$

$$\hat{w}(s) = n_w(-s)\ m_w^{-1}(-s)$$

with

$$n_v(s) = \left[A_v^T, -P_v^{-1}A_{12}^T, C_vP_v + B_1^TA_{12}, B_1^T\right]$$

$$n_w(s) = \left[A_w^T, -P_w^{-1}A_{21}, C_wP_w + C_1A_{21}^T, C_1\right]$$

where P_v and P_w are the uniquely negative definite solutions to the Lyapunov equations, respectively,

$$A_vP_v + P_vA_v^T + A_{12}^TA_{12} = 0$$

$$A_wP_w + P_wA_w^T + A_{21}A_{21}^T = 0$$

It is important to notice that $n_v(-s)$ and $n_w(-s)$ are minimum-phase inner and, hence, have stable left inverse. With these notations, we have the following theorem.

Theorem 4.1: $\begin{bmatrix}h & H\end{bmatrix} \in \mathbf{RH}_{-,k}^{\infty,p\times q}\hat{V}(-s)$ if and only if

$$H \in \hat{G}_1 n_v l_{n_v}\hat{V}_\perp(-s) + \mathbf{RH}_{-,k}^{\infty,p\times(q-1)} := \mathcal{H}$$

where $l_{n_v}(-s)$ is any stable left inverse of $n_v(-s)$. □

Now by Theorem 4.1, it follows that H and H_1 can be parametrized as

$$H = \hat{G}_1 n_v l_{n_v} \hat{V}_\perp(-s) + R$$

$$H_1 = \hat{G}_1 n_v l_{n_v} \hat{V}_\perp(-s) + R_1$$

for some R and R_1 in $\mathbf{RH}_{-,k}^{\infty, p\times(q-1)}$. In other words, R_1 can be computed as follows

$$R_1 = \hat{G}_1 (I - n_v l_{n_v}) \hat{V}_\perp(-s)$$

wherein the function $V_2 := (I - n_v l_{n_v}) \hat{V}_\perp(-s)$ has at most $n-1$ states and is antistable. This can be verified by letting

$$l_{n_v}(s) = \left[A_v^T, C_v^T, A_{12}(I - P_v^{-1} Q_v^{-1}), B_1\right]$$

according to Lemma 2.2. Then a series of state-space calculations yields that

$$V_2(s) = \left[A_v^T, P_v^{-1} Q_v^{-1} B_2 B_\perp^T, C_v P_v + B_1^T A_{12}, B_\perp^T\right]$$

As R_1 is found, Condition (C2) is equivalent to $n_w^T(R - R_1) = 0$. The following theorem characterize all such function R.

Theorem 4.2: $R \in \mathbf{RH}_{-,k}^{\infty, p\times(q-1)}$ and $n_w^T(R - R_1) = 0$ if and only if

$$R \in l_{n_w}^T n_w^T R_1 + \hat{W}_\perp \mathbf{RH}_{-,k}^{\infty,(p-1)\times(q-1)}$$

where $l_{n_w}(-s)$ is any stable left inverse of $n_w(-s)$. □

Hence by theorem 4.2, there exist functions $\hat{Q}$ and Q_1 in $\mathbf{RH}_{-,k}^{\infty,(p-1)\times(q-1)}$ such that

$$R = l_{n_w}^T n_w^T R_1 + \hat{W}_\perp \hat{Q}$$

$$R_1 = l_{n_w}^T n_w^T R_1 + \hat{W}_\perp Q_1$$

and, in other words

$$Q_1 = \hat{W}_\perp^\sim (I - l_{n_w}^T n_w^T) \hat{G}_1 V_2$$

To compute the function $W_2 := \hat{W}_\perp^\sim (I - l_{n_w}^T n_w^T)$, we choose

$$l_{n_w} = \left[A_w^T, C_w^T, A_{21}^T (I - P_w^{-1} Q_w^{-1}), C_1^T\right]$$

according to Lemma 2.3. then a series of state-space calculations gives that

$$W_2 = \left[A_w, P_w C_w^T + A_{21} C_1^T, C_\perp^T C_2 Q_w^{-1} P_w^{-1}, C_\perp^T\right]$$

which is also antistable and has McMillan degree no more than $n-1$. Finally, the function which need to be minimized in (C3) becomes

$$F_1 - \hat{W}_\perp^\sim (H - H_1) = F_1 - \hat{W}_\perp^\sim (R - R_1) = F_1 + Q_1 - \hat{Q}$$

Define

$$Q := F_1 + Q_1 = \hat{W}_\perp^\sim (G - \hat{G}_1) \hat{V}_\perp(-s) + W_2 \hat{G}_1 V_2 \tag{4.1}$$

Then Condition (C3) is reduced to the problem that finds a $\hat{Q}$ in $\mathbf{RH}_{-,k}^{\infty,(p-1)\times(q-1)}$ such that the sequence

$$s_1^{\infty}(Q-\hat{Q}), s_2^{\infty}(Q-\hat{Q}), \cdots$$

is minimized lexicographically. This is just the same super-optimal problem but with one less dimension. This dimension peeling process can be recursively invoked until the row or column dimension is reduced to one, wherein the solution can be uniquely determined. By induction, the super-optimal solution is clearly unique. And once the super-optimal solution $\hat{Q}$ is found, $\hat{G}$ can be recovered as follows

$$\hat{G}(s) = \hat{G}_1 + \hat{W}_{\perp}(\hat{Q} - Q_1)\hat{V}_{\perp}^T \tag{4.2}$$

Although this algorithm is conceptually workable, the computation should not follow (4.1) and (4.2) directly. The reason is that the sizes of the A-matrices of Q and $\hat{G}$ will blow up very rapidly if pure state-space additions and multiplications are used, and a further computation is required in order to get the minimal realizations of Q and $\hat{G}$. Owing to this observation, the pole-zero cancellations occur in (4.1) and (4.2) should be analyzed in detail.

It can be proved that Q is stable and can be realized with $n-1$ states. Then, the super-optimal solution $\hat{G}$ will require no more than $\deg(\hat{Q}) + n - 1$ states, where $\hat{Q}$ is the super-optimal Hankel-norm approximation of Q. As a result, the required computation time in each recursive step will gradually decrease rather than increase. Since our intension is to apply the theory to large-scale multivariable systems, it is clear that a feasible computer program can be setup only when the full analysis of the pole-zero cancellations is carried out. Thus, the result of this section is valuable from the pratical consideration. The proof requires a series of tedious state-space calculations and is omitted.

Theorem 4.3: A realization of Q is given by

$$Q(s) = \left[\begin{array}{c|c} -A_v & Q_v^{-1}B_2B_{\perp}^T \\ \hline \sigma^2 C_{\perp}^T C_2 Q_w^{-1}(X_w X_v - I) & 0 \end{array}\right]$$

which is stable and minimal. Moreover, Let $\hat{Q} = [\check{A}, \check{B}, \check{C}, \check{D}] \in \mathbf{RH}_{-,k}^{\infty,(p-1)\times(q-1)}$ be any optimal Hankel-norm approximation to Q with L^{∞} error $\sigma_{k+1}(Q)$. Then a realization for $\hat{G}$ is

$$\hat{G}(s) = \left[\begin{array}{cc|c} A_w & -Q_w^{-1}C_2^T C_{\perp}\check{C} & Q_w^{-1}(\Sigma B_2 - \sigma C_2^T C_1 B_1 - C_2^T C_{\perp}\check{D}B_{\perp}) \\ 0 & \check{A} & \check{B}B_{\perp} + P_{12}^T C_v^T \\ \hline C_w & C_{\perp}\check{C} & D + \sigma C_1 B_1 + C_{\perp}\check{D}B_{\perp} \end{array}\right]$$

where P_{12} satisfies $A_v P_{12} - P_{12}\check{A}^T = Q_v^{-1}B_2 B_{\perp}^T \check{B}^T$. □

V. Concluding Remarks

Throughout this paper, we have concentrated on the computation of super-optimal Hankel-norm approximations. The existence and uniqueness of the super-optimal solution are proved by use of simple state-space calculations. The approach is unlike the work in [10], which based on conceptual operator-theoretic constructions. In addition, we have given a detailed analysis of pole-zero cancellations in the algorithm and a bound on the McMillan degree of the super-optimal solution, which generalize the results in [13].

References

[1] S. Y. Kung and D. W. Lin, "Optimal Hankel-norm model reductions: Multivariable systems," *IEEE Trans. Automat. Contr.*, vol. AC-26, pp. 832–852, 1981.

[2] K. Glover, "All optimal Hankel-norm approximations of linear multivariable system and their L^∞ error bounds," *Int. J. Contr.*, vol. 39, pp. 1115–1193, 1984.

[3] J. A. Ball and A. C. M. Ran, "Optimal Hankel-norm model reductions and Wiener-Hopf factorizations II: the noncanonical case," *Integral Eqn. Operator Theory*, vol. 10, pp. 416–436, 1987.

[4] V. M. Adamjan, D. Z. Arov, and M. G. Krein, "Analytic properties of Schmidt pairs for a Hankel Operator and the generalized Schur–Takagi problem," *Math. of the USSR: Sborink*, vol. 15, pp. 31–73, 1971.

[5] V. M. Adamjan, D. Z. Arov, and M. G. Krein, "Infinite block Hankel matrices and related extension problems," *AMS Transl.*, ser. 2, vol. 111, pp. 133–156, 1978.

[6] B. A. Francis, *A Course in H^∞ Control Theory*. New York: Springer-Verlag, 1987.

[7] J. C. Doyle, "Lecture Notes in Advances in Multivariable Control," *ONR / Honeywell Workshop*, Minneapolis, MN, 1984.

[8] F. B. Yeh and L. F. Wei, "Inner–outer factorizations of right-invertible real-rational matrices," *Syst. Contr. Lett.*, vol. 14, pp. 31–36, 1990.

[9] N. J. Young, "The Nevanlinna-Pick problem for matrix-valued functions," *J. Operator Theory*, vol. 15, pp. 239–265, 1986.

[10] N. J. Young, "Super-optimal Hankel-norm approximations," in *Modeling Robustness and Sensitivity Reduction in Control Systems*, R. F. Curtain, Ed. New York: Springer-Verlag, 1987.

[11] F. B. Yeh and T. S. Hwang, "A computational algorithm for the super-optimal solution of the model matching problem," *Syst. Contr. Lett.*, vol. 11, pp. 203–211, 1988.

[12] M. C. Tsai, D. W. Gu, and I. Postlethwaite, "A state-space approach to super-optimal H^∞ control problems," *IEEE Trans. Automat. Contr.*, vol. AC-33, pp. 833–843, 1988.

[13] D. J. N. Limebeer, G. D. Halikias, and K. Glover, "State-space algorithm for the computation of superoptimal matrix interpolating functions," *Int. J. Contr.*, vol. 50, pp. 2431–2466, 1989.

ROBUST CONTROL AND APPROXIMATION IN THE CHORDAL METRIC

Jonathan R. Partington

School of Mathematics, University of Leeds, Leeds LS2 9JT, U.K.

E-mail: PMT6JRP@UK.AC.LEEDS.CMS1 Fax: +44 532 429925.

Abstract

The chordal metric on SISO transfer functions (possibly unstable and infinite-dimensional), which generates the graph topology, is considered as a measure of robustness for stabilizing controllers. Connections with approximation are also explored.

1. Introduction.

One aim of this paper is to express the margin of robust stabilization of control systems in terms of the chordal metric on the Riemann sphere. This metric has been studied by function theorists (see e.g. Hayman [6]), since it is a natural analogue of the H_∞ distance between bounded analytic functions which can be used for functions with poles in the domain in question. Some results in this direction have been given by El-Sakkary [2,3], and it is clear that this is a fruitful area of enquiry.

In section 2 we shall study the continuity of the feedback process with respect to the topology given by the chordal metric (which, as we shall show, coincides with the topology given by the graph and gap metrics when we consider *all* stabilizable systems – this removes a technical restriction which was necessary in [10]). It will thus be possible to consider the stability margin in the chordal metric and we shall give an explicit expression for this. An application to approximation is also given.

Finally section 3 discusses some examples, including one in which the optimally robust constant feedback for an unstable delay system is calculated explicitly.

All our transfer functions will be single input, single output – it seems that a widely-used definition of the chordal metric for multivariable systems has still to be given.

2. The chordal metric and robustness.

In this section we relate the chordal metric between two (possibly unstable) plants to the gap and graph metrics. We shall see that it yields the same topology, and that it can therefore be used as a measure of robustness.

For convenience we shall throughout this paper use the symbols $c = \|C\|_\infty$ and $g = \|G\|_\infty$ to denote the H_∞ norms of functions in which we are particularly interested.

Three natural metrics can be defined to measure the closeness of one meromorphic function to another. The first two assume that two plants P_1 and P_2 have normalized coprime factorizations $P_k = N_k/D_k$ $(k = 1,2)$ with $N_k^* N_k + D_k^* D_k = 1$ on $\mathbb{T}$, and with X_k, Y_k in H_∞ such that $X_k N_k + Y_k D_k = 1$. See Vidyasagar [13] for details. We write $\hat{G}_k = \begin{bmatrix} N_k \\ D_k \end{bmatrix}$, for $k = 1,2$.

The *graph* metric $d(P_1, P_2)$ is defined by

$$d(P_1,P_2) = \max\left\{\inf_{Q\in H_\infty, \|Q\|\le 1} \|\hat{G}_1 - \hat{G}_2 Q\|_\infty, \inf_{Q\in H_\infty, \|Q\|\le 1} \|\hat{G}_2 - \hat{G}_1 Q\|_\infty\right\} \tag{2.1}$$

(see Georgiou [4], Vidyasagar [13]).

Likewise the *gap* metric $\delta(P_1, P_2)$ can be defined by

$$\delta(P_1,P_2) = \max\left\{\inf_{Q\in H_\infty} \|\hat{G}_1 - \hat{G}_2 Q\|_\infty, \inf_{Q\in H_\infty} \|\hat{G}_2 - \hat{G}_1 Q\|_\infty\right\} \tag{2.2}$$

(see Georgiou [4]). Since $\delta(P_1,P_2) \le d(P_1,P_2) \le 2\delta(P_1,P_2)$ these define the same topology, which is the natural topology for considering robust stabilization of systems. Meyer [9] has shown recently that the gap topology on systems of a fixed degree n is quotient Euclidean, though in general it is more complicated.

A third metric, the *chordal* metric $\kappa(P_1,P_2)$, may be defined as follows. For two complex numbers w_1 and w_2 the chordal distance between them on the Riemann sphere is

$$\kappa(w_1,w_2) = \frac{|w_1 - w_2|}{\sqrt{(1+|w_1|^2)(1+|w_2|^2)}} \tag{2.3}$$

with $\kappa(w,\infty) = 1/\sqrt{1+|w|^2}$.

Any function of the form

$$\tau(z) = e^{i\theta}\frac{az+b}{-\bar{b}z+\bar{a}},$$

with a, $b \in \mathbf{C}$, and $\theta \in \mathbf{R}$ is a conformal isometry of the Riemann sphere, that is, it satisfies $\kappa(\tau(w_1),\tau(w_2)) = \kappa(w_1,w_2)$ for all w_1, w_2, and corresponds to a rotation of the sphere. It is not hard to see that any point w_1 can be mapped to any other point w_2 by using these maps, an observation that will be useful later.

For two meromorphic functions P_1, P_2 in the open right half plane we write

$$\kappa(P_1,P_2) = \sup\{\kappa(P_1(s),P_2(s)) \ : \ \operatorname{Re} s > 0\}. \tag{2.4}$$

More about this metric can be found in Hayman [6].

El-Sakkary [2] showed that the chordal metric coincides with the gap metric when restricted to systems with no unstable zeroes or poles. The following inequality (see [6]) will be useful to us:

$$\kappa(w_1,w_2) \le \min\left(|w_1 - w_2|, \left|\frac{1}{w_1} - \frac{1}{w_2}\right|\right). \tag{2.5}$$

In some ways, this metric is a more natural one to consider than the graph and gap metrics: one can see at once that two functions P and P' are close if at each point either $|P(s) - P'(s)|$ is small or $|1/P(s) - 1/P'(s)|$ is small – this latter case handles poles of P and P'. In order to obtain a robustness result, we shall need a lower bound for κ, which acts as a partial converse to inequality (2.5).

Lemma 2.1. *For any two complex numbers w_1 and w_2, and for any a with $0 < a < 1$,*

$$\kappa(w_1, w_2) \geq \min\{a^2|w_1 - w_2|, a^2|1/w_1 - 1/w_2|, 1 - a^2\}/(1 + a^2).$$

Proof Consider the following three cases, which are collectively exhaustive:

(a) $|w_1| \leq 1/a$, and $|w_2| \leq 1/a$;

(b) $|w_1| \geq a$, and $|w_2| \geq a$;

(c) Either $|w_1| \leq a$ and $|w_2| \geq 1/a$, or vice-versa.

In case (a), the formula for κ shows that $\kappa(w_1, w_2) \geq |w_1 - w_2|/(1 + 1/a^2)$; in case (b) the same applies with w_1 and w_2 replaced by $1/w_1$ and $1/w_2$; and in case (c) $\kappa(w_1, w_2) \geq \kappa(a, 1/a) = (1 - a^2)/(1 + a^2)$.

□

El-Sakkary [3] has given some robustness results for the chordal metric, of which the following is typical: if $1/(1 + P)$ is stable, and

$$\kappa(P(s), P'(s)) < \frac{1}{3(1 + \|[1 + P(s)]\|^{-1}|^2)^{1/2}},$$

then $1/(1 + P')$ is also stable (and there is a bound on $\left|\frac{1}{1+P(s)} - \frac{1}{1+P'(s)}\right|$.)

Using Lemma 2.1, we can now prove some new robustness results in the chordal metric.

Theorem 2.2. *Suppose that $P(s)$ and $P'(s)$ are SISO transfer functions and $C(s) \in H_\infty$ is a stable controller such that the closed loop transfer function*

$$G(s) = P(s)/(1 + C(s)P(s))$$

is BIBO stable, i.e. $G(s) \in H_\infty$. If

$$\kappa(P, P') \leq (1/3)\min\{1, g^{-1}, c^{-1}(1 + cg)^{-1}\},$$

then P' is also stabilized by C, i.e. $G'(s) = P'(s)/(1 + C(s)P'(s)) \in H_\infty$.

Proof Suppose that s is a point such that $G'(s) = \infty$ but $|G(s)| \leq g = \|G\|_\infty < \infty$. We may estimate the chordal distance of $P(s)$ from $P'(s)$ using Lemma 2.1. Let a be any number with $0 < a < 1$.

(a) $|P(s) - P'(s)| = |1/(C(s)(1 - C(s)G(s)))| \geq 1/(c(1 + cg))$; and

(b) $|1/P(s) - 1/P'(s)| = |1/G(s)| \geq 1/g$.

Thus

$$\kappa(P, P') \geq \min\{a^2/(c(1 + cg)), a^2/g, 1 - a^2\}/(1 + a^2).$$

An easy limiting argument shows that the same is true if G' is merely unbounded, rather than having a pole in the right half plane.

The result now follows on taking $a = 1/\sqrt{2}$.

□

Given $c = \|C\|$ and $g = \|G\|$ it is in general possible to choose a in order to obtain tighter estimates, but we do not do this here.

We can improve upon the above result and show that if P is close to P' then G is close to G' in H_∞ norm, as follows.

Theorem 2.3. *Under the hypotheses of Theorem 2.2, let $\epsilon > 0$ be given. Then, if*

$$\kappa(P, P') < (1/3)\min\{1, \epsilon/((1+cg)(1+c(g+\epsilon))), \epsilon/(g(g+\epsilon))\},$$

then $\|G - G'\|_\infty < \epsilon$.

Proof Again, we use Lemma 2.1, and the estimates

$$|P(s) - P'(s)| = \frac{|G(s) - G'(s)|}{|1 - C(s)G(s)|\,|1 - C(s)G'(s)|},$$

and

$$|1/P(s) - 1/P'(s)| = \frac{|G(s) - G'(s)|}{|G(s)|\,|G'(s)|}$$

to bound κ from below, under the hypothesis that $|G(s) - G'(s)| = \epsilon$.

□

We can now give an alternative proof of the result of [10] that the chordal metric gives the graph topology; indeed we can now treat the case when P is any stabilizable plant (and hence when all nearby plants in the chordal metric are stabilizable by the same controller). Note that all stabilizable plants have coprime factorizations over H_∞, by [12].

Lemma 2.4. *Suppose that $P(s)$ and $P'(s)$ have coprime factorizations over H_∞, namely $P(s) = E(s)/F(s)$ and $P'(s) = E'(s)/F'(s)$. Then*

$$\kappa(P, P') \le A\max(\|E - E'\|, \|F - F'\|),$$

where A is a constant that depends only on P, not on P'.

Proof Since E and F are coprime, they satisfy a Bezout identity $X(s)E(s) + Y(s)F(s) = 1$ with $X, Y \in H_\infty$. Let $x = \|X\|_\infty$ and $y = \|Y\|_\infty$. Then certainly for each s we have that $x|E(s)| + y|F(s)| \ge 1$, and hence if $|E(s)| \le 1/2x$ then $|F(s)| \ge 1/2y$. We can thus estimate $\kappa(E/F, E'/F')$: when $|F(s)| \ge 1/2y$,

$$\kappa(E/F, E'/F') \le |E/F - E'/F'| \le \frac{|E||F' - F| + |F||E - E'|}{|F||F'|}$$

which will be at most a constant times $\max(\|E - E'\|_\infty, \|F - F'\|_\infty)$ provided that $\|F - F'\|_\infty < 1/4y$; alternatively, when $|F(s)| < 1/2y$ and $|E(s)| \ge 1/2x$,

$$\kappa(E/F, E'/F') \le |F/E - F'/E'| \le \frac{|F||E' - E| + |E||F - F'|}{|E||E'|},$$

using (2.5) again, and this is bounded by a constant times $\max(\|E-E'\|_\infty, \|F-F'\|_\infty)$ provided that $\|E-E'\| < 1/4x$.

Corollary 2.5. *With the notation above, and with G, C, P fixed and G', P' varying, the following conditions are equivalent:*

$$(a)\ \|G'-G\| \to 0; \qquad (b)\ \kappa(P,P') \to 0; \qquad (c)\ \delta(P,P') \to 0.$$

Hence the chordal metric determines the graph topology on the set of stabilizable systems.

Proof By Theorem 2.3 (b) implies (a); by Lemma 2.4 and the definition of δ, (c) implies (b); finally that (a) implies (c) is well-known using the results of [13], since $P = G/(1-CG)$ and G and $1-CG$ are coprime, similarly $P' = G'/(1-CG')$; then $\|G-G'\| \to 0$ and $\|(1-CG)-(1-CG')\| \to 0$, which implies (c).

□

We give now an explicit formula for the optimal robust stability margin for a stable controller measured in the chordal metric. This formula shows a similarity to many of the H_∞ optimization problems encountered in other robustness questions.

Theorem 2.6. *Let $P(s)$ be stabilizable, and let $C(s)$ be a stabilizing controller. Then the chordal distance from P to the set of plants not stabilizable by C is $\sqrt{1-\eta^2}$, where*

$$\eta = \kappa(C,\overline{P}) = \sup_s\{\kappa(C(s),\overline{P(s)})\}. \tag{2.6}$$

Hence any C minimizing (2.6) is optimal.

Proof Note that $P'/(1+CP')$ is unstable precisely when

$$\inf_s\{\kappa(P'(s), -1/C(s))\} = 0.$$

Hence if $\kappa(P(s),-1/C(s)) \geq \delta$ for all s then $\kappa(P,P') \geq \delta$ for any P' not stabilized by C; and if $\kappa(P(s),-1/C(s)) = \delta$ for some s then there is a plant P' with $\kappa(P,P') = \delta$ and $P'(s) = -1/C(s)$, namely $P'(s) = \tau(P(s))$ for some rotation τ of the Riemann sphere such that $\tau(P(s)) = -1/C(s)$. Hence the stability margin of C is given by

$$\epsilon = \inf_s\{\kappa(P(s),-1/C(s))\} = \inf_s\{\kappa(C(s),-1/P(s))\}.$$

Since $(\kappa(w_1,w_2))^2 + (\kappa(w_1,-1/\bar{w}_2))^2 = 1$ for any w_1 and w_2, it follows that $\epsilon = \sqrt{1-\eta^2}$, where

$$\eta = \sup_s \kappa(C(s),\overline{P(s)}) = \kappa(C,\overline{P}).$$

□

For robust stabilization using normalized coprime factors, as in [8] and [1], the plant P is given as $N(s)/D(s)$ where N and D are normalized and coprime. A perturbed model is $(N+\Delta N)/(D+\Delta D)$, where ΔN and ΔD are in H_∞ with small norm. We write $\Delta P(s) = [\Delta N(s), \Delta D(s)]$. The following simple result estimates such a stability margin using the chordal metric.

Theorem 2.7. *For N, D, ΔN, ΔD as above,*

$$\kappa(N/D,(N+\Delta N)/(D+\Delta D)) \leq \frac{\|\Delta P\|}{1/2 - \|\Delta P\|/\sqrt{2}}.$$

Proof Using (2.5), we have that (at a point s)

$$\kappa(N/D,(N+\Delta N)/(D+\Delta D)) \leq \min\left\{\left|\frac{D\Delta N - N\Delta D}{D(D+\Delta D)}\right|, \left|\frac{N\Delta D - D\Delta N}{N(N+\Delta N)}\right|\right\}.$$

Now either $|N(s)| \geq 1/\sqrt{2}$ or $|D(s)| \geq 1/\sqrt{2}$, and also

$$|D\Delta N - N\Delta D| \leq \|[N,D]\|\,\|[\Delta D, \Delta N]\| = \|\Delta P\|,$$

which gives the result on taking the supremum over s.

□

Finally we give an application to approximation of systems. Let P be a given system, and C a stabilizing controller, such that C and $G = P/(1+CP)$ are both in H_∞. Let G' be an H_∞ approximation to G. This yields an approximation $P' = G'/(1-CG')$ to P and our final result estimates the distance of P' from P in the chordal metric.

Theorem 2.8. *Let P be a possibly unstable open-loop transfer function and $C \in H_\infty$ a stable controller such that the closed-loop transfer function $G = P/(1+CP) \in H_\infty$ is stable; let G' be an approximation to the closed loop transfer function G, with $\|G - G'\| = \alpha$ and $P' = G'/(1-CG')$ the corresponding approximation to the open-loop transfer function P. Then the following error bound holds for the chordal metric $\kappa(P,P')$.*

$$\kappa(P,P') \leq \frac{\alpha(1+c)^2}{1-\alpha(1+c)\max(1,c)}.$$

The same inequalities hold if the roles of G and P and their primed equivalents are reversed.

Proof Note that $|C(s)||G(s)| + |1 - C(s)G(s)| \geq 1$ at each point.

We consider the cases when (a) $|G|/|1-CG| \leq 1$ and $|1-CG| \geq 1/(1+c)$ and (b) $|G|/|1-CG| \geq 1$ and $|G| \geq 1/(1+c)$.

In the first case,

$$P - P' = \frac{(G-G')}{(1-CG)(1-CG')}.$$

Hence

$$|P - P'| \leq \frac{\alpha}{(1/(1+c))(1/(1+c) - c\alpha)}.$$

In the second case,

$$1/P - 1/P' = \frac{G'-G}{GG'}.$$

Hence

$$|1/P - 1/P'| \leq \frac{\alpha}{(1/(1+c))(1/(1+c) - \alpha)}.$$

Finally, taking the supremum over all points s, we obtain

$$\kappa(P,P') \leq \frac{\alpha(1+c)^2}{1-\alpha(1+c)\max(1,c)}$$

using (2.5) again.

□

An application to identification is given in [7]. Further results on approximation of unstable systems in the chordal metric may be found in [10].

3. Examples.

We begin with an elementary example. Consider the plant $P(s) = 1/s$. By Theorem 2.6, an optimally robust controller C can be found by minimizing

$$\eta = \kappa(C(s), 1/\bar{s}).$$

This is the same as

$$\kappa\left(\frac{1-C(s)}{1+C(s)}, \frac{\bar{s}-1}{\bar{s}+1}\right),$$

and transforming to the disc, with $z = (1-s)/(1+s)$, and noting that $C(s) \neq -1$ at an optimum, we have to minimize $\kappa(D(z), \bar{z})$ for $D = (1-C)/(1+C) \in H_\infty$. Winding number considerations show that, as in the H_∞ case, the optimum D is identically zero, so that $C(s) = 1$. The stability margin is $(1-\kappa(1,1/\bar{s})^2)^{1/2} = 1/\sqrt{2}$. A function P' with $\kappa(1/s, P') = 1/\sqrt{2}$, that is not stabilized by C, is $\tau \circ P$, where τ is a suitable rotation of the Riemann sphere. Indeed $P'(s) = (1+s)/(1-s)$ will do.

Now consider the continuous-time delay system with transfer function $P(s) = e^{-s}/s$. It is well known that, with a constant controller C, the closed-loop transfer function $G(s) = e^{-s}/(s + Ce^{-s})$ is stable for $0 < C < \pi/2$. Consider now this restricted problem of robust stabilization - to determine which value of C gives the maximal stability margin in the chordal metric.

We can re-express the results of Theorem 2.6 as

$$\epsilon = \max_{0<C<\pi/2} \inf_s \{\kappa(1/P(s), -C)\} = \max_{0<C<\pi/2} \inf_s \{\kappa(se^s, -C)\}.$$

For each C this infimum, $\epsilon(C)$ say, will be attained at some s_0 and one can then use the isometries of the Euclidean sphere to obtain a function P' such that $P'(s_0) = -1/C$ and $\kappa(se^s, P') \leq \epsilon(C)$ for all s (geometrically this is just $P' = \tau \circ P$ for some τ rotating the sphere.)

The robustness margin $\epsilon(C)$ is plotted below for the relevant values of C; its maximum value is 0.397, achieved at $C = 0.46$. Thus any function P' whose chordal distance from e^{-s}/s is at most 0.397 is stabilized by $C = 0.46$, and there is a function at this distance which is not stabilized by C.

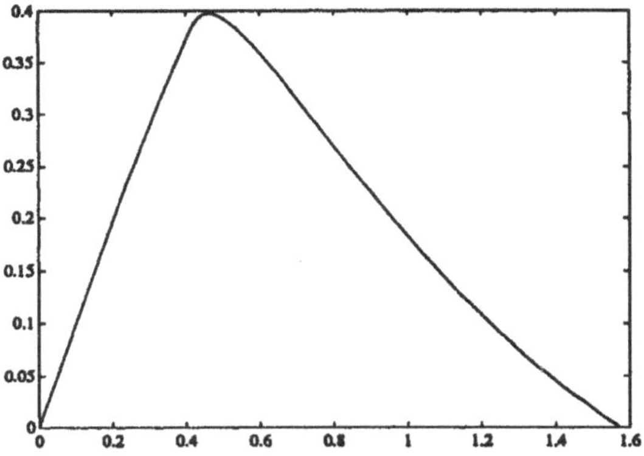

Plot of chordal robustness margin $\epsilon(C)$ against C

References.

[1] R.F. Curtain, "Robust stabilizability of normalized coprime factors: the infinite-dimensional case," *Int. Journal of Control* 51 (1990), 1173-1190.

[2] A.K. El-Sakkary, "The gap metric: robustness of stabilization of feedback systems," *I.E.E.E. Trans. Autom. Control* 30 (1985), 240-247.

[3] A.K. El-Sakkary, "Robustness on the Riemann sphere," *Int. Journal of Control* 49 (1989), 561-567.

[4] T.T. Georgiou, "On the computation of the gap metric," *Systems and Control Letters* 11 (1988), 253-257.

[5] T.T. Georgiou and M.C. Smith, "Optimal robustness in the gap metric," *I.E.E.E. Trans. Autom. Control* 35 (1990), 673-686.

[6] W.K. Hayman, *Meromorphic functions*, Oxford University Press (1975).

[7] P.M. Mäkilä and J.R. Partington, "Robust identification of stabilizable systems," Report 91-9, University of Leeds (1991).

[8] D.C. McFarlane and K. Glover, *Robust controller design using normalized coprime factor plant descriptions*, Springer-Verlag (1990).

[9] D.G. Meyer, "The graph topology on nth order systems is quotient Euclidean," *I.E.E.E. Trans. Autom. Control* 36 (1991), 338-340.

[10] J.R. Partington, "Approximation of unstable infinite-dimensional systems using coprime factors," *Systems and Control Letters* 16 (1991), 89-96.

[11] J.R. Partington and K. Glover, "Robust stabilization of delay systems by approximation of coprime factors," *Systems and Control Letters* 14 (1990), 325-331.

[12] M.C. Smith, "On stabilization and the existence of coprime factors," *I.E.E.E. Trans. Autom. Control* 34 (1989), 1005-1007.

[13] M. Vidyasagar, *Control System Synthesis: a factorization approach*, MIT Press, 1985.

FINITE-DIMENSIONAL ROBUST CONTROLLER DESIGNS FOR DISTRIBUTED PARAMETER SYSTEMS: A SURVEY

R.F. Curtain
Mathematics Institute, University of Groningen
P.O. Box 800, 9700 AV GRONINGEN, The Netherlands

1. Introduction

Although there has been much research on the stabilization of distributed parameter systems, the theory usually produces infinite-dimensional controllers. For example, the theory of linear quadratic control produces a state-feedback law in which both the state and the gain operator are infinite-dimensional. To produce an implementable finite-dimensional controller entails approximating both this gain operator and the state which is a complicated numerical procedure and sometimes (in the case of unbounded inputs) difficult to justify theoretically. The continuing popularity of this approach is somewhat surprising, since there exist several direct methods of designing finite-dimensional controllers for infinite-dimensional systems. The first finite-dimensional compensator design was a state-space approach in 1981 and since then there have been many others. In particular, there exist various frequency domain theories for designing finite-dimensional controllers which are robust to certain types of uncertainties or which satisfy other additional performance objectives. This talk will be a survey and comparison of ways of designing finite-dimensional controllers for infinite-dimensional systems with the emphasis on robust controllers.

2. Approximation of infinite-dimensional deterministic LQG controllers: Numerical approximation of infinite-dimensional Riccati equations

One of the early successes of infinite-dimensional systems theory was the solution of the linear quadratic control problem [Lions, 1971] and this topic dominated the literature for decades. In particular, the case of boundary control for p.d.e. systems and delayed control for delay systems presented difficult mathematical problems and was the subject of many treatises. The solution yields a state-feedback controller whose gain is the solution of an infinite-dimensional Riccati equation. The numerical approximation of infinite-dimensional Riccati equations has been the subject of many papers, most of which assume bounded input operators. A recent survey of infinite-dimensional Riccati equations, including those with unbounded input operators can be found in [Lasiecka and Triggiani, 1991]. Although there are fewer results on numerical approximations with unbounded inputs, the underlying idea is the same. One approximates the underlying system operator (p.d.e. or delay type) by a suitable finite-dimensional scheme and one gives sufficient conditions on the approximations to obtain strong convergence of the finite-rank approximating solutions to the infinite-dimensional operator. Recently, more compact sufficient conditions have been given to ensure convergence ([Kappel and Salomon], [Itô], but the choice of the approximations still depends strongly on the type of system one is dealing with (parabolic, hyperbolic or delay). However, by now there is a lot of knowledge about the performance of different approximation schemes, at least for the case of bounded inputs.

Of course, even if one has a sequence of finite-rank gains which converge strongly to the optimal gain, implemention of the *LQ* control law requires knowledge of the whole state. While there is a stochastic *LQG* theory [Curtain and Pritchard, 1978], most attempts at designing finite-dimensional compensators have been deterministic and there have been very few of these. In [Itô, 1990] general conditions are given for the convergence of finite-dimensional approximation schemes to infinite-dimensional compensators. These infinite-dimensional compensators will have been developed in a way so that the closed loop system is exponentally stable. The classic application is to the combination of a *LQ* control law with a state-observer – a deterministic *LQG* design approach. In principle, this reduces to two Riccati approximation schemes: one for the control Riccati equation and the other for the "filter" Riccati equation. However, a more careful analysis of a specific application to flexible systems in [Gibson and Adamarian] reveals that if one wishes to do a realistic engineering filter design, it might be difficult to satisfy all of the assumptions one needs to guarantee convergence. In this example, a physically meaningful choice of the state noise weights did not satisfy their theoretical assumptions and so they could not prove convergence. However, their numerical results indicated that things did converge nicely.

Summarizing, we can make the following, general comments on finite-dimensional compensator design via approximation of infinite-dimensional, deterministic *LQG* compensators. At least for the case of bounded control and measurement, there is a fairly complete theory of convergence of the approximations and sufficient knowledge about the appropriate choice of the approximations for parabolic, hyperbolic and delay systems. The theory for the case of unbounded inputs and outputs is somewhat less developed and one may have to exercise some care in the choice of weights to satisfy the theoretical conditions. This notwithstanding, this design method is not recommended for non experts (see [Burns et al, 1988], and research is still continuing into important properties of the approximations, such as the stabilizability and detectability properties of the approximations [Burns].

3. Direct state-space design of infinite-dimensional compensators

Surprisingly enough, it took a long time to realize that it was possible to directly design a finite-dimensional compensator for an infinite-dimensional system without approximating anything. The first design in [Schumacher, 1981, 1983] was a geometric approach and was numerically simple. The assumptions were basically that the inputs and output operators B and C were bounded, the system operator A should generate a C_0-semigroup, (A,B,C) should be stabilizable and detectable and the eigenvectors of A

should span the state-space. It was this last assumption which was restrictive. While most p.d.e. systems will satisfy this assumptions, many retarded systems will not. The finite-dimensional compensator designs in [Sakawa, 1984] and [Curtain, 1984] are similar to each other and use a type of modal approach in designing the compensator; they also need the assumption the the eigenvectors span the state-space. A completely different approach was taken in [Bernstein and Hyland, 1986] who presented a theory for designing a fixed order, finite-dimensional compensator. They do not need the assumptions on the eigenvectors, but they do need to assume the existence of solutions of nonlinear "optimal projection equations" and they need to solve these numerically. This is a nontrivial task, in contrast to the other three design approaches which can be carried out in using PCMATLAB routines.

4. Finite-dimensional compensator designs based on reduced order models

Basic engineering methodology is to design a compensator based on approximate model of the true system: the reduced order model. In our application the true system is infinite-dimensional, the approximation is finite-dimensional and the compensator will be finite-dimensional. However, we apply the finite-dimensional controller to the infinite-dimensional system, hoping that it will stabilize, although we are only guaranteed that it will stabilize the reduced order model. Usually it works fine, but it can happen that it fails to stabilize the original infinite-dimensional systems. This phenomenon was demonstrated in [Balas, 1978] and christened as the "spillover" problem. In [Balas, 1980] he suggested various choices of reduced modal on Galerkin approximations of the infinite-dimensional state-space and a list of sufficient conditions which were rather complicated to check. In [Bontsema and Curtain, 1988], an explanation of and a remedy for the spillover problem were given in terms of robustness properties of the plant. In fact, it is a simple corollary of the theory of additively robust controllers which we will discuss in section 5.

We shall use the following notation:

H^∞ The class of functions of a complex variable which are bounded and holomorphic in $Re(s)>0$.

H^∞_- The class of functions of a complex variable which are bounded and holomorphic in $Re(s)>-\varepsilon$, for some $\varepsilon>0$.

$H^\infty_-+\mathbb{R}(s)$ The class of transfer functions which are the sum of one in H^∞_- and a proper rational one.

$M(H^\infty)$ Transfer matrices of any size whose components are elements of H^∞.

H^∞/H^∞ The quotient space of H^∞.

The class of systems considered comprises plants with the decomposition $G=G_s+G_u$, where $G_s\in M(H^\infty_-)$ and G_u is a strictly proper rational transfer matrix. In particular, this means that the plant has only finitely many unstable poles in $Re(s)\geq 0$.

The appropriate concept of stability is defined next.

Definition 4.1 The feedback system (G,K) with $G,K\in M(H^\infty_-+\mathbb{R}(s))$ is said to be *input–output stable* if and only if (a) and (b) are satisfied.

(a) $\inf_{\mathbb{C}^+} |det(I-GK)(s)| > 0$

(b) $S=(I-GK)^{-1}$, KS, SG, $I-KSG \in M(H_-^{\infty})$ ∎

The following Lemma is an easy consequence of the theory of robust stabilization under additive perturbations which we shall discuss in more detail in section 5.

Lemma 4.2 Suppose the plant G is strictly proper and has a decomposition $G=G_u+G_s$, where $G_s \in M(H_-^{\infty})$ and G_u is a strictly proper, rational transfer matrix with all its poles in $Re(s)>0$. Let $G_f=G_u+G_a$, where G_a is an L_∞-approximation for G_s. If K_f stabilizes the reduced order model G_f, then K_f also stabilizes G provided that

$$\|G-G_f\|_\infty = \|G_a-G_s\|_\infty < \|K_f(I-G_fK_f)^{-1}\|_\infty^{-1}$$

Furthermore, a stabilizing finite-dimensional compensator K_f for G_f and G exists provided that

$$\|G-G_f\|_\infty < \sigma_{min}(G_u^{\dagger}).$$

($\sigma_{min}(G_u^{\dagger})$ is the smallest Hankel singular value of the stable rational transfer function $G_u^{\dagger}(s)=G_u^{\dagger}(-s)$. It gives an upper bound to the L_∞-approximation error.) ∎

The class of infinite-dimensional systems covered by this lemma is very large. It includes systems with a state-space representation (A,B,C,D) in which B and C may be unbounded. The system should be "well-posed" in the sense that it should have well-defined controllability, observability map and input-output map, but these are very mid requirements (see Curtain and Weiss, 1989). The essential restriction is that the operator A should have only finitely many unstable eigenvalues, but this is also necessary for the LQG theory in the case of bounded, finite-rank inputs and outputs. In addition, this class includes systems which do not have nice state-space representation and for which Riccati equations are not well-posed. For example, it includes the important class of systems of the type $G(s)=e^{-\alpha s}G_f(s)$, where $G_f(s)$ is a proper, rational transfer function. Lemma 4.2 suggests the following

Reduced order model design approach 4.3

Step 1 Find a reduced order model G_f so that $\|G_s-G_f\|_\infty<\varepsilon$

Step 2 Design your favorite compensator K_f(e.g. LQG) for G_f and calculate its robustness margin $\|K_f(I-G_fK_f)^{-1}\|_\infty^{-1}$.

Step 3 Check whether $\|G_s-G_f\|_\infty < \|K_f(I-G_fK_f)^{-1}\|_\infty^{-1}$. If not, go to step 1 and obtain a more accurate reduced-order model G_f and repeat the loop.

Step 4 If the inequality checks out, K_f stabilizes G.

The only numerical approximation occurs in step 1 and there are many excellent techniques for these approximations, [Glover, Curtain and Partington, 1988], [Hover, Lam and Partington, 1990, 1991], [Gu, Khargoneka, Lee, 1989], [Zwart, Curtain, Partington and Glover, 1988] and [Partington, Glover, Zwart and Curtain, 1988]. This is a numerically simpler design approach than that approximating solution of infinite-dimensional Riccati equations which we discussed in section 1. In addition, the number $\sigma_{min}(G_u^{\dagger})$ gives a good indication of the order of the reduced-order model required. The L_∞-approximation error should be less than $\sigma_{min}(G_u^\dagger)$. A small value of $\sigma_{min}(G_u^\dagger)$ indicates, a priori, that a low order reduced order model could lead to spillover problems.

Finally, a word about exponential stability of the closed loop system and stability in the sense of Lemma 3.1: in general, this type of input-output stability is weaker than exponential stability. However, for large classes of state-space realizations it will imply exponential stability of the closed loop system (see Curtain, 1988).

5. Robust controller designs with a-priori robustness bounds

Since the infinite-dimensional model is, at best, an approximate one, it is highly desirable that the controllers will also stabilize perturbations of the nominal model, the *robustness property*. Depending on the class of perturbations chosen and the topology chosen to measure the size of the perturbations, one obtains different definitions and measurements of robustness. A minimal requirement is that of robustness with respect to the graph topology, [Vidyasagar, chapter 7].

Recently, in [Morris, 1991], it was shown that under the usual assumptions made in approximating infinite-dimensional system in the design of infinite-dimensional deterministic *LQG* controllers, the approximations converged in the graph topology. This can be interpreted to show that the deterministic *LQG* controller design discussed in section 2 is robust to perturbations which are "small" in the graph topology. Reassuring as this result is, it gives no information as to the size on nature of perturbations which are allowed. There are no results on the robustness of the direct state-space design compensators discussed in section 3, although numerical results carried out on an example using the design in [Curtain, 1984] were disappointing. On the other hand, it is clear that Lemma 4.1 is a type of robustness result. It says that there is robustness with respect to stable additive perturbations in the plant and it also gives an indication of the robustness tolerance in $\|K_f(I-G_fK_f)^{-1}\|_\infty^{-1}$.

In this section, we discuss two simple finite-dimensional robust controller designs for infinite-dimensional systems with finitely many unstable poles, which give information about the size and nature of perturbations which retain stability.

The first is that of *robust stabilization of a plant G under additive perturbations Δ*, G and Δ are assumed to satisfy the following assumptions

(5.1) G is regarded as the nominal plant and is assumed to be a strictly proper transfer matrix in $M(H_-^\infty+\mathbb{R}(s)$ and moreover, it is assumed to have no poles on the imaginary axis.

(5.2) The permissible perturbations Δ are also in $M(H_-^\infty+\mathbb{R}(s))$ and they are such that Δ and $\Delta+G$ have the same number of unstable poles in $Re(s)\geq 0$.

Definition 5.1 The feedback system (G,K) with $G,K\in M(H_-^\infty+\mathbb{R}(s))$ is said to be *additively robustly stable* with *robustness margin ε* if the feedback system is input-output stable for all G and Δ satisfying (5.1) and (5.2) and $\|\Delta\|_\infty<\varepsilon$. ∎

The assumption (5.1) and (5.2) are needed to apply a Nyquist argument and the following result from [Logemann, 1990] is a generalization of first version for infinite-dimensional systems in [Curtain and Glover, 1986].

Theorem 5.2 If $G_s \in M(H_-^\infty + \mathrm{R}(s))$ and (5.1) is satisfied, given an $\varepsilon > 0$ there exists a compensator K which stabilizes $G+\Delta$ for all Δ satisfying (5.2) and $\|\Delta\|_\infty < \varepsilon$ if and only if

$$\sigma_{min}(G_u^\dagger) \geq \varepsilon \tag{5.4}$$

where $\sigma_{min}(G_u^\dagger)$ is as in Lemma 4.2. ∎

In addition to this existence result, there are also explicit formulas for a controller K which achieves the maximum robustness margin $\sigma_{min}(G_u^\dagger)$. These depend on the stable part of G, G_s, and so this K is *infinite-dimensional*, which is undesirable for applications. However, is one replaces G_s by a finite-dimensional L_∞-approximation, G_s^k, say, then one can design a finite-dimensional controller (of the order of k plus the McMillan degree of G_u) which has a robustness margin of at least $\sigma_{min}(G_u^\dagger) - \|G_s - G_s^k\|_\infty$. So in combination with a theory for good L_∞-approximations, we have a practical design technique for additively robust finite-dimensional controllers for a large class of infinite-dimensional systems.

Reduced-order model design approach 5.3

Step 1 Calculate $\sigma_{min}(G_u^\dagger)$.

Step 2 Approximate G by $G_s^k + G_u$ so that $\|G_s^k - G_s\|_\infty << \sigma_{min}(G_u^\dagger)$.

Step 3 Find the finite-dimensional additively robustly stabilizing compensator K_f which stabilizes $G_s^k + G_u$.

Step 4 K_f additively robustly stabilizes G with a robustness margin of at least $\sigma_{min}(G_u^\dagger) - \|G_s^k - G_s\|_\infty$. ∎

The same remarks made in section 4 concerning the L_∞-approximation techniques and the class of systems covered apply here too. The advantage of this approach over the reduced order model design approach 4.3 is that we can give an a priori robustness margin of $\sigma_{min}(G_u^\dagger) - \|G_s - G_s^k\|_\infty$ and design a finite-dimensional controller to achieve this. Applications of this technique to design robustly stabilizing finite-dimensional controllers for a class of flexible systems can be found in [Bontsema, 1989].

The second robust controller design is applicable to a wider class of systems, including those with infinitely many unstable poles.

Robustness stabilization under normalized coprime factor perturbations

The plant G and the controller are allowed to be in $M(H^\infty/H^\infty)$; the elements of the transfer matrix are quotients of elements in H^∞. In addition, it is assumed that G has a normalized coprime factorization.

Definition 5.4 If $G \in M(H^\infty/H^\infty)$, then $\tilde{M}^{-1}\tilde{N}$ is called a *left coprime factorization* of G if and only if

(*i*) $\tilde{M}$ and $\tilde{N} \in M(H^\infty)$

(*ii*) $G = \tilde{M}^{-1}\tilde{N}$

(*iii*) $\tilde{M}$ and $\tilde{N}$ are left coprime in the sense that there exist X and Y in $M(H^\infty)$ such that

$$\tilde{N}Y - \tilde{M}X + I \tag{5.3}$$

$\tilde{M}^{-1}\tilde{N}$ is called a *normalized* left coprime factorization of G if it is a left coprime factorization and

$$\tilde{N}(j\omega)\tilde{N}^t(-j\omega) + \tilde{M}(j\omega)\tilde{M}^t(-j\omega) = I \text{ for all real } \omega \tag{5.4}$$

The definition of a right coprime factorization is analogous.

It is known that $G \in M(H^{\infty}/H^{\infty})$ has a normalized left–coprime factorization if and only if G is stabilizable in the sense of Definition 4.1 (see [Smith, 1989]. (This definition can be extended in the obvious way to G, $K \in M(H^{\infty}/H^{\infty})$).

We consider the robustness of stability with respect to perturbations of the form

$$G_{\Delta} = (\tilde{M} + \Delta_N)^{-1}(\tilde{N} + \Delta_N) \tag{5.5}$$

where $G = \tilde{M}^{-1}\tilde{N}$ is a left coprime factorization of G and Δ_M, $\Delta_N \in M(H^{\infty})$.

The robust design objective is to find a feedback controller $K \in M(H^{\infty}/H^{\infty})$ which stabilizes not only the nominal system G, but also the family of perturbed systems defined by

$$G_{\epsilon} = \{G_{\Delta} = (\tilde{M} + \Delta_M)^{-1}(\tilde{N} + \Delta_N) \text{ such that } \Delta_M, \Delta_N \in M(H^{\infty}) \tag{5.6}$$
$$\text{and } \|[\Delta_M, \Delta_N]\|_{\infty} < \epsilon\} \text{ (see Figure 5.2)}$$

This leads to the following definition of factor robust stability.

Definition 5.5 Suppose that the system $G \in M(H^{\infty}/H^{\infty})$ has the left normalized coprime factorization of Definition 5.3. Then the feedback system (G,K) is *factor robustly stable* if and only if (G_{Δ},K) is input–output stable for all $G_{\Delta} \in G_{\epsilon}$. If there exists a K such that (G,K) is factor robustly stable, then (G,ϵ) is said to be factor robustly stabilizable with *factor robustness margin ϵ*. ∎

This problem has an elegant solution.

Theorem 5.6 [Georgiou and Smith, 1990] If $G \in M(H^{\infty}/H^{\infty})$ has a left normalized coprime factorization, then $(\tilde{M},\tilde{N},K,\epsilon)$ is factor robustly stable if and only if

$$\epsilon \le (1 - \|[\tilde{N},\tilde{M}]\|_H)^{\frac{1}{2}}$$

where $\|\cdot\|_H$ represents the Hankel norm of the system $(\tilde{M},\tilde{N})$. The maximum factor robustness margin is

$$\epsilon_{max} \le (1 - \|[\tilde{N},\tilde{M}]\|_H)^{\frac{1}{2}}$$ ∎

In general, explicit formulas for normalized coprime factorizations are not available. However, for proper rational transfer functions and a class of infinite–dimensional systems they are given in terms of the solutions of algebraic Riccati equations (see [Glover and McFarlaine, 1988, 1990], [Curtain, 1990] and [Curtain and Van Keulen, 1990]). Moreover, explicit formulas for robustly stabilizing controllers are also known for these classes of systems. While it is nice to have an elegant mathematical solution for infinite–dimensional systems in terms of algebraic Riccati equations, these are not a good basis for designing finite–dimensional robustly stabilizing controllers. In [Bontsema and Curtain, 1989] the following approximation result was proved for plants in $G \in M(H^{\infty}_{-}+R(s))$.

Lemma 5.7 Suppose that $G \in M(H^{\infty}_{-}+R(s))$ has the decomposition $G = G_f + G_s$, where $G_s \in M(H^{\infty}_{-})$ and G_f is a rational transfer function. G_f is a reduced order model for G such that $\|G - G_f\|_{\infty} < \mu < \epsilon_{max}$. If K_f stabilizes G_f with a factor robustness margin of $\epsilon > \mu$, it stabilizes G with the factor robustness margin of at least $\epsilon - \mu$. ∎

This yields the following algorithm for designing a robust finite–dimensional compensator.

Reduced order model design approach 5.8

Step 1 Find a reduced-order model G_f for G with an L_∞-error $\|G-G_f\|_\infty<\mu$.

Step 2 Obtain a minimum realisation for G_f and calculate the maximal factor robustness margin for G_f (This involves solving Riccati equations), say ε_{max}.

Step 3 Compare ε_{max} and μ. If $\varepsilon_{max}-\mu$ is acceptable as a robustness margin, go to step 4. Otherwise, return to step 1 and find a reduced-order model with a smaller L_∞-error.

Step 4 Find a factor robustly stabilizing compensator K_f for G_f with robustness margin $\varepsilon>>\mu$.

Step 5 K_f factor robustly stabilizes G with robustness margin $\varepsilon-\mu$.

This is analogous to the reduced-order model design 5.3 for additive perturbations. It is applicable to a wider class of systems, since (5.1) and (5.2) do not need to be satisfied here. In particular, G can have finitely many poles on the imaginary axis and G and G_Δ do not need to have the same number of unstable poles. The design approach 5.8 was applied to a flexible beam model with parameter uncertainty in [Bontsema and Curtain, 1989]. This study was a follow-up of that on additively robustly stabilizing compensators in [Bontsema, 1989]. Both the additive and the factor theories allow one to measure the perturbations allowed (in the L_∞-norm and in the gap metric) and this was done for parameter variations in a flexible beam model. The reduced order model design approach 5.8 allowed the largest range of parameter variations in the model, both in the theoretically guaranteed robustness margin and in the actual robustness margin. This is to be expected as it admits a larger class of perturbations; in particular, G and G_Δ do not need to have the same number of unstable poles.

6. Conclusions

We have discussed five different approaches to designing finite-dimensional compensators for infinite-dimensional systems. The infinite-dimensional *LQG* approach in section 1 finds an infinite-dimensional compensator first and then approximates to obtain a finite-dimensional compensator, whereas the state-space approaches in section 2 lead in one step to a finite-dimensional compensator. The remaining approaches following the philosophy of approximating the irrational transfer function first by a reduced-order model and designing a finite-dimensional compensator for it. Coupled with good L_∞-approximation techniques, these offer computational simplicity. The reduced order model design approach 4.3 has long been standard engineering practice. The two robust controller designs discussed in section 5 have the extra advantage of guaranteeing robustness to uncertainties in the model. The robustness margin can be calculated a priori and it corresponds to L_∞- or gap-metric perturbations in the nominal infinite-dimensional model.

References: *These have been omitted due to space limitations.*

A complete version of this paper is available on request.

ROBUST COVARIANCE CONTROL

J.-H. Xu and R.E. Skelton
School of Aeronautics & Astronautics
Purdue University
West Lafayette, IN 47907

Abstract
This paper extends first the covariance assignment theory for continuous time systems to the assignment of a covariance upper bound in the presence of structured perturbations. A systematic way is given to construct an assignable covariance upper bound for the purpose of robust design.

1. Introduction

The main idea of the covariance control theory developed in [1-4] is to choose a state covariance X_o according to the different requirements on the system performance and then to design a controller so that the specified state covariance is assigned to the nominal closed loop system. However, because of modeling error, the real system may differ from the nominal system for the design. As a result, the state covariance X of the real system will not be equal to the desired X_o.

Reference [5] provides an upper bound for the perturbed state covariance of a linear system when all system perturbations are in a given structured set. The purpose of this paper is to present a way for designing a robust covariance controller which guarantees that the state covariance X of the perturbed system is upper bounded by a specified matrix $\overline{X}$ as long as the plant perturbations are in a given structured set.

The description of the paper is organized into 5 sections. Section 2 reviews briefly the needed robustness analysis result in [5]. Section 3 extends the covariance assignment theory to the assignment of a covariance upper bound when the plant matrix is perturbed in a structured way. In Section 4, the problem of designing a robust covariance controller is solved by constructing an assignable covariance upper bound. The paper is concluded in Section 5 where directions for further research are discussed.

2. Upper Covariance Bound of Perturbed Systems

Consider the following time-invariant feedback system

$$\dot{x}(t) = A_o x(t) + Dw(t) , \tag{2.1}$$

where A_o is stable, w(t) is a white noise signal of unit intensity. The state covariance X_o defined as

$$X_o \triangleq \lim_{t \to \infty} E\,[x(t)x^*(t)] \tag{2.2}$$

is the unique solution to the following Lyapunov equation [6]

$$A_o X_o + X_o A_o^* + DD^* = 0\,. \tag{2.3}$$

In practice, perturbations in the system matrix A_o are inevitable. As a result, the real system matrix is actually $A_o + \Delta A$, instead of A_o, where the perturbation ΔA describes the uncertainty of the system and can be characterized in certain cases by a specified set Ω as follows

$$\Omega \triangleq \{\Delta A \;:\; \Delta A \Delta A^* \le \bar{A}\,,\ \text{and } (A_o + \Delta A\,,\ D) \text{ is controllable}\}\,. \tag{2.4}$$

The controllability condition is guaranteed if D is nonsingular. In this case, the state covariance of the perturbed system $X = X^* \ge 0$ is the unique solution of the following equation

$$(A_o + \Delta A)X + X(A_o + \Delta A)^* + DD^* = 0\,, \tag{2.5}$$

as long as $A_o + \Delta A$ remains stable.

Obviously, it is impossible to know X exactly because ΔA is uncertain. However, with the help of the following theorem we can find an upper bound on the perturbed state covariance X.

Theorem 2.1

Suppose there exist a real $\beta > 0$ *and an* $\bar{X} > 0$ *satisfying the following algebraic Riccati equation*

$$A_o \bar{X} + \bar{X} A_o^* + \frac{\bar{X}\bar{X}}{\beta} + \beta \bar{A} + DD^* = 0\,, \tag{2.6a}$$

then for the nominal feedback system we have the H_∞ *bound*

$$\|(sI - A_o)^{-1} D\|_\infty < \sqrt{\beta}\,, \tag{2.6b}$$

and for all $\Delta A \in \Omega$, $A_o + \Delta A$ *is asymptotically stable and* $X \le \bar{X}$, *where X satisfies (2.5).*

□

Proof (Theorem 2.1)

By using the inequality

$$\frac{\bar{X}^2}{\beta} + \beta \bar{A} \ge \frac{\bar{X}^2}{\beta} + \beta \Delta A \Delta A^* \ge \Delta A \bar{X} + \bar{X} \Delta A^*\,,\ \forall \beta > 0\,,$$

we have

$$(A_o + \Delta A)\overline{X} + \overline{X}(A_o + \Delta A)^* + DD^* \le A_o\overline{X} + \overline{X}A_o^* + \frac{\overline{X}\overline{X}}{\beta} + \beta\overline{A} + DD^* ,$$

i.e. from (2.6a),

$$(A_o + \Delta A)\overline{X} + \overline{X}(A_o + \Delta A)^* + DD^* \le 0 , \tag{2.7}$$

which implies that $A_o + \Delta A$ is stable for all $\Delta A \in \Omega$ since $\overline{X} > 0$ and $(A_o + \Delta A , D)$ is controllable.

By subtracting (2.5) from (2.7), we have

$$(A_o + \Delta A)(\overline{X} - X) + (\overline{X} - X)(A_o + \Delta A)^* \le 0 .$$

Obviously, $X \le \overline{X}$ for all $\Delta A \in \Omega$, because $(A_o + \Delta A)$ is stable. The proof of (2.6b) can be found in [5] and hence is omitted. Q.E.D.

3. Upper Covariance Bound Assignment Theory

Consider the following linear time-invariant plant model

$$\dot{x}_p(t) = A_p x_p(t) + B_p u(t) + D_p w(t) , \; z(t) = M_p x_p(t) , \; y(t) = C_p x_p(t) , \tag{3.1}$$

where $w \in R^{n_w}$, $x_p \in R^{n_x}$, $u \in R^{n_u}$, $z \in R^{n_z}$ and $y \in R^{n_y}$ are external input, state, control input, measurement and output vectors, respectively. The dynamic controller to be designed is of specified order n_c:

$$\dot{x}_c(t) = A_c x_c(t) + B_c z(t) , \; u(t) = C_c x_c(t) + D_c z(t) . \tag{3.2}$$

By using Theorem 2.1 and the following definitions

$$A \triangleq \begin{bmatrix} A_p & 0 \\ 0 & 0 \end{bmatrix}, B \triangleq \begin{bmatrix} B_p & 0 \\ 0 & I_{n_c} \end{bmatrix}, D \triangleq \begin{bmatrix} D_p \\ 0 \end{bmatrix}, M \triangleq \begin{bmatrix} M_p & 0 \\ 0 & I_{n_c} \end{bmatrix}, C \triangleq \begin{bmatrix} C_p & 0 \end{bmatrix}, G \triangleq \begin{bmatrix} D_c & C_c \\ B_c & A_c \end{bmatrix}, \tag{3.3}$$

an upper covariance bound

$$\overline{X} = \begin{bmatrix} \overline{X}_p & \overline{X}_{pc} \\ \overline{X}_{pc}^* & \overline{X}_c \end{bmatrix} \tag{3.4}$$

can be found by solving the algebraic Riccati equation (2.6a),

$$(A + BGM)\overline{X} + \overline{X}(A + BGM)^* + \frac{\overline{X}\overline{X}}{\beta} + \beta\overline{A} + DD^* = 0 \tag{3.5}$$

for a given G and β. Note that $A_o = A + BGM$.

For a given β, a specified $\overline{X}$ is said to be *assignable* as an upper bound if a controller G exists such that $\overline{X}$ solves (3.5). In this case, we say that this controller *assigns* $\overline{X}$ as a covariance upper bound for the system (3.1). Now, the upper covariance bound assignment problem can be mathematically formulated as follows:

For a given $\overline{X}$, what are the necessary and sufficient conditions for the solvability of G from (3.5)? If these conditions are satisfied, what are all solutions G for this equation?

By considering only the perturbation ΔA_p in A_p, we have

$$\overline{A} = \begin{bmatrix} \overline{A}_p & 0 \\ 0 & 0 \end{bmatrix}, \quad \overline{X} = \begin{bmatrix} \overline{X}_p & \overline{X}_{pc} \\ \overline{X}_{pc}^* & \overline{X}_c \end{bmatrix},$$

where $\overline{A}_p$ is an upper bound for $\Delta A_p \Delta A_p^*$. Then by defining

$$\hat{X}_p = \overline{X}_p - \overline{X}_{pc}\overline{X}_c^{-1}\overline{X}_{pc}^*, \tag{3.6a}$$

$$Q_p = \overline{X}_p A_p^* + A_p \overline{X}_p + D_p D_p^* + \beta \overline{A}_p + \frac{\overline{X}_p^2 + \overline{X}_{pc}\overline{X}_{pc}^*}{\beta}, \tag{3.6b}$$

$$\overline{Q}_p = \hat{X}_p^{-1}(\hat{X}_p A_p^* + A_p \hat{X}_p + D_p D_p^* + \beta \overline{A}_p + \frac{\hat{X}_p^2}{\beta})\hat{X}_p^{-1}, \tag{3.6c}$$

we can present the upper covariance bound assignability conditions in the following theorem.

Theorem 3.1

A specified upper bound $\overline{X} > 0$ is assignable, iff

$$\text{(i)} \quad (I - B_p B_p^+) Q_p (I - B_p B_p^+) = 0, \tag{3.7a}$$

$$\text{(ii)} \quad (I - M_p^+ M_p) \overline{Q}_p (I - M_p^+ M_p) = 0, \tag{3.7b}$$

where $[\cdot]^+$ denotes the Moore-Penrose inverse. □

Proof (Theorem 3.1)

It has been proved in [7] that there exists a solution G to (3.5) for a given $\overline{X}$ if and only if

$$(I - BB^+)(A\overline{X} + \overline{X}A^* + \frac{\overline{X}\overline{X}}{\beta} + \beta\overline{A} + DD^*)(I - BB^+) = 0,$$

$$(I - M^+M)\overline{X}^{-1}(A\overline{X} + \overline{X}A^* + \frac{\overline{X}\overline{X}}{\beta} + \beta\overline{A} + DD^*)\overline{X}^{-1}(I - M^+M) = 0.$$

By considering that

$$I - BB^+ = \begin{bmatrix} I - B_p B_p^+ & 0 \\ 0 & 0 \end{bmatrix}, \quad I - M^+M = \begin{bmatrix} I - M_p^+ M_p & 0 \\ 0 & 0 \end{bmatrix},$$

and

$$\bar{A} = \begin{bmatrix} \bar{A}_p & 0 \\ 0 & 0 \end{bmatrix},$$

where $\bar{A}_p$ is an upper bound for the perturbation ΔA_p: $\Delta A_p \Delta A_p^* \le \bar{A}_p$, we obtain (3.7).

Q.E.D.

For parameterizing the whole set of controllers that assign this $\bar{X}$ to the system, we need the following definitions.

$$Q \triangleq \bar{X}A^* + A\bar{X} + DD^* + \beta\bar{A} + \frac{\bar{X}^2}{\beta}, \quad \bar{Q} \triangleq \bar{X}^{-1}Q\bar{X}^{-1}, \tag{3.8a}$$

$$\Gamma \triangleq M^+M\bar{X}(I - BB^+), \tag{3.8b}$$

$$\Phi \triangleq 2\,\bar{Q}\bar{X}(I - BB^+)\Gamma^+ + \Gamma\Gamma^+\bar{Q}[I - \bar{X}(I - BB^+)\Gamma^+]. \tag{3.8c}$$

Theorem 3.2

Suppose a specified $\bar{X} > 0$ is assignable for a given β, then all controllers that assign this $\bar{X}$ to the system (3.1) are parameterized by $Z \in R^{(n_u+n_c)\times(n_z+n_c)}$ and $S = -S^* \in R^{(n_z+n_c)\times(n_z+n_c)}$ as follows

$$G = -\frac{1}{2}B^+\bar{X}(2I - M^+M)\bar{Q}M^+ + \frac{1}{2}B^+\bar{X}M^+M(\Phi^* - \Phi)M^+$$
$$+ B^+\bar{X}M^+M(I - \Gamma\Gamma^+)S(I - \Gamma\Gamma^+)M^+ + Z - B^+BZMM^+. \tag{3.9}$$

□

The proof of this theorem follows along the same lines as that in the standard covariance control theory [7] and hence is omitted.

Remark 3.1

For convenience we assume that there are no redundant inputs and measurements (B_p and M_p are of full rank). In this case, the freedom Z disappears from G in (3.9).

4. Design of Robust Covariance Controller

In this section, we are to use the upper covariance bound assignment result to construct first an assignable $\bar{X}$ and then to design a controller which assigns this upper bound $\bar{X}$ to the system (3.1). For constructing an assignable $\bar{X}$ we need the following theorem.

Theorem 4.1

A specified $\bar{X} > 0$ with the constraint $\bar{X}_c = \alpha I$ is assignable, if there exist a

$$Z_m \triangleq \begin{bmatrix} Z_{m11} & Z_{m12} \\ Z_{m12}^* & 0 \end{bmatrix}, \text{ with } Z_{m11} = Z_{m11}^* \in R^{n_s x n_s}, \ Z_{m12} \in R^{n_s x(n_x - n_s)}$$

and a

$$Z_b \triangleq \begin{bmatrix} Z_{b11} & Z_{b12} \\ Z_{b12}^* & 0 \end{bmatrix}, \text{ with } Z_{b11} = Z_{b11}^* \in R^{n_s x n_s}, \ Z_{b12} \in R^{n_s x(n_x - n_s)},$$

such that

$$\text{(i)} \quad \hat{X}_p A_p^* + A_p \hat{X}_p + D_p D_p^* + \beta \bar{A}_p + \hat{X}_p (\frac{I}{\beta} + V_m Z_m V_m^*) \hat{X}_p = 0, \tag{4.1a}$$

$$\text{(ii)} \quad \bar{X}_p A_p^* + A_p \bar{X}_p + D_p D_p^* + \beta \bar{A}_p + \frac{\bar{X}_p^2 + \alpha \bar{X}_p - \alpha \hat{X}_p}{\beta} + U_b Z_b U_b^* = 0, \tag{4.1b}$$

where V_m and U_b come from the singular value decomposition of M_p and B_p respectively

$$M_p = U_m \begin{bmatrix} \Lambda_m & 0 \\ 0 & 0 \end{bmatrix} V_m^*, \quad B_p = U_b \begin{bmatrix} \Lambda_b & 0 \\ 0 & 0 \end{bmatrix} V_b^*.$$

□

Proof (Theorem 4.1)

For the specified $\bar{X}$, the first assignability condition in Theorem 3.1 is equivalent to

$$\exists Z = Z^*, \text{ such that } Q_p = Z - (I - B_p B_p^+) Z (I - B_p B_p^{+}), \tag{4.2}$$

which, by using (3.6) and the assumption $\bar{X}_c = \alpha I$, is equivalent to (4.1b). The second assignability condition in Theorem 3.1 is equivalent to

$$\exists Z = Z^*, \text{ such that } \bar{Q}_p = Z - (I - M_p^+ M_p) Z (I - M_p^+ M_p), \tag{4.3}$$

which again, by using (3.6), is equivalent to (4.1a). Q.E.D.

Remark 4.1

Both (4.1a) and (4.1b) are algebraic Riccati equations, and (4.1b) can be written in the following standard form

$$\bar{X}_p (A_p + \frac{\alpha I}{2\beta})^* + (A_p + \frac{\alpha I}{2\beta}) \bar{X}_p + \frac{\bar{X}_p^2}{\beta} + (D_p D_p^* + \beta \bar{A}_p + U_b Z_b U_b^* - \frac{\alpha \hat{X}_p}{\beta}) = 0.$$

With the help of Theorem 4.1, a systematic way for constructing an assignable $\bar{X}$ can be stated as follows.

step 1 Construct $\overline{A}_p$, such that $\Delta A_p \Delta A_p^* \le \overline{A}_p$.

step 2 Choose β according to the requirement on the H_∞ norm of the closed-loop transference (2.6b).

step 3 Solve the algebraic Riccati equations (4.1a) and (4.1b) by adjusting Z_m and Z_b such that

$$\hat{X}_p > 0\,,\ \overline{X}_p - \hat{X}_p \ge 0\,,\ \mathrm{rank}(\overline{X}_p - \hat{X}_p) = n_c\,,$$

step 4 Construct $\overline{X}_c$ and $\overline{X}_{pc}$ by using (3.6a) as follows

$$\overline{X}_p - \hat{X}_p = [U_1 \quad U_2] \begin{bmatrix} \Lambda & 0 \\ 0 & 0 \end{bmatrix} \begin{bmatrix} U_1^* \\ U_2^* \end{bmatrix} \qquad \text{(SVD)}$$

$$= \sqrt{\alpha}\, U_1 \Lambda^{1/2} \cdot \frac{I}{\alpha} \cdot \sqrt{\alpha}\, \Lambda^{1/2} U_1^*$$

$$\overline{X}_{pc} \triangleq \sqrt{\alpha}\, U_1 \Lambda^{1/2}\,,\quad \overline{X}_c \triangleq \alpha I\,,$$

and obtain an assignable upper bound

$$\overline{X} = \begin{bmatrix} \overline{X}_p & \overline{X}_{pc} \\ \overline{X}_{pc}^* & \alpha I \end{bmatrix}.$$

step 5 Obtain a controller which assigns this $\overline{X}$ to the system (3.1) by setting any $S = -S^*$ in (3.9).

Remark 4.2

The conditions for obtaining positive definite solutions to the algebraic Riccati equations (4.1a) and (4.1b) can be found in [8].

Remark 4.3

It is reasonable to construct $\overline{X}_c$ as a scaled identity matrix because, as stated in [9-10], the computational errors introduced via controller realization in computers are explicitly related to the properties of the controller covariance matrix $\overline{X}_c$, and the best $\overline{X}_c$ in this sense should have all its diagonal elements equal to α, where α is a scaling factor.

Theorem 4.2

For all $\Delta A \in \Omega$ defined in (2.4), the closed-loop system with the perturbed plant (3.1) and the controller obtained in the step 5 has the following properties:

(i) $A_o + \Delta A$ *is asymptotically stable, where* $A_o = A + BGM$;

(ii) $X \le \overline{X}$, *where* X *satisfies* (2.5);

(iii) $\|(sI - A_o)^{-1} D\|_\infty < \sqrt{\beta}$.

□

Proof (Theorem 4.2)

The $\overline{X}$ constructed above in step 5 is positive definite and assignable since both $\overline{X}_c$ and $\hat{X}_p = \overline{X}_p - \overline{X}_{pc}\overline{X}_c^{-1}\overline{X}_{pc}^*$ are positive definite and $\hat{X}_p$ and $\overline{X}_p$ satisfy (4.1a) and (4.1b) respectively for some chosen Z_m and Z_b. Hence, we can construct a controller of order n_c by using Theorem 3.2 which stabilizes the closed loop system (see Theorem 2.1) and assigns this $\overline{X}$ to the system (3.1). As a result of Theorem 2.1, the closed loop system has the above three properties. Q.E.D.

5. Conclusion

A systematic way to construct an assignable covariance upper bound is presented. By constructing such an upper bound, a robust covariance controller can be designed to tolerate structured plant perturbations in the sense that the perturbed closed system covariance remains beneath this bound for all plant perturbations in the given set.

Further research is needed on how to construct a "smallest" covariance upper bound which is assignable to the given system.

References

[1] A. Hotz and R. Skelton, "Covariance Control Theory", *Int. J. Control* Vol. 46, No. 1, pp. 13-32, 1987.

[2] E. Collins and R. Skelton, "A Theory of State Covariance Assignment for Discrete Systems", *IEEE Trans. Auto. Control*, Vol. AC-32, No. 1, pp. 35-41, Jan. 1987.

[3] R. Skelton and M. Ikeda, "Covariance Controllers for Linear Continuous Time Systems", *Int. J. Control*, Vol. 49, No. 5, pp. 1773-1785, 1989.

[4] C. Hsieh and R. Skelton, "All Discrete-Time Covariance Controllers", *Proceedings ACC*, Pittsburgh, 1989.

[5] J.-H. Xu, R.E. Skelton and G. Zhu, "Upper and Lower Covariance Bounds for Perturbed Linear Systems", *IEEE Trans. Auto. Control*, vol. AC-35, no. 8, pp. 944-948, Aug. 1990.

[6] H. Kwakernaak and R. Sivan, *Linear Optimal Control Systems*, Wiley, New York, 1972.

[7] K. Yasuda, R.E. Skelton and K. Grigoriadis, "Covariance Controllers: A New Parameterization of the Class of All Stabilizing Controllers," submitted for publication in *Automatica*, 1990.

[8] J.C. Doyle, K. Glover, P.P. Khargonekar and B.A, Francis, "State-Space Solution to Standard H_2 and H_∞ Control Problems", *IEEE Trans. Auto. Control*, vol. AC-34, no. 8, pp. 831-847, Aug. 1989.

[9] D. Williamson, "Roundoff Noise Minimization and Pole-Zero Sensitivity in Fixed Point Digital Filters Using Residue Feedback," *IEEE Trans. Acoustics, Speech and Signal Processing*, vol. ASSP-34, no. 4, pp. 1013-1016, Aug. 1986.

[10] D. Williamson and R.E. Skelton, "Optimal q-Markov Cover for Finite Wordlength Implementation," *Proc. ACC*, Atlanta, 1988.

An Inverse LQ Based Approach to the Design of Robust Tracking System with Quadratic Stability

T.Fujii and T.Tsujino

Dept. of Control Engineering and Science, Kyushu Institute of Technology, Japan

ABSTRACT

A tracking problem for linear uncertain systems is considered. A robust feedback controller is designed such that the closed loop system is quadratically stable and its output tracks a command step input. One such controller is obtained by applying the quadratic stabilization method from the viewpoint of Inverse LQ problem. Attractive features of the design method proposed lie in its simplicity in several senses from the practical viewpoint.

1 INTRODUCTION

This paper addresses the problem of designing robust tracking controllers for linear uncertain systems and command step inputs. Similar problems have already been considered in [1],[2] for linear uncertain systems. In [1], however, desired tracking is guaranteed essentially for small parameter uncertainty, whereas in [2] it is guaranteed for large parameter uncertainty but under a strong assumption of matching condition. On the contrary, we consider here a more realistic design problem in that 1) the linear system considered here has time-invariant uncertain parameters with no such assumptions, 2) both robust stability and robust tracking are guaranteed for all allowable uncertainty specified in advance. In the area of robust stability for such systems above, much progress has been made recently on quadratic stability [3]. In particular, design methods for quadratically stabilizing controllers have been obtained in a similar form as in the LQ design[4]-[6]. Although these methods can be applied directly to the above-mentioned design problem, the resulting design method involves inherent practical difficulties in the choice of design parameters similar to those well known in LQ design. In order to overcome these difficulties, a new design method for optimal servo systems, called *Inverse LQ design method*, has been proposed recently by the first author from the viewpoint of the Inverse LQ problem[7],[8]. The most practical features of the method lie in its design objective of output response specification as well as analytical expressions of the gain matrices as a function of design parameters and the system matrices.

In view of the similarity between LQ and quadratically stabilizing controls as stated above, we apply here the quadratic stabilization method to the servo problem above from the viewpoint of Inverse LQ problem. Our objective here is thus to develop an Inverse LQ based design method of robust tracking controllers such that the closed-loop system is quadratically stable and its output tracks a command step input for all allowable uncertainties. We have already developed such a design method in [10] for linear systems with uncertain parameters entering purely into the state matrix. In this paper we consider the case where uncertain parameters enter purely into the input matrix or both into the state matrix and the input matrix, and obtain a unified design method for these two cases. In particular, we design such controllers in the form of both state feedback and observer based output feedback.

2 PROBLEM FORMULATION

Consider a linear system with structured uncertainty described by

$$\begin{aligned} \dot{x}(t) &= [A + DFE_a]x(t) + [B + DFE_b]u(t) \\ y(t) &= Cx(t) \end{aligned} \tag{1}$$

Here $x(t) \in \boldsymbol{R}^n$ is the state, $u(t) \in \boldsymbol{R}^m$ is the control input and $y(t) \in \boldsymbol{R}^m$ is the output; A, B, and C are the nominal system matrices with rank $B = m$; D, E_a, and E_b are known real matrices characterizing the structure of the uncertainty with $E_b \neq 0$. Moreover F is a matrix of uncertain parameters with its maximum singular value bounded by unity, i.e.,

$$F \in \boldsymbol{F} = \{F : \|F\| \leqq 1\} \tag{2}$$

For this uncertain system we consider a design problem of robust servo systems tracking a step reference input $r(t)$ such that

1) the closed loop system is quadratically stable under the parameter uncertainty described above (*robust stability*),

2) its output $y(t)$ approaches $r(t)$ asymptotically as $t \to \infty$ for all allowable F (*robust tracking*),

by use of state feedback or observer based output feedback controllers.

To solve this servo problem, we first consider a familiar augmented system used often in a design problem of servo systems for a step reference input $r(t)$:

$$\begin{aligned} \dot{\xi}_e &= \begin{bmatrix} A + DFE_a & 0 \\ C & 0 \end{bmatrix} \xi_e + \begin{bmatrix} B + DFE_b \\ 0 \end{bmatrix} u \\ &= [A_\xi + D_\xi F E_{a\xi}]\xi_e + [B_\xi + D_\xi F E_b]u \end{aligned}$$

$$y = [C \quad 0]\xi_e, \tag{3a}$$

where

$$A_\xi = \begin{bmatrix} A & 0 \\ C & 0 \end{bmatrix}, B_\xi = \begin{bmatrix} B \\ 0 \end{bmatrix}, \tag{3b}$$

$$D_\xi = \begin{bmatrix} D \\ 0 \end{bmatrix}, E_{a\xi} = \begin{bmatrix} E_a & 0 \end{bmatrix} \tag{3c}$$

and we make the following assumption as is usual in this type of servo problems considered here.

$$\det \begin{bmatrix} A & B \\ C & 0 \end{bmatrix} \neq 0 \tag{4}$$

Then the above servo problem can be reduced to *a design problem of quadratically stabilizing controllers for the augmented system (3).*

3 PRELIMINARY RESULTS

The notion of robust stability of our interest here is the so-called *quadratic stabilty* for linear uncertain systems as defined below.

Definition 1 [3] *The unforced system (1) with $u = 0$ is said to be quadratically stable if there exists an $n \times n$ real symmetric matrix $P > 0$ and a constant $\alpha > 0$ such that for any admissible uncertainty F, the Lyapunov function $V(x) = x^T Px$ satisfies*

$$L(x,t) := \dot{V} = 2x^T P[A + DFE_a]x \leq -\alpha \|x\|^2$$
$$\textit{for all pairs } (x,t) \in \boldsymbol{R}^n \times \boldsymbol{R}^n. \tag{5}$$

We first describe a quadratic stabilization problem(QSP) for the system (3), and then consider its inverse problem.

Definition 2 (QSP) *Determine first whether the system (3) is quadratically stabilizable or not, and if so, construct a quadratically stabilizing control(QSC).*

The solution to QSP is stated below:

Fact 1 [6] *The system (3) is quadratically stabilizable if and only if there exists $\epsilon > 0$ such that the Riccati equation*

$$\begin{aligned} [A_\xi - B_\xi \Xi E_b^T E_{a\xi}]^T P + P[A_\xi - B_\xi \Xi E_b^T E_{a\xi}] \\ + P[D_\xi D_\xi^T - B_\xi R^{-1} B_\xi^T]P \\ + E_{a\xi}^T\{I - E_b \Xi E_b^T\}E_{a\xi} + \epsilon I = 0 \end{aligned} \tag{6}$$

has a real symmetric solution $P > 0$. Moreover, a QSC is given by

$$u = -(R^{-1}B_\xi^T P + \Xi E_b^T E_{a\xi})\xi_e \qquad (7)$$

where R and Ξ are defined by

$$R = (V_1 J^{-2} V_1 + \frac{1}{\epsilon} V_2 V_2^T)^{-1},\ \Xi = V_1 J^2 V_1^T \qquad (8)$$

based on the singular value decomposition of E_b:

$$E_b = [U_1 \quad U_2] \begin{bmatrix} J & 0 \\ 0 & 0 \end{bmatrix} \begin{bmatrix} V_1^T \\ V_2^T \end{bmatrix} \qquad (9)$$

In view of Fact 1, we transform the system (3) first by the feedback transformation

$$u = v - \Xi E_b^T E_{a\xi}\xi_e \qquad (10)$$

into

$$\begin{aligned} \dot{\xi}_e &= \begin{bmatrix} A_F + DFU_2U_2^T E_a & 0 \\ C & 0 \end{bmatrix} \xi_e \\ &+ \begin{bmatrix} B + DFE_b \\ 0 \end{bmatrix} v \qquad (11a) \\ y &= [C \quad 0]\xi_e, \qquad (11b) \end{aligned}$$

where

$$A_F = A - B\Xi E_b^T E_a$$

and then by the state transformation $\xi_e = \Gamma x_e$ into

$$\begin{aligned} \dot{x}_e &= (A_e + D_eFE_e)x_e + (B_e + D_eFE_b)v \\ y &= C_e x_e \qquad (12a) \end{aligned}$$

where

$$\begin{aligned} \Gamma &= \begin{bmatrix} A_F & B \\ C & 0 \end{bmatrix} \quad (\det \Gamma \neq 0) \qquad (12b) \\ A_e &= \begin{bmatrix} A_F & B \\ 0 & 0 \end{bmatrix}, \quad B_e = \begin{bmatrix} 0 \\ I \end{bmatrix} \\ C_e &= \begin{bmatrix} C & 0 \end{bmatrix} \qquad (12c) \\ D_e &= \Gamma^{-1} \begin{bmatrix} D \\ 0 \end{bmatrix}, E_e = \begin{bmatrix} U_2U_2^T E_a & 0 \end{bmatrix} \Gamma \qquad (12d) \end{aligned}$$

As a result of these transformations the QSP for the system (3) can be reduced to that for the system (12). Furthermore, by Fact 1, a QSC for the system (12) is given by

$$v = -K_e x_e,\ K_e = R^{-1}B_e^T P_e \qquad (13a)$$

where $P_e > 0$ is a solution to the following Riccati equation for some $\epsilon > 0$

$$\begin{aligned} &P_eA_e + A_e^T P_e - P_e(D_eD_e^T - B_eR^{-1}B_e^T)P_e \\ &+E_e^T(I - E_b\Xi E_b^T)E_e + \epsilon\Gamma^T\Gamma = 0 \qquad (13b) \end{aligned}$$

According to the design theory of servo systems, with the QSC (13) for the augmented system (12) we can construct a desired robust servo-system as stated in Section 2 by the following state feedback plus integral control:

$$u(t) = -K_F x(t) - K_I \int_0^t (r(\tau) - y(\tau))d\tau \qquad (14a)$$

where

$$[K_F \quad K_I] := K_e\Gamma^{-1} + \Xi E_b^T E_{a\xi} \qquad (14b)$$

or by the following type of observer based output feedback control:

$$\begin{aligned} \dot{\xi}(t) &= (A - LC)\xi(t) + Ly(t) + Bu(t) \qquad (15a) \\ u(t) &= K_F\xi(t) + K_I \int_0^t (r(\tau) - y(\tau))d\tau \qquad (15b) \end{aligned}$$

4 APPLICATION OF QUADRATIC STABILIZATION METHOD FROM THE VIEWPOINT OF INVERSE PROBLEM

We are now ready to define two types of inverse quadratic stabilization problems.

Definition 3 (Inverse QSP of K-type) *Given a linear feedback control $v(t) = -K_e x_e(t)$ for the system (12), find conditions such that it is a QSC of a particular type, say K-type, that is the one given by (13).*

Definition 4 (General Inverse QSP) *Given a linear feedback control $v(t) = -K_e x_e(t)$ for the system (12), find conditions such that it is a QSC of a general type.*

We then state below some pertinent results associated with these inverse problems.

Fact 2 [9] *A feedback control $v = -K_e x_e$ is a quadratically stabilizing control of K-type for the system (12) if and only if there* *exists $R > 0$ of the form (8) and $P_e > 0$ that satisfy the following relations:*

$$RK_e = B_e^T P_e \tag{16a}$$

$$P_e(\frac{1}{2}B_eK_e - A_e) + (\frac{1}{2}B_eK_e - A_e)^T P_e$$
$$-P_eD_eD_e^T P_e - (U_2^T E_e)^T (U_2^T E_e) > 0 \tag{16b}$$

Fact 3 [7] *Let B_e be given by (12c), and K_e by (13a) for some $R > 0$ and $P_e > 0$. Then*

1. K_e can be expressed as

$$K_e = V^{-1}\Sigma V[K \quad I] \tag{17}$$

for some nonsingular matrix $V \in R^{m\times m}$, some positive definite diagonal matrix $\Sigma = \mathrm{diag}\{\sigma_i\} \in R^{m\times m}$, and some real matrix $K \in R^{m\times m}$.

2. The matrices R and P_e in (13a) are expressed by

$$P_e = (VK_e)^T \Lambda \Sigma^{-1}(VK_e) + \text{block-diag}(Y, 0) \tag{18}$$

$$R = V^T \Lambda V \tag{19}$$

for the matrix V in (17), some positve definite diagonal matrix $\Lambda \in R^{m\times m}$ and some positive definite matrix $Y \in R^{n\times n}$.

Fact 4 *The Riccati inequality (16b) is equivalent to the following linear matrix inequality.*

$$\begin{bmatrix} P_e\Psi_e + \Psi_e^T P_e - (U_2^T E_e)^T (U_2^T E_e) & P_eD_e \\ D_e^T P_e & I \end{bmatrix} > 0 \tag{20}$$

$$\Psi_e := \frac{1}{2}B_eK_e - A_e \tag{21}$$

Fact 5 [6] *A feedabck control $v(t) = -K_e x_e(t)$ is a QSC for the system (12) if and only if K_e satisfies the following conditions.*

1. $\Phi_e := A_e - B_eK_e$ is stable (22a)

2. $\mu_e(K_e) := \|G_e(s)\|_\infty < 1,$ (22b)

$$G_e(s) := (E_e - E_bK_e)(sI - \Phi_e)^{-1}D_e$$

5 MAIN RESULTS

To design a desired robust tracking controller stated in Section 2, we first derive a parameterization of QSC from the Inverse LQ viewpoint, and then obtain conditions for QSC to be satisfied by the associated parameters, which leads finally to the desired design algorithm for QSC.

5.1 Parameterization of QSC

The key idea of our design method of QSC is to parameterize a QSC law K_e in the form of (17) based on Facts 1 and 2, and then determine the associated parameter matrices V, K, and Σ based on Facts 1 to 3, in such a way that the K_e so parameterized is a QSC law for the system (12). Note by Fact 1 that the R as in (13a) is restricted in the form of (8), or equivalently, by

$$R = \begin{bmatrix} V_1 & V_2 \end{bmatrix} \begin{bmatrix} J^2 & 0 \\ 0 & \epsilon I \end{bmatrix} \begin{bmatrix} V_1^T \\ V_2^T \end{bmatrix}$$

and hence by 2 of Fact 3 we can determine V as

$$V = [V_1 \quad V_2]^T \tag{23}$$

5.2 Conditions for QSC and design algorithms

We then show necessary and sufficient conditions for QSC of the state feedback control (13) and the observer based output feedback control (15) associated with the parameterized gain K_e, which lead to determination of the remaining parameter matrices K and Σ.

5.2.1 State feedback case

Theorem 1 *Set V as in (23). Then the state feedback (13) associated with K_e given by (17) is a QSC law of K-type for some Σ only if K satisfies the following conditions.*

1. $A_K := A_F - BK$ *is stable.* (24a)
2. $\mu_s(K) := \|U_2^T G_s(s)\|_\infty < 1$ (24b)

where

$$G_s(s) := U_2U_2^T E_s(sI - A_K)^{-1}(I - BB^-)D$$

$$B^- := (CA_K^{-1}B)^{-1}CA_K^{-1}$$

The proof of this theorem is shown in Appendix.

Theorem 2 [9] *Set V as in (17). Then the matrix $G_e(s)$ given by (22b) has the following asymptotic property as $\{\sigma_i\} \to \infty$.*

$$G_e(s) \quad \to \quad \bar{G}_e(s) := G_a(s) - G_b(s)$$

where

$$G_b(s) := E_b[K(sI - A_K)^{-1}(I - BB^-) + B^-]D$$

These results lead to the following basic design algorithm of QSC.

Design algorithm 1

Step 1 Set $V = [V_1 \quad V_2]^T$.

Step 2 First obtain a matrix K satisfying (24a) by the following pole assignment method.

1. First specify stable poles $s_1 \sim s_n$ and n dimensional vectors $g_1 \sim g_n$ and then find a nonsingular T that satisfies the following equation.

$$TS - A_F T = BG \tag{25}$$

where

$$S = diag\{s_1, s_2, \cdots, s_n\}$$
$$G = [g_1, g_2, \cdots, g_n]$$

2. Determine K by the following equation.

$$K = -GT^{-1} \tag{26}$$

Then check whether it satisfies the condition $\|\bar{G}_e(s)\|_\infty < 1$ or not (see Theorem 2). If it satisfies this condition, proceed to Step 3; otherwise check whether it satisfies the condition (24b). If it satisfies this condition, then proceed to Step 3; otherwise obtain a different K similarly and repeat this step.

Step 3 First choose some positive values of $\sigma_1 \sim \sigma_m$ and then check whether the resulting control law K_e given by (17) satisfies the condition (22). If it satisfies this condition, then proceed to Step 4; otherwise choose different $\{\sigma_i\}$ again and repeat this step.

Step 4 With the parameter matrices V,K, and Σ obtained above, the QSC law K_e can be obtained by (17), from which the desired gain matrices K_F,K_I in (14) of the servo system can be obtained by

$$[K_F \quad K_I] = V^{-1}\Sigma V[K \quad I]\Gamma^{-1} + \Xi E_b^T E_{e\xi} \tag{27}$$

Remark

1)When K satisfies the condition $\|\bar{G}_e(s)\|_\infty < 1$ in Step 2, Theorem 2 suggests us to choose $\{\sigma_i\}$ large enough in Step 3 so that the resulting K_e satisfies the condition (22). In fact, validity of such a choice of $\{\sigma_i\}$ is ensured by noting the following asymptotic property of the eigenvalues of Φ_e [7]:

$$\{\lambda_i(\Phi_e)\} \rightarrow \{s_i\} \cup \{-\sigma_i\} \;\; as \; \{\sigma_i\} \rightarrow \infty$$

2)If we cannot find a desired K in Step 2 or desired $\{\sigma_i\}$ in Step 3, then we conclude that this system cannot be quadratically stabilizable.

4.2.2 Obserser based output feedback case

Theorem 3 [10] *The closed loop system (1),(15) is quadratically stable if and only if the following conditions are satisfied.*

1. $$\Phi_{oe} := \begin{bmatrix} \Phi_e & \Gamma^{-1}\begin{bmatrix} -BK_F \\ 0 \end{bmatrix} \\ 0 & A - LC \end{bmatrix}$$ *is stable.* (28a)
2. $\|G_{oe}(s)\|_\infty < 1$ (28b)

$$G_{oe}(s) := [E_e - E_b K_e \; - E_b K_F](sI - \Phi_{oe})^{-1}\begin{bmatrix} D_e \\ -D \end{bmatrix}$$

and furthermore

$$G_{oe}(s) \rightarrow G_e(s) - G_b(s) \quad as \quad \{\sigma_i\} \rightarrow \infty$$

We thus obtain a desired design algorithm for this case from Design algorithm 1 by replacing the condition (22) in Step 3 with (28).

5 ILQ DESIGN ALGORITHM

In this section we show that ILQ design algorithm[8] can be used to determine K and Σ in such a way as stated in the preceding design algorithms described in Section 5.

5.1 ILQ design method [7],[8]

This design method for servo systems tracking a step input has been derived from the viewpoint of inverse LQ problem on the basis of a parameterization of the form (17) of LQ gains, and has several attractive features as shown below from the practical viewpoint.

1)The primary design computation is the determination of K in (17) by the pole assignment method as stated in Design alogorithm 1, which is obviously much simpler than solving Riccati equations in the usual LQ design.

2)The step output response can be specified by suitable choice of design parameters $\{s_i\}$ and $\{g_i\}$ for the pole assignment method above.

3)The design parameters $\{\sigma_i\}$ as in (17) can be used as the tradeoff parameters between the magnitude of control inputs and the tracking property of output responses.

4)The resulting feedback matrices can be expressed explicitly in terms of the system matrices and the design parameters selected.

Fig.1 shows the configulation of ILQ servo system associated with the nominal augmented system (12) which has the state feedback gain $K_F = V^{-1}\Sigma V K_F^0 + \Xi E_b^T E_a$ and the integral gain $K_I = V^{-1}\Sigma V K_I^0$ as in (7). Here K_F^0 and K_I^0 are determined as the ILQ principal gains for the system S_F indicated by the dotted line in Fig.1 so that the closed loop response approaches a specified responce as $\Sigma \to \infty$. Note that the above definition of K_F, K_I together with (7), (17) yields

$$[K_F^0 \quad K_I^0] = [K \quad I]\Gamma^{-1} \tag{29}$$

Denote the ith row of C by $c_i (1 \leq i \leq m)$ and define the following indices d_i,d and matrix M:

$$\begin{aligned} d_i &:= min\{k | c_i A_F^k B \neq 0\} \\ &= min\{k | c_i A^k B \neq 0\} \quad (1 \leq i \leq m) \end{aligned} \tag{30}$$

$$d := d_1 + d_2 + \cdots + d_m \tag{31}$$

$$M := \begin{bmatrix} c_1 A_F^{d_1} B \\ \vdots \\ c_m A_F^{d_m} B \end{bmatrix} = \begin{bmatrix} c_1 A^{d_1} B \\ \vdots \\ c_m A^{d_m} B \end{bmatrix} \tag{32}$$

and make the following assumption.

Assumption The nominal system (1) is minimum phase with det $M \neq 0$.

Let K_i be a set of integers with $d_i + 1$ elements such that

$$\{k\}_{k=1}^{d+m} = K_1 \cup K_2 \cup \cdots \cup K_m \tag{33}$$

and define two polynomials for each K_i,

$$\phi_i(s) := \prod_{k \in K_i} (s - s_k) \quad 1 \leq i \leq m \tag{34}$$

$$\psi_i(s) := \{\phi_i(s) - \phi_i(0)\}/s \quad 1 \leq i \leq m \tag{35}$$

where $\{s_k, k \in K_i\}_{i=1}^m$ are those stable poles specified freely in the pole assignment and the remaining poles should be specified by all the system zeros. With these assumption and definitions, we can state the analytical expression of the ILQ principal gains K_F^0 and K_I^0 as well as the pole assignment gain K as follows.

$$K = M^{-1}N \tag{36}$$

$$K_F^0 = M^{-1}N_0, \quad K_I^0 = M^{-1}M_0 \tag{37}$$

where

$$\begin{aligned} N &:= \begin{bmatrix} c_1\phi_1(A_F) \\ \vdots \\ c_m\phi_m(A_F) \end{bmatrix} \\ N_0 &:= \begin{bmatrix} c_1\psi_1(A_F) \\ \vdots \\ c_m\psi_m(A_F) \end{bmatrix} = \begin{bmatrix} c_1\psi_1(A) \\ \vdots \\ c_m\psi_m(A) \end{bmatrix} \\ M_0 &:= \mathrm{diag}\{\phi_1(0), \cdots, \phi_m(0)\} \end{aligned} \tag{38}$$

This expression yields the following result on which the second features of ILQ design method is based.

Theorem 4 *Under the assumption, the step response of the nominal closed loop system shown in Fig.1 approaches that of a system with the transfer function $G_d(s)$ as $\{\sigma_i\} \to \infty$,where*

$$G_d(s) := \mathrm{diag}\left\{\frac{\phi_i(0)}{\phi_i(s)}\right\} \tag{39}$$

Furthermore, this property also holds even if we use the observer based output feedback (15) instead of the state feedback (14).

This result suggests us to use $\{s_k, k \in K_i\}$ in (34) as design parameters for specifying the i-th output response, and $\{\sigma_k\}$ in (17) as those tradeoff parameters mentioned earlier of an ILQ servo system shown in Fig.1.

In connection of the choice of these ILQ design parameters $\{s_k\}$ and $\{\sigma_i\}$ with Design algorithm 1, the selection of $\{s_k\}$

should be made in order for the resulting pole assignment gain K to satisfy the quadratic stability conditions $\|\bar{G}_e(s)\|_\infty < 1$ or (24b) in Step 2, and that of $\{\sigma_i\}$ should be made in order for another quadratic stability condition (22) in Step 3 to be satisfied by the resulting K_e.

6 CONCLUSION

We have developed here a unified design method of robust tracking controllers for the case where uncertain parameters enter purely into the input matrix or both into the state matrix and the input matrix. Although the method proposed are similar to those proposed in [10] for the case with parameter uncertainty only in the state matrix, there exists an essential difference between them in that for the latter case a high gain robust controller can be always designed, whenever a robust controller exists, but not for the former case as treated here. This may indicate the difficulty of designing robust tracking controllers under parameter uncertainty particularly in the input matrix. Clarifying this difference from the system theoretical point of view is left as a future study.

REFERENCE

[1] E.J.Davison(1976), "The robust control of a servomechanism problem for linear time-invariant multivariable systems," IEEE Trans. Auto. Control, vol. AC-21.

[2] W. E. Schmitendorf and B. R. Barmish(1986), "Robust Asymptotic Tracking for Linear Systems with Unknown Parameters," Automatica, vol.22, 355-360.

[3] B. R. Barmish(1985), "Necessary and sufficient conditions for quadratic stabilizability of an uncertain linear system," J. Optim Theory. Appl., vol. 46, 399-408.

[4] I. R. Petersen(1987), "A stabilization algorithm for a class of uncertain linear systems," Systems & Control Lett., 8, 351-357.

[5] I. R. Petersen(1988), "Stabilization of an uncertain linear system in which uncertain linear parameters enter into the input matrix," SIAM J. Control Optim., vol. 26, 1257-1264.

[6] P. P. Khargonekar, I. R. Petersen, and K. Zhou (1990), "Robust stabilization of uncertain linear systems: quadratic stabilizability and H_∞ control theory," IEEE Trans. Auto. Control, vol.35, 356-361.

[7] T.Fujii(1987), "A new approach to LQ design from the viewpoint of the inverse regulator problem," IEEE Trans. Auto. Control, vol.AC-32. 995-1004.

[8] T. Fujii, et al.(1987), "A Practical Approach to LQ design and its application to engine control," Proc. 10th IFAC World Congress, Munich.

[9] T.Fujii, T.Tsujino, and H.Uematsu, "The design of robust servo systems based on the quadratic stability theory," Trans. of ISCIE, to be published (in Japanese).

[10] T.Fujii and T.Tsujino(1991), "An Inverse LQ Based Approach to the Design of Robust Tracking Controllers for linear uncertain systems," Proc. of MTNS, Kobe.

[11] K.Zhou and P.P.Khargonekar(1988), "An algebraic Riccati equation approach to H_∞ optimization," Systems & Control Lett., 11, 85-91.

Appendix

We note by Facts 2 and 4 that the control (14) is quadratically stabilizing control of K-type for the system (12) only if there exists a $P_e > 0$ in the form of (18) that satisfies (20). Hence we show below that the existence of such a P_e implies the condition (24). First we substitute (18) into (20) and make the following equivalent transformation.

$$\begin{bmatrix} T_e^T & 0 \\ 0 & I \end{bmatrix} \times \begin{bmatrix} P_e\Psi_e + \Psi_e^T P_e - (U_2^T E_e)^T(U_2^T E_e) & P_e D_e \\ D_e^T P_e & I \end{bmatrix} \times \begin{bmatrix} T_e & 0 \\ 0 & I \end{bmatrix} > 0$$

This inequality can be rewritten as

$$\begin{bmatrix} X_e H_e + H_e^T X_e - (U_2^T E_e T_e)^T(U_2^T E_e T_e) & X_e \tilde{D}_e \\ \tilde{D}_e^T P_e & I \end{bmatrix} > 0 \tag{40}$$

where

$$T_e := \begin{bmatrix} T & 0 \\ G & V^{-1} \end{bmatrix} \tag{41a}$$

$$X_e := T_e^T P_e T_e = \begin{bmatrix} T^T Y T & 0 \\ 0 & \Sigma\Lambda \end{bmatrix} \tag{41b}$$

$$H_e := T_e^{-1}\Psi_e T_e = \begin{bmatrix} 0 & 0 \\ 0 & \frac{1}{2}\Sigma \end{bmatrix} - \begin{bmatrix} S & T^{-1}BV^{-1} \\ -VGS & VFBV^{-1} \end{bmatrix} \quad (41c)$$

$$\tilde{D}_e := T_e^{-1}D_e = \begin{bmatrix} Z_{11} \\ Z_{21} \end{bmatrix} D \quad (41d)$$

$$\begin{bmatrix} Z_{11} \\ Z_{21} \end{bmatrix} := (\Gamma T_e)^{-1}\begin{bmatrix} I \\ 0 \end{bmatrix} \quad (41e)$$

and T,S and G are those matrices satisfying the following relations.

$$TST^{-1} = A_F - BK := A_K, \ \det T \neq 0$$
$$G = -KT \quad (41f)$$

Furthermore (40) can be transformed equivalently as follows

$$(40) \Leftrightarrow X_e H_e + H_e^T X_e^T - (U_2^T E_e T_e)^T (U_2^T E_e T_e) - (X_e\tilde{D}_e)(X_e\tilde{D}_e)^T := \begin{bmatrix} L_{11} & L_{12} \\ L_{12}^T & L_{22} \end{bmatrix} > 0$$
$$\Leftrightarrow \begin{cases} L_{11} > 0 \\ L_{22} - L_{12}^T L_{11}^{-1} L_{12} > 0 \end{cases} \quad (42)$$

$$L_{11} := -T^T YTS - (T^T YTS) - (U_2^T E_a TS)^T (U_2^T E_a TS) - (T^T YTZ_{11}D)(T^T YTZ_{11}D)^T \quad (43)$$

In the following we show that the inequality $L_{11} > 0$ implies (24). We first note that $L_{11} > 0$ is equivalent to

$$Y(TST^{-1}) + (TST^{-1})^T Y + \{U_2^T E_a(TST^{-1})\}^T \times \{U_2^T E_a(TST^{-1})\} + Y(TZ_{11}D)(TZ_{11}D)^T Y < 0$$

which can be rewritten by (41f) as

$$YA_K + A_K^T Y + (U_2^T E_a A_K)^T (U_2^T E_a A_K) + Y(TZ_{11}D)(TZ_{11}D)^T Y < 0 \quad (44)$$

or equivalently

$$Y_K A_K + A_K^T Y_K + (U_2^T E_a)^T (U_2^T E_a) + Y_K(A_K TZ_{11}D)(A_K TZ_{11}D)^T Y_K < 0 \quad (45)$$

$$Y_K := (A_K^{-1})^T Y A_K^{-1} \quad (46)$$

By Lemma 2.2 of [11] the inequality (44) is equivalent to

$$A_K \textit{ is stable} \quad (47a)$$
$$\|U_2^T E_a(sI - A_K)^{-1} A_K TZ_{11}D\|_\infty < 1 \quad (47b)$$

To derive (24b) from (47b), we premultiply (41e) by the following term.

$$\begin{bmatrix} T & 0 \end{bmatrix} = \begin{bmatrix} I & 0 \end{bmatrix} N^{-1} T_e \quad (48)$$

$$N := \begin{bmatrix} I & 0 \\ -K & I \end{bmatrix} \quad (49)$$

Then we have

$$TZ_{11} = \begin{bmatrix} I & 0 \end{bmatrix} (\Gamma N)^{-1} \begin{bmatrix} I \\ 0 \end{bmatrix} = A_K^{-1} + A_K^{-1}B(-CA_K^{-1}B)^{-1}CA_K^{-1} \quad (50)$$

Substituting this equation into the transfer function of (47b) yields

$$U_2^T E_a(sI - A_K)^{-1} A_K TZ_{11}D = U_2^T E_a(sI - A_K)^{-1} \times \{I - B(CA_K^{-1}B)^{-1}CA_K^{-1}\}D \quad (51)$$

Hence from (47) we obtain

$$A_K \textit{ is stable} \quad (52a)$$
$$\|U_2^T E_a(sI - A_K)^{-1}(I - BB^-)D\|_\infty < 1 \quad (52b)$$

Q.E.D.

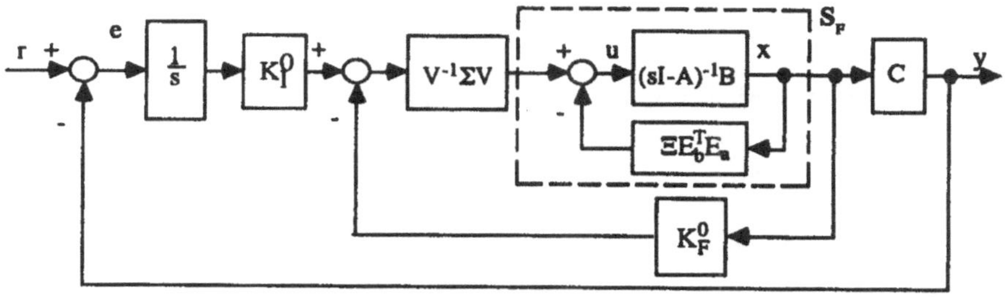

Fig. 1 Configulation of ILQ servo systems

Linear systems and robustness: a graph point of view [1]

Tryphon T. Georgiou [2] and Malcolm C. Smith [3]

Abstract

This paper presents a framework for modelling and robust control of linear systems based on a *graph* point of view. Basic concepts such as linearity, causality, stabilizability can be defined as properties of the graph of a system. Uncertainty is naturally thought of in terms of perturbations of the graph. Modelling and approximation are also fundamentally related to the graph of the system. Robustness will be quantified in the graph setting using the gap metric. Necessary and sufficient conditions for robustness will be explained geometrically in terms of the graphs of the plant and controller. A flexible structure design example will be discussed.

1 Introduction

The importance of a "graph" viewpoint in systems was highlighted in the early 1980's by the work of Vidysagar et al. [18], [16], in a fundamental study of the graph topology, and Zames and El-Sakkary [19], who introduced the gap metric into the control literature. The later work of Vidyasagar and Kimura [17] and Glover and McFarlane [9], [12] was the basis for a substantial quantitative theory which is still rapidly developing. This paper presents an overview of several recent results of the authors and others concerning the gap metric as well as a discussion of a control system design example and some related topics.

2 Basics of Linear Systems

In this paper we consider a *system* to be defined mathematically as an operator which maps an input space to an output space. In Fig. 1 the *plant* is a system $\mathbf{P} : \mathcal{D}(\mathbf{P}) \subset \mathcal{U} \rightarrow \mathcal{Y}$ where $\mathcal{U} := L_2^m[0,\infty)$ (resp. $\mathcal{Y} := L_2^p[0,\infty)$) is the input (resp. output) space and $\mathcal{D}(\mathbf{P})$ is the domain. The *graph* of $\mathbf{P}$ is defined by

[1]Supported by the National Science Foundation, U.S.A. and the Fellowship of Engineering, U.K.
[2]Department of Electrical Engineering, University of Minnesota, Minneapolis, MN 55455, U.S.A.
[3]University of Cambridge, Department of Engineering, Cambridge CB2 1PZ, U.K.

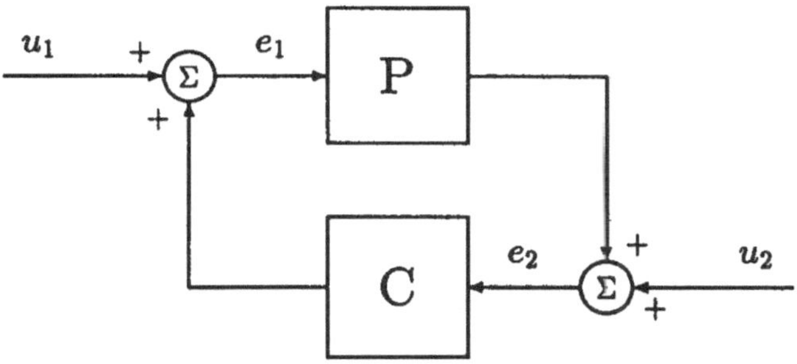

Figure 1: Standard feedback configuration.

$$\mathcal{G}(\mathbf{P}) := \begin{pmatrix} \mathbf{I} \\ \mathbf{P} \end{pmatrix} \mathcal{D}(\mathbf{P}).$$

In general, a system $\mathbf{P}$ is said to be *linear* if $\mathcal{G}(\mathbf{P})$ is a linear subspace of $\mathcal{L} := \mathcal{U} \oplus \mathcal{Y}$. In an input-output setting the study of linear systems is the study of linear subspaces $\mathcal{K} \subset \mathcal{L}$ which are graphs, i.e. such that $\binom{0}{y} \in \mathcal{K} \Rightarrow y = 0$. An advantage of the graph point of view is that the basic definitions and theorems can be compactly stated without assuming the systems are stable.

A linear system $\mathbf{P}$ is said to be *shift-invariant* if $\mathcal{G}(\mathbf{P})$ is a shift-invariant subspace of $\mathcal{L}$, i.e. $\mathbf{S}_\tau \mathcal{G}(\mathbf{P}) \subset \mathcal{G}(\mathbf{P})$ for all $\tau > 0$, where $\mathbf{S}_\tau x(t) = x(t+\tau)$. The system $\mathbf{P}$ is said to be *causal* if $\mathbf{T}_\tau \mathcal{G}(\mathbf{P})$ is a graph for all $\tau > 0$, where $\mathbf{T}_\tau x(t) = x(t)$ for $t < \tau$ and zero otherwise. In Fig. 1 the *controller* is a system $\mathbf{C} : \mathcal{D}(\mathbf{C}) \subset \mathcal{Y} \to \mathcal{U}$. The *inverse graph* of $\mathbf{C}$ is defined by

$$\mathcal{G}'(\mathbf{C}) := \begin{pmatrix} \mathbf{C} \\ \mathbf{I} \end{pmatrix} \mathcal{D}(\mathbf{C}).$$

The feedback configuration of Fig. 1, denoted by $[\mathbf{P}, \mathbf{C}]$, is defined to be *stable* if the operators mapping $u_i \to e_j$ are bounded for $i, j = 1, 2$. It can be shown that $[\mathbf{P}, \mathbf{C}]$ is stable if and only if $\mathcal{G}(\mathbf{P})$ and $\mathcal{G}'(\mathbf{C})$ are closed in $\mathcal{L}$,

$$\begin{aligned} \mathcal{G}(\mathbf{P}) \cap \mathcal{G}'(\mathbf{C}) &= \{0\} \quad \text{and} \\ \mathcal{G}(\mathbf{P}) + \mathcal{G}'(\mathbf{C}) &= \mathcal{U} \oplus \mathcal{Y}. \end{aligned}$$

The theorems below give necessary and sufficient conditions, in the frequency domain, for shift-invariance, causality and stabilizability. For $x \in L_2^n[0,\infty)$ we denote by $\hat{x} \in H_2^n$ the Fourier transform of x. The operator $\hat{\mathbf{P}} : H_2^m \to H_2^p$ is defined by the relation $\hat{\mathbf{P}}\hat{x} = \widehat{(\mathbf{P}x)}$ for all $x \in \mathcal{D}(\mathbf{P})$. For $U \in H_\infty^{n \times r}$ with $r \leq n$ we introduce the notation $d_U :=$ {gcd of all highest order minors of U} (see [15] for a proof of existence of gcd's in H_∞).

Theorem 1. Consider any shift-invariant linear system **P** such that $\mathcal{G}(\mathbf{P})$ is a closed subspace of $\mathcal{L}$. Then there exists $r \leq m$, $M \in H_\infty^{m\times r}$ of rank r, and $N \in H_\infty^{p\times r}$ such that

$$\mathcal{G}(\hat{\mathbf{P}}) = \begin{pmatrix} M \\ N \end{pmatrix} H_2^r =: GH_2^r. \tag{1}$$

and $G^*G = I$. Moreover **P** is causal if and only if $\gcd(d, e^{-s}) = 1$ where $d := d_M/d_G \in H_\infty$.

Proof. The first part is a well-known consequence of the Beurling-Lax theorem [11]. The condition for causality is established in [8]. □

Theorem 2. Consider any shift-invariant linear system **P**. Then there exists a $\mathbf{C} : \mathcal{Y} \to \mathcal{U}$ (linear, shift-invariant) such that $[\mathbf{P}, \mathbf{C}]$ is stable if and only if (1) holds with $r = m$ and G is left invertible over H_∞.

Proof. A more detailed version of this proof can be found in [8]. Let $[\mathbf{P}, \mathbf{C}]$ be stable. Then (1) holds for $r \leq m$, $M \in H_\infty^{m\times r}$ and $N \in H_\infty^{p\times r}$. Also, by Theorem 1, we can take $\mathcal{G}'(\hat{\mathbf{C}}) = \binom{U}{V} H_2^q$ with $U \in H_\infty^{m\times q}$, $V \in H_\infty^{p\times q}$ and $q \leq p$. Define

$$T := \begin{pmatrix} M & U \\ N & V \end{pmatrix} \in H_\infty^{(p+m)\times(r+q)}. \tag{2}$$

Then $[\mathbf{P}, \mathbf{C}]$ stable implies that $TH_2^{r+q} = H_2^{m+p}$. This requires $r = m$, $q = p$ and T is invertible over H_∞. The first m rows of T^{-1} now provide a left inverse of G. For the converse direction assume that $r = m$ and that G has a left inverse over H_∞. By Tolokonnikov's Lemma [13, p. 293] there exists $U \in H_\infty^{m\times p}$ and $V \in H_\infty^{p\times p}$ such that T as defined in (2) is invertible over H_∞. Moreover this can be done with a non-singular V, in which case we can define a stabilizing **C** in terms of the graph $\mathcal{G}(\hat{\mathbf{C}}) = \binom{V}{U} H_2^p$. □

3 Uncertainty

An intrinsic requirement in robust control design is a suitable measure of modelling uncertainty. In an input-output setting it is natural to measure uncertainty by perturbations of the graph. Such a measure can be quantified using the *gap metric*. Let $\mathbf{P}_i$ have graph $\mathcal{G}_i := G_i H_2^m$ where $G_i^* G_i = I$ and $G_i \in H_\infty^{(p+m)\times m}$ for $i := 1, 2$. Then the *gap* between $\mathbf{P}_1$ and $\mathbf{P}_2$ is defined to be $\delta(\mathbf{P}_1, \mathbf{P}_2) := \|\Pi_{\mathcal{G}_1} - \Pi_{\mathcal{G}_2}\|$ ([10], [19]) where $\Pi_{\mathcal{K}}$ denotes the orthogonal projection onto $\mathcal{K}$. It can be shown [4] that the following formula holds for the gap: $\delta(\mathbf{P}_1, \mathbf{P}_2) = \max\left\{\vec{\delta}(\mathbf{P}_1, \mathbf{P}_2), \vec{\delta}(\mathbf{P}_2, \mathbf{P}_1)\right\}$ where $\vec{\delta}(\mathbf{P}_1, \mathbf{P}_2) = \inf_{Q \in H_\infty} \|G_1 - G_2 Q\|_\infty$.

A natural generalization of the gap metric that includes a frequency domain "scaling" of the spaces of inputs and outputs can be defined as follows [6]. Let $\mathbf{W}_i, \mathbf{W}_o$ of appropriate dimension be bounded and have bounded inverses. Define the *weighted gap*

$$\delta(\mathbf{P}_1, \mathbf{P}_2; \mathbf{W}_o, \mathbf{W}_i) := \delta(\mathbf{W}_o \mathbf{P}_1 \mathbf{W}_i, \mathbf{W}_o \mathbf{P}_2 \mathbf{W}_i).$$

It follows from the definition that $\delta(\cdot,\cdot\,;\mathbf{W}_o,\mathbf{W}_i)$ is a metric. Moreover, the weighted gap inherits the property $0 \le \delta(\cdot,\cdot\,;\mathbf{W}_o,\mathbf{W}_i) \le 1$ from the gap metric. In Section 6 we will discuss how the weighted gap metric can be incorporated into a design procedure.

4 Modelling and Identification

Broadly speaking, any identification experiment on the plant $\mathbf{P}$ involves measurements (or estimates) $\tilde{u}$, $\tilde{y}$ of the input and output signals. The vector $\binom{\tilde{u}}{\tilde{y}}$ can be thought of as an approximate element of the graph of $\mathbf{P}$ with the uncertainty being in the graph topology. Identification of the graph of a system—as opposed to the transfer function—is particularly natural in the case of closed loop experiments, as in adaptive control or in the identification of unstable systems. Error bounds are naturally expressed in terms of an appropriate metric such as the gap.

A closely related topic is that of approximation: given a plant $\mathbf{P}$ of order n, find the closest approximant of order $k < n$ in the gap metric. This problem is currently unsolved, though there exist upper and lower bounds on the achievable error in terms of the singular values of two Hankel operators [7]. That is

$$\sigma_{k+1}\left(\Pi_{H_2^\perp} G G^* |_{H_2}\right) \le \inf_{\tilde{\mathbf{P}}} \delta(\mathbf{P},\hat{\mathbf{P}}) \le \sum_{i=k+1}^{n} \sigma_i\left(\Pi_{H_2^\perp} G^* |_{H_2}\right) \tag{3}$$

where the infimum is taken over all $\tilde{\mathbf{P}}$ with degree at most $k < n$. In the matrix case, the upper bound in (3) was derived under a non-restrictive assumption that it is less than a certain quantity $\lambda(\mathbf{P})$—see [5], [7].

5 Robustness

In this section we determine the radius of the largest uncertainty ball in the gap metric which a feedback system can tolerate. Let $\mathcal{M} := \mathcal{G}(\mathbf{P})$ and $\mathcal{N} := \mathcal{G}'(\mathbf{C})$. Then, it is straightforward to show that $[\mathbf{P},\mathbf{C}]$ is stable if and only if $\mathbf{A}_{\mathcal{M},\mathcal{N}} := \Pi_{\mathcal{N}^\perp}|_{\mathcal{M}}$ is invertible (see [3]). When $\mathbf{A}_{\mathcal{M},\mathcal{N}}$ is invertible, then $\mathbf{Q}_{\mathcal{M},\mathcal{N}} := \mathbf{A}_{\mathcal{M},\mathcal{N}}^{-1}\Pi_{\mathcal{N}^\perp}$, is the *parallel projection onto* $\mathcal{M}$ *along* $\mathcal{N}$, i.e., $\mathbf{Q}_{\mathcal{M},\mathcal{N}}^2 = \mathbf{Q}_{\mathcal{M},\mathcal{N}}$, range$(\mathbf{Q}_{\mathcal{M},\mathcal{N}}) = \mathcal{M}$ and kernel$(\mathbf{Q}_{\mathcal{M},\mathcal{N}}) = \mathcal{N}$. Note that $\mathbf{Q}_{\mathcal{M},\mathcal{N}}$ can be expressed in terms of $\mathbf{P}$ and $\mathbf{C}$ as follows:

$$\mathbf{Q}_{\mathcal{M},\mathcal{N}} = \begin{pmatrix}\mathbf{I}\\ \mathbf{P}\end{pmatrix}(\mathbf{I}-\mathbf{CP})^{-1}(\mathbf{I},-\mathbf{C}). \tag{4}$$

When $[\mathbf{P},\mathbf{C}]$ is stable define $b_{P,C} := \|\mathbf{Q}_{\mathcal{M},\mathcal{N}}\|^{-1} = \sqrt{1-\delta(\mathcal{M},\mathcal{N}^\perp)^2}$ and observe that $0 < b_{P,C} \le 1$.

Theorem 3. Suppose $[\mathbf{P},\mathbf{C}]$ is stable. Then $[\mathbf{P}_1,\mathbf{C}]$ is stable for all $\mathbf{P}_1$ with $\delta(\mathbf{P},\mathbf{P}_1) < b$ if and only if $b \leq b_{P,C}$.

Proof. From a purely geometric point of view the sufficiency part of the theorem follows from the equation:

$$\mathbf{A}_{\mathcal{M}_1,\mathcal{N}} = \mathbf{A}_{\mathcal{M},\mathcal{N}}(\mathbf{I}_{\mathcal{M}} + \mathbf{X})(\mathbf{\Pi}_{\mathcal{M}_1}|_{\mathcal{M}})^{-1}$$

where $\mathbf{X} = \mathbf{Q}_{\mathcal{M},\mathcal{N}}(\mathbf{\Pi}_{\mathcal{M}_1} - \mathbf{\Pi}_{\mathcal{M}})$. To see this, note that if $\|\mathbf{\Pi}_{\mathcal{M}_1} - \mathbf{\Pi}_{\mathcal{M}}\| < b_{P,C}$ then $\|\mathbf{X}\| < 1$. This implies that $\mathbf{A}_{\mathcal{M}_1,\mathcal{N}}$ is invertible. A construction for the necessity part in the case of time-varying systems is given in [3]. See also [5] for a frequency domain approach to the shift-invariant case. □

A dual result to Theorem 3 involves interchanging the roles of **P** and **C**. It is interesting to note that $b_{P,C} = b_{C,P}$ (see [5]) so that the robustness margin for plant perturbations is the same as for controller perturbations. An equivalent fact is $\|\mathbf{Q}_{\mathcal{M},\mathcal{N}}\| = \|\mathbf{Q}_{\mathcal{N},\mathcal{M}}\|$ (see [3]) where $\mathbf{Q}_{\mathcal{N},\mathcal{M}}$ is the parallel projection onto $\mathcal{N}$ along $\mathcal{M}$. For combined plant and controller uncertainty, the following sharp result has been obtained by Qiu and Davison [14].

Theorem 4. Suppose $[\mathbf{P},\mathbf{C}]$ is stable. Then $[\mathbf{P}_1,\mathbf{C}_1]$ is stable for all $\mathbf{P}_1$ with $\delta(\mathbf{P},\mathbf{P}_1) < b_1$ and all $\mathbf{C}_1$ with $\delta(\mathbf{C},\mathbf{C}_1) < b_2$ if and only if

$$b_1^2 + b_2^2 + 2b_1b_2\sqrt{1-b_{P,C}^2} \leq b_{P,C}^2. \quad \Box$$

6 Control System Design

The problem of weighted gap optimization amounts to solving the following H_∞-minimization problem:

$$b_{\mathrm{opt}}(\mathbf{P}) := \left(\inf_{\mathbf{C}\ \mathrm{stblz}} \left\| \begin{pmatrix} \mathbf{W}_i^{-1} \\ \mathbf{W}_o\mathbf{P} \end{pmatrix} (\mathbf{I}-\mathbf{CP})^{-1}(\mathbf{W}_i, -\mathbf{C}\mathbf{W}_o^{-1}) \right\|_\infty \right)^{-1}. \tag{5}$$

This is a well-posed problem [2] (unlike several popular choices of H_∞ design problems such as S/CS or S/T optimization) which offers interesting possibilities for controller design. We remark that the robustness results of the previous section apply to the weighted gap when $\delta(\mathbf{P},\mathbf{P}_1)$ is replaced by $\delta(\mathbf{P},\mathbf{P}_1;\mathbf{W}_o,\mathbf{W}_i)$, $\delta(\mathbf{C},\mathbf{C}_1)$ by $\delta(\mathbf{C},\mathbf{C}_1;\mathbf{W}_i^{-1},\mathbf{W}_o^{-1})$ and $b_{P,C}$ by the corresponding weighted expression for the parallel projection as in (5).

In case $\mathbf{W}_i$ is the identity matrix and $\mathbf{W}_o$ is a scalar weighting function then (5) reduces to finding:

$$b_{\mathrm{opt}}(\mathbf{P}) := \left(\inf_{\mathbf{C}\ \mathrm{stblz}} \left\| \begin{pmatrix} (\mathbf{I}-\mathbf{CP})^{-1} & -(\mathbf{I}-\mathbf{CP})^{-1}\mathbf{C}\mathbf{W}_o^{-1} \\ \mathbf{W}_o\mathbf{P}(\mathbf{I}-\mathbf{CP})^{-1} & -\mathbf{P}(\mathbf{I}-\mathbf{CP})^{-1}\mathbf{C} \end{pmatrix} \right\|_\infty \right)^{-1}. \tag{6}$$

The choice of the weighting function $\mathbf{W}_o$ in (6) is interpreted here in a closed-loop sense as a trade-off between "control action", as represented by the transfer function $(\mathbf{I}-\mathbf{CP})^{-1}\mathbf{C}$, and "disturbance attenuation", represented by $\mathbf{P}(\mathbf{I}-\mathbf{CP})^{-1}$. An upper bound equal to $b_{opt}(\mathbf{P})^{-1}$ is also guaranteed on $\|(\mathbf{I}-\mathbf{CP})^{-1}\|_\infty$ and $\|\mathbf{P}(\mathbf{I}-\mathbf{CP})^{-1}\mathbf{C}\|_\infty$. This interpretation, as well as the natural one of robustness in the weighted gap, will be exploited in the design example below.

In some recent work [12], McFarlane and Glover proposed a certain "loop shaping" design procedure. This procedure begins with an augmented plant transfer function $\mathbf{P}_{aug} := \mathbf{W}_o\mathbf{P}\mathbf{W}_i$ containing input and output "weighting" functions $\mathbf{W}_o, \mathbf{W}_i$. The weighting functions are selected on the basis of a desired loop shape for the control system. The controller is designed to give optimal robustness with respect to normalized coprime fraction perturbations of $\mathbf{P}_{aug}$. We point out that this choice of design synthesis problem is precisely the same as (5).

7 Flexible Structure Example

The weighted gap design approach was applied to a 12-meter vertical cantilever truss structure at the Wright-Patterson Air Force Base (Dayton, Ohio). A 4-input/4-output/14-state nominal model was selected for the design and a higher order 42-state model was used for robustness analysis (see Fig. 2). An initial choice of $\mathbf{W}_o = 0.65$ led to an appropriate level of control action for the lower frequency modes. The weight was modified

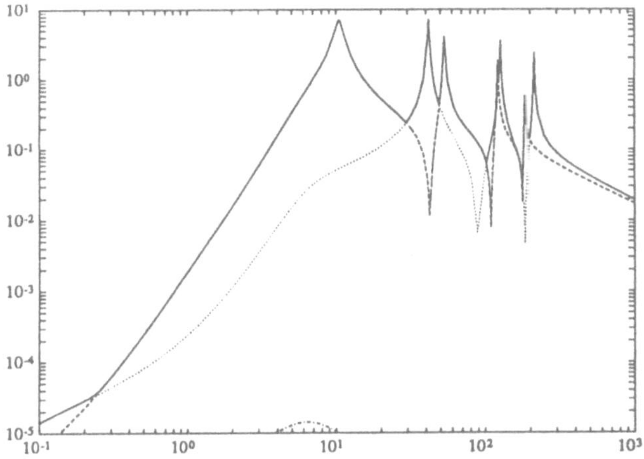

Figure 2: Open loop singular values for 42-state model.

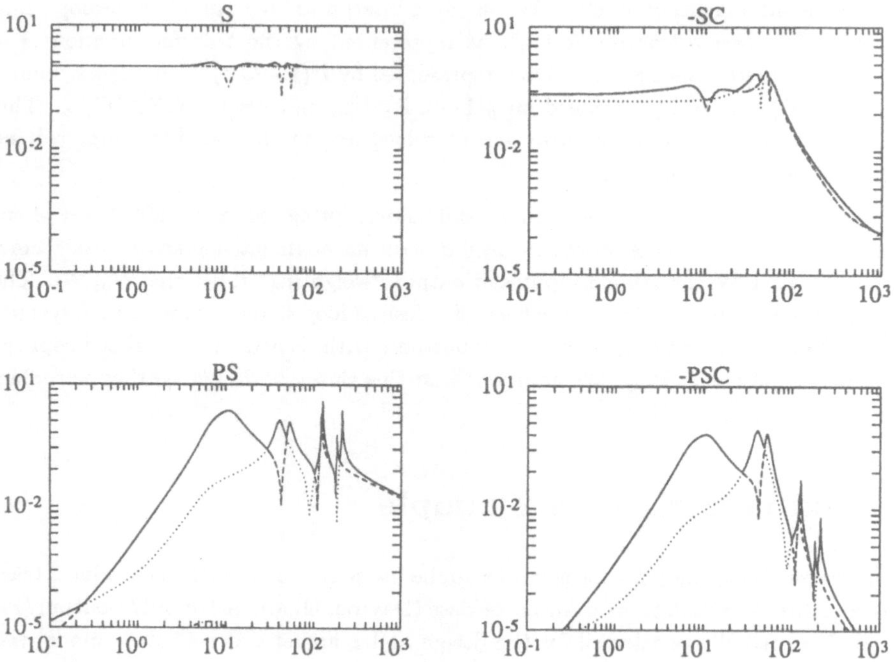

Figure 3: Singular values of entries of parallel projection.

to $\mathbf{W}_o = c(0.1s + \alpha)^2/(s^2 + 0.5\alpha s + \alpha^2)$ to give, additionally, increased disturbance attenuation over the middle frequency range and robustness to neglected high frequency dynamics. This choice of weighting function gave the value $b_{opt}(\mathbf{P}) = 0.655$ for $c = 0.65$ and $\alpha = 52.0$. The weighted gap between the 14-state and 42-state models was 0.446 which guarantees robustness by the weighted version of Theorems 3 and 4. A suboptimal (16-state) controller was designed for the associated one-block H_∞ optimization problem. A plot of the singular values for each 4×4-block entry of (4) for the 42-state plant and 16-state controller is shown in Fig. 2. This controller was implemented on the structure and gave disturbance attenuation close to the anticipated levels. A detailed report on the implementations is given in [1].

References

[1] S. Buddie, T.T. Georgiou, Ü. Özgüner and M.C. Smith, Flexible structure experiments at JPL and WPAFB, report in preparation.

[2] M.J. Englehart and M.C. Smith, A four-block problem for H_∞ design: properties and applications, *Automatica*, to appear.

[3] C. Foias, T.T. Georgiou, and M.C. Smith, Geometric techniques for robust stabilization of linear time-varying systems, preprint, February 1990; and in the Proceedings of the 1990 IEEE Conference on Decision and Control, pp. 2868–2873, December 1990.

[4] T.T. Georgiou, On the computation of the gap metric, *Systems and Control Letters*, **11**, 253–257, 1988.

[5] T.T. Georgiou and M.C. Smith, Optimal robustness in the gap metric, *IEEE Trans. on Automat. Control*, **35**, 673–686, 1990.

[6] T.T. Georgiou and M.C. Smith, Robust control of feedback systems with combined plant and controller uncertainty, Proceedings of the 1990 American Control Conference, pp. 2009–2013, May 1990.

[7] T.T. Georgiou and M.C. Smith, Upper and lower bounds for approximation in the gap metric, preprint, February 1991.

[8] T.T. Georgiou and M.C. Smith, Graphs, causality and stabilizability: linear, shift-invariant systems on $L_2[0,\infty)$, report in preparation.

[9] K. Glover and D.C. McFarlane, Robust stabilization of normalized coprime factor plant descriptions with H_∞-bounded uncertainty, *IEEE Trans. on Automat. Contr.*, **34**, 821–830, 1989.

[10] T. Kato, *Perturbation Theory for Linear Operators*, Springer-Verlag: New York, 1966.

[11] P.D. Lax, Translation invariant subspaces, *Acta Math.*, **101**, 163–178, 1959.

[12] D.C. McFarlane and K. Glover, *Robust controller design using normalized coprime factor plant descriptions*, Lecture notes in Control and Information Sciences, Springer-Verlag, 1990.

[13] N.K. Nikol'skiĭ, *Treatise on the Shift Operator*, Springer-Verlag: Berlin, 1986.

[14] L. Qiu and E.J. Davison, Feedback stability under simultaneous gap metric uncertainties in plant and controller, preprint, 1991.

[15] M.C. Smith, On stabilization and the existence of coprime factorizations, *IEEE Trans. on Automat. Contr.*, **34**, 1005-1007, 1989.

[16] M. Vidyasagar, The graph metric for unstable plants and robustness estimates for feedback stability, *IEEE Trans. on Automat. Contr.*, **29**, 403–418, 1984.

[17] M. Vidyasagar and H. Kimura, Robust controllers for uncertain linear multivariable systems, *Automatica*, **22**, 85–94, 1986.

[18] M. Vidyasagar, H. Schneider and B. Francis, Algebraic and topological aspects of feedback stabilization, *IEEE Trans. on Automat. Contr.*, **27**, 880-894, 1982.

[19] G. Zames and A.K. El-Sakkary, Unstable systems and feedback: The gap metric, Proceedings of the Allerton Conference, pp. 380–385, October 1980.

Experimental Evaluation of H^∞ Control for a Flexible Beam Magnetic Suspension System

Masayuki Fujita†, Fumio Matsumura†, and Kenko Uchida‡

† Department of Electrical and Computer Engineering, Kanazawa University, Kodatsuno 2-40-20, Kanazawa 920, Japan

‡ Department of Electrical Engineering, Waseda University, Okubo 3-4-1, Shinjuku, Tokyo 169, Japan

Abstract

An H^∞ robust controller is experimentally demonstrated on a magnetic suspension system with a flexible beam. The experimental apparatus utilized in this study is a simplified model of magnetic bearings with an elastic rotor. We first derive a suitable mathematical model of the plant by taking account of various model uncertainties. Next we setup the standard problem, where the generalized plant is constructed with frequency weighting functions. An iterative computing environment MATLAB is then employed to calculate the central controller. The techniques devised for the selection of design parameters involve both the game theoretic characterizations of the central controller in the time domain and the all-pass property of the closed loop transfer function in the frequency domain. Finally, the digital controller is implemented using a DSP μPD77230 where the control algorithm is written in the assembly language. Several experiments are carried out in order to evaluate the robustness and the performance of this H^∞ design. These experimental results confirm us that the flexible beam magnetic suspension system is robustly stable against various real parameter changes and uncertainties.

1. Introduction

The problem of state-space formulae for the H^∞ control has been recently settled (see [1], [2], and the references therein). A direct consequence of the above results is that the associated complexity of computation is fairly reduced. Further, there has been a substantial development in the computer aided control systems design packages. Accordingly, one could readily practice this powerful methodology. We believe that one of the most challenging issues in this present situation is an application oriented study of the H^∞ control [3] - [7]. In this paper, an H^∞ controller is experimentally demonstrated on a magnetic suspension system with a flexible beam. Based on several experimental results, we will show that the magnetic suspension system designed by the H^∞ methodology, is robustly stable against various real parameter changes and uncertainties.

2. Modeling

We consider a magnetic suspension system with a flexible beam shown in Fig. 1. The apparatus utilized in this study is a simplified model of magnetic bearings with an elastic rotor. The experimental configuration consists of a flexible aluminum beam with an electromagnet and a gap sensor. The beam is supported by a hinge at the left side. Mass M is attached at the center of the beam and mass m is attached at the right side. An U-shaped electromagnet is located as an actuator at the right side. As a gap sensor, a standard induction probe of eddy-current type is placed at the same position in the right side.

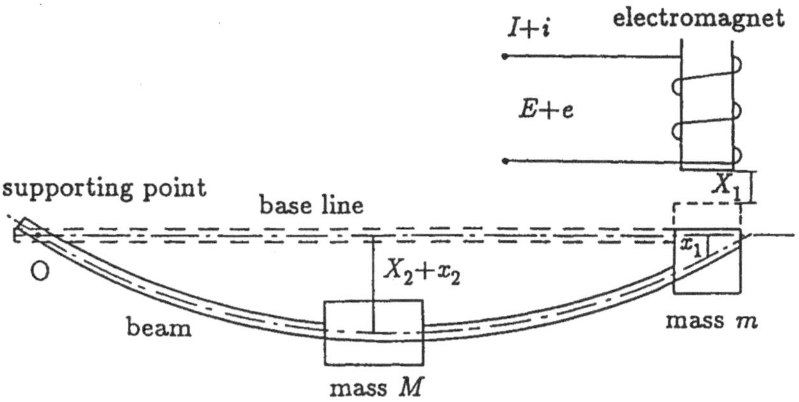

Fig. 1. Flexible beam magnetic suspension system.

A mathematical model of this experimental apparatus has been derived in [6], [7]. The state-space representation is of the following form

$$\dot{x}_g = A_g x_g + B_g u + D_g v_0, \qquad y = C_g x_g + w_0 \tag{1}$$

where $x_g := [x_1 \;\; x_2 \;\; \dot{x}_1 \;\; \dot{x}_2 \;\; i]^T$, $u := e$, and

$$A_g = \begin{bmatrix} 0.0 & 0.0 & 1.0 & 0.0 & 0.0 \\ 0.0 & 0.0 & 0.0 & 1.0 & 0.0 \\ 7070 & 712 & -0.327 & 0.654 & -41.9 \\ 399 & -797 & 0.654 & -1.31 & 0.0 \\ 0.0 & 0.0 & 0.0 & 0.0 & -18.0 \end{bmatrix}, \; B_g = \begin{bmatrix} 0.0 \\ 0.0 \\ 0.0 \\ 0.0 \\ 0.317 \end{bmatrix}, \; D_g = \begin{bmatrix} 0.0 & 0.0 & 0.0 \\ 0.0 & 0.0 & 0.0 \\ 0.172 & 0.0 & 0.0 \\ 0.0 & 0.0965 & 0.0 \\ 0.0 & 0.0 & 0.317 \end{bmatrix}$$

$$C_g = \begin{bmatrix} 1 & 0 & 0 & 0 & 0 \end{bmatrix}. \tag{2}$$

Here (A_g, B_g) and (A_g, D_g) are controllable, and (A_g, C_g) is observable. It should be noted [6], [7] that the disturbance terms (v_0, w_0) in (1) come from various model uncertainties such as a) the errors associated with the model parameters, b) the neglected higher order terms in the Taylor series expansions in the linearization procedure, c) the unmodeled higher order vibrational dynamics of the beam, d) the effect of the eddy-current in the electromagnet, e) the real physical disturbances acting on the beam, and f) the sensor noises. The transfer function of the nominal plant $G(s) := C_g(sI-A_g)^{-1}B_g$ is then given by

$$G(s) = \frac{-13.3(s+0.654-j28.2)(s+0.654+j28.2)}{(s+84.4)(s-84.1)(s+18.0)(s+0.697-j28.8)(s+0.697+j28.8)}. \tag{3}$$

3. Controller Design

For the magnetic suspension system described in the previous section, our principal control objective is its robust stabilization against various real parameter changes and uncertainties. To this end, we will setup the problem within the framework of the H^∞ control theory.

First let us consider the disturbances v_0 and w_0. In practice, these disturbances v_0 and w_0 mainly act on the plant over a low frequency range and a high frequency range, respectively. Hence it is helpful to introduce frequency weighing functions. Let v_0 and w_0 be of the form

$$v_0(s) = W_1(s)v(s), \qquad w_0(s) = W_2(s)w(s) \tag{4}$$

where $W_1(s) =: C_{W1}(sI-A_{W1})^{-1}B_{W1} + D_{W1}$ and $W_2(s) =: C_{W2}(sI-A_{W2})^{-1}B_{W2} + D_{W2}$. These functions, as yet unspecified, can be regarded as free design parameters. Next we consider the variables which we want to regulate. In this study, since our main concern is in the stabilization of the beam, it is natural to choose as

$$z_g := \begin{bmatrix} x_1 \\ x_2 \\ \dot{x}_1 \\ \dot{x}_2 \end{bmatrix} = H_g x_g, \quad H_g := \begin{bmatrix} 1\,0\,0\,0\,0 \\ 0\,1\,0\,0\,0 \\ 0\,0\,1\,0\,0 \\ 0\,0\,0\,1\,0 \end{bmatrix}. \tag{5}$$

Then, as the error vector, let us define as

$$z := \begin{bmatrix} z_1 \\ z_2 \end{bmatrix} = \begin{bmatrix} \Theta z_g \\ \rho u \end{bmatrix}, \quad \Theta := diag\begin{bmatrix} \theta_1\ \theta_2\ \theta_3\ \theta_4 \end{bmatrix} \tag{6}$$

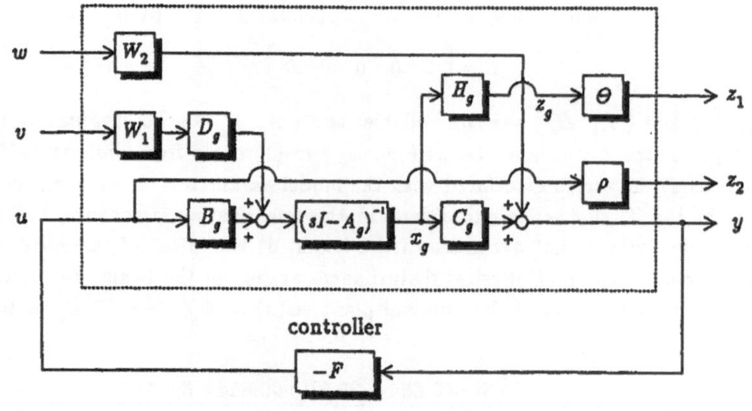

Fig. 2. Control problem setup.

where Θ is a weighting matrix on the regulated variables z_g and ρ is a scalar weighting factor on the control input u. These values, as yet unspecified, are also free design parameters. Now, we will setup the generalized plant as in Fig. 2.

The state-space representation of the generalized plant is

$$\dot{x} = Ax + B_1 d + B_2 u, \quad z = C_1 x + D_{11} d + D_{12} u, \quad y = C_2 x + D_{21} d + D_{22} u \tag{7}$$

where

$$x := \begin{bmatrix} x_g \\ x_{W1} \\ x_{W2} \end{bmatrix}, \quad d := \begin{bmatrix} v \\ w \end{bmatrix} \tag{8}$$

with the states x_{W1} and x_{W2} in the frequency weightings $W_1(s)$ and $W_2(s)$, respectively, and

$$A := \begin{bmatrix} A_g & D_g C_{W1} & 0 \\ 0 & A_{W1} & 0 \\ 0 & 0 & A_{W2} \end{bmatrix}, \quad B_1 := \begin{bmatrix} D_g D_{W1} & 0 \\ B_{W1} & 0 \\ 0 & B_{W2} \end{bmatrix}, \quad B_2 := \begin{bmatrix} B_g \\ 0 \\ 0 \end{bmatrix} \tag{9a}$$

$$C_1 := \begin{bmatrix} \Theta H_g & 0 & 0 \\ 0 & 0 & 0 \end{bmatrix}, \quad D_{11} := \begin{bmatrix} 0 & 0 \\ 0 & 0 \end{bmatrix}, \quad D_{12} := \begin{bmatrix} 0 \\ \rho I \end{bmatrix} \tag{9b}$$

$$C_2 := \begin{bmatrix} C_g & 0 & C_{W2} \end{bmatrix}, \quad D_{21} := \begin{bmatrix} 0 & D_{W2} \end{bmatrix}, \quad D_{22} := 0. \tag{9c}$$

Substituting $u(s) = -F(s)y(s)$ yields

$$\begin{bmatrix} z_1 \\ z_2 \end{bmatrix} = \begin{bmatrix} T_{z_1 v} & T_{z_1 w} \\ T_{z_2 v} & T_{z_2 w} \end{bmatrix} \begin{bmatrix} v \\ w \end{bmatrix} \tag{10}$$

where $\Phi(s) := (sI - A_g)^{-1}$ and

$$\begin{bmatrix} T_{z_1 v} \\ T_{z_2 v} \end{bmatrix} := \begin{bmatrix} \Theta H_g \\ -\rho F C_g \end{bmatrix} \Phi [I + B_g F C_g \Phi]^{-1} D_g W_1, \quad \begin{bmatrix} T_{z_1 w} \\ T_{z_2 w} \end{bmatrix} := -\begin{bmatrix} \Theta H_g \Phi B_g \\ \rho I \end{bmatrix} F [I + GF]^{-1} W_2. \tag{11}$$

Now our control problem is: find a controller $F(s)$ such that the closed-loop system is internally stable and the closed-loop transfer function satisfies the following H^∞ condition:

$$\left\| \begin{matrix} T_{z_1 v} & T_{z_1 w} \\ T_{z_2 v} & T_{z_2 w} \end{matrix} \right\|_\infty < 1. \tag{12}$$

The design has been carried out using the computer aided design package MATLAB. The frequency weightings are chosen as (see, Fig. 3 and 4)

$$W_1(s) = \frac{9.76\times10^3}{(1+s/(2\pi\cdot0.016))(1+s/(2\pi\cdot0.5))}\begin{bmatrix}1\\1\\1\end{bmatrix} \tag{13}$$

$$W_2(s) = \frac{5.21\times10^{-7}(1+s/(2\pi\cdot0.01))(1+s/(2\pi\cdot0.05))(1+s/(2\pi\cdot100))}{(1+s/(2\pi\cdot3.0))(1+s/(2\pi\cdot5.0))(1+s/(2\pi\cdot8.0))}. \tag{14}$$

While, the weightings Θ and ρ are chosen as

$$\theta_1 = 66.0,\ \theta_2 = 3.5\times10^{-1},\ \theta_3 = 9.1\times10^{-1},\ \theta_4 = 7.0\times10^{-1},\ \rho = 8.0\times10^{-5}. \tag{15}$$

For the selection of the frequency weightings $W_1(s)$ and $W_2(s)$, the well known "all-pass" property of the H^∞ control at optimality is very helpful. Besides, the LQ differential game theoretic characterizations of the H^∞ control enable us to choose the weightings Θ and ρ effectively. Direct calculations yield the central controller [1]

$$F(s) = \frac{-3.64\times10^{10}(s+84.4)(s+50.3)(s+31.4)(s+18.9)(s+18.1)}{(s+3650)(s+615-j480)(s+615+j480)(s+14.8)(s+3.13)(s+0.0758)}$$
$$\times\frac{(s+2.40-j2.17)(s+2.40+j2.17)(s+0.687-j28.9)(s+0.687+j28.9)}{(s+53.1-j31.0)(s+53.1+j31.0)(s+0.412-j29.3)(s+0.412+j29.3)}. \tag{16}$$

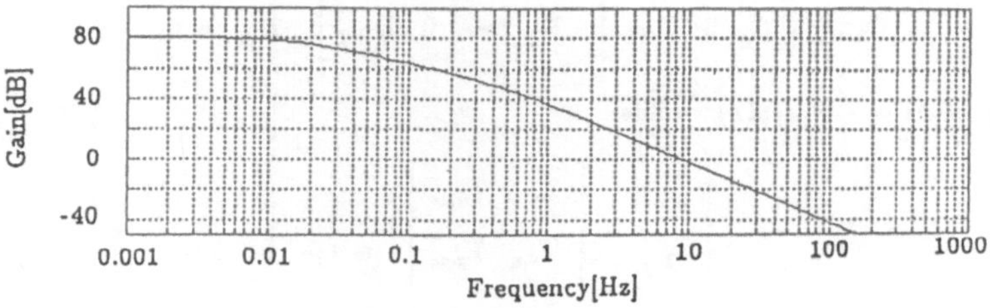

Fig. 3. Frequency weighting W_1.

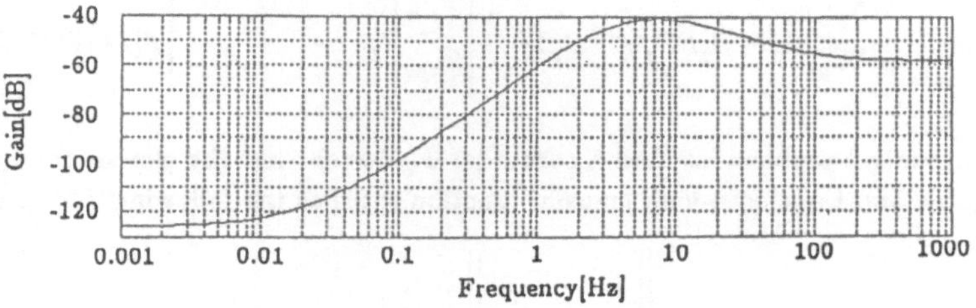

Fig. 4. Frequency weighting W_2.

4. Experimental Results

The obtained continuous-time controller (16) is discretized via the Tustin transform at the sampling rate of 70 μs. A digital signal processor (DSP)-based real-time controller is implemented using NEC μPD77230 on a special board, which can execute one instruction in 150 ns with 32-bit floating point arithmetic (see, Fig. 5). The control algorithm is written in the assembly language for the DSP and the software development is assisted by the host personal computer NEC PC9801. The data acquisition boards consist of a 12-bit A/D converter module DATEL ADC-B500 with the maximum conversion speed of 0.8 μs and a 12-bit D/A converter DATEL DAC-HK12 with the maximum conversion speed of 3 μs.

We first evaluate the nominal stability of the designed control system with time-responses for a step-type disturbance. The disturbance is added to the experimental system as an applied voltage in the electromagnet, which in fact amounts to about 20 % of the steady-state force. Fig. 6 shows the displacements of x_1 and x_2. From this result, we can see that the nominal stabilization is certainly achieved. Since our concerns are also in the robust stability of the system against various model uncertainties, we further continue the same experiments with the values of mass M, mass m, and the resistance of the electromagnet R changed. The parameters have been changed in the following ways:

(i) M = 9.63 kg, which amounts to 7 % decrease for its nominal value of 10.36 kg,

(ii) M = 8.06 kg, which amounts to 22 % decrease for its nominal value of 10.36 kg.

(iii) M = 11.51 kg, which amounts to 11 % increase for its nominal value of 10.36 kg.

(iv) m = 6.82 kg, which amounts to 18 % increase for its nominal value of 5.8 kg,

(v) R = 62.0 Ω, which amounts to 9 % increase for its nominal value of 57.0 Ω,

The results are shown in Fig. 7 - 11. In any case, it can be seen that the beam is still suspended stably even if the step-type disturbance is added. Therefore, these experimental results confirm us that the magnetic suspension system designed by the H^∞ control theory is robustly stable against various uncertainties.

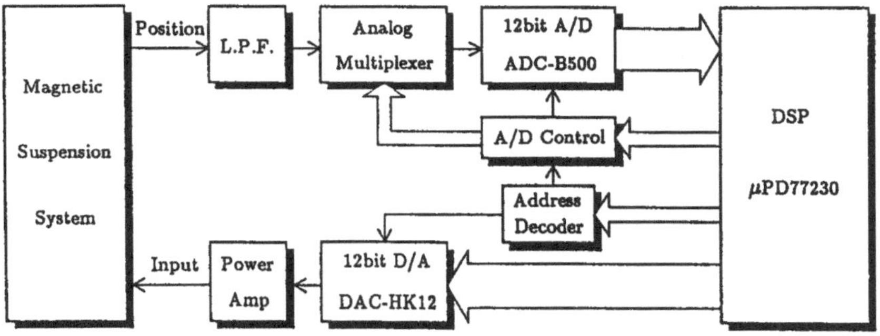

Fig. 5. DSP-based controller.

5. Conclusions

In this paper, an H^∞ controller has been experimentally demonstrated on a magnetic suspension system with a flexible beam. Several experimental results showed that the magnetic suspension system is robustly stable against various model uncertainties.

References

[1] K. Glover and J.C. Doyle, "State-space formulae for all stabilizing controllers that satisfy an H_∞-norm bound and relations to risk sensitivity", *Syst. Contr. Lett.*, vol. 11, no. 3, pp. 167-172, 1988.

[2] J.C. Doyle, K. Glover, P.P. Khargonekar, and B.A. Francis, "State-space solutions to standard H_2 and H_∞ control problems," *IEEE Trans. Automat. Contr.*, vol. 34, no. 8, pp. 831-847, 1989.

[3] M. Fujita, F. Matsumura, and M. Shimizu, "H^∞ robust control design for a magnetic suspension system," in *Proc. 2nd Int. Symp. Magnetic Bearings*, Tokyo, Japan, 1990.

[4] M. Fujita and F. Matsumura, "H^∞ control of a magnetic suspension system (in Japanese)," *J. IEE of Japan*, vol. 110, no. 8, pp. 661-664, 1990.

[5] F. Matsumura, M. Fujita, and M. Shimizu, "Robust stabilization of a magnetic suspension system using H^∞ control theory (in Japanese)," *Trans. IEE of Japan*, vol. 110-D, no. 10, pp. 1051-1057, 1990.

[6] M. Fujita, F. Matsumura, and K. Uchida, "Experiments on the H^∞ disturbance attenuation control of a magnetic suspension system," in *Proc. 29th IEEE Conf. Decision Contr.*, Honolulu, Hawaii, 1990.

[7] M. Fujita, F. Matsumura, and K. Uchida, "H^∞ robust control of a flexible beam magnetic suspension system (in Japanese)," *J. SICE*, vol. 30, no. 8, 1991.

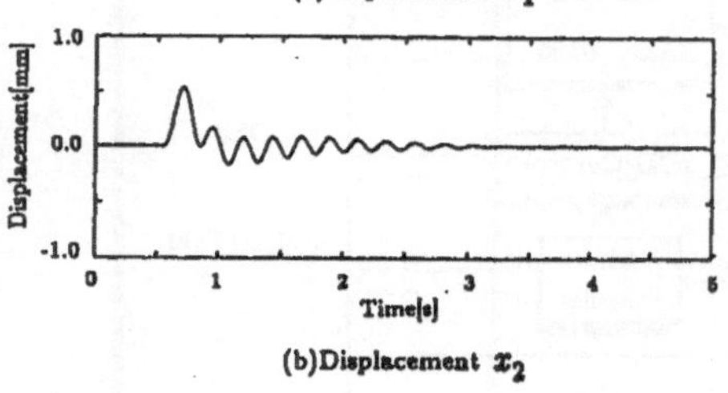

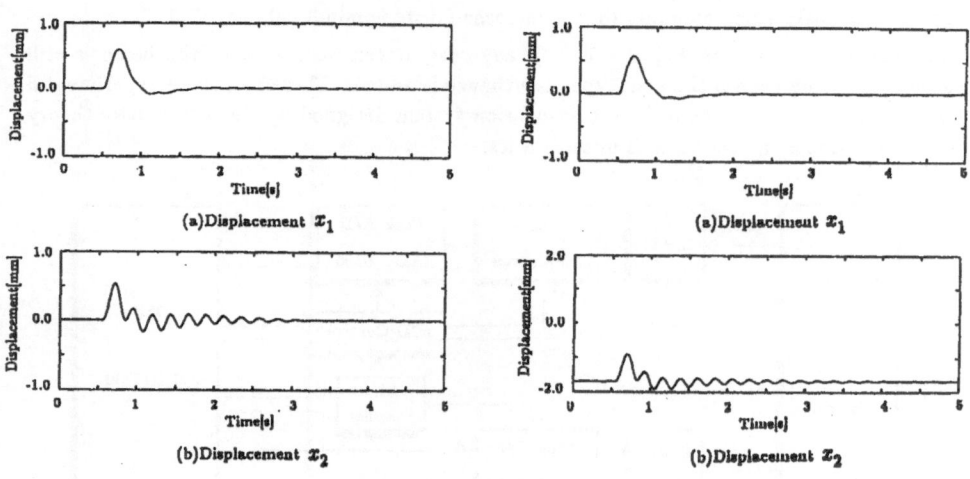

Fig. 6. Responses for step-type disturbance (nominal).

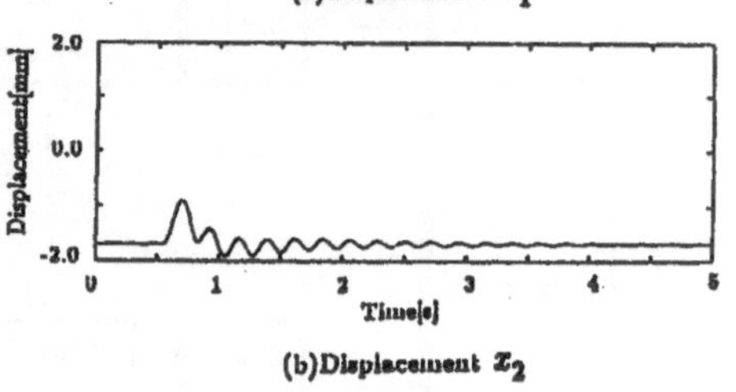

Fig. 7. Responses for step-type disturbance ($M = 9.63$ kg).

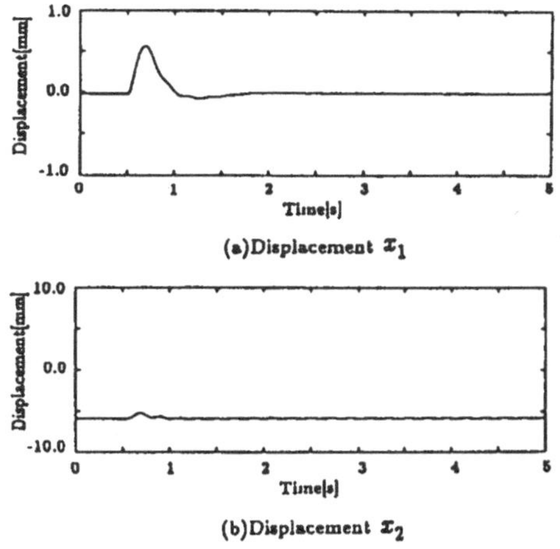

Fig. 8. Responses for step-type disturbance (M = 8.06 kg).

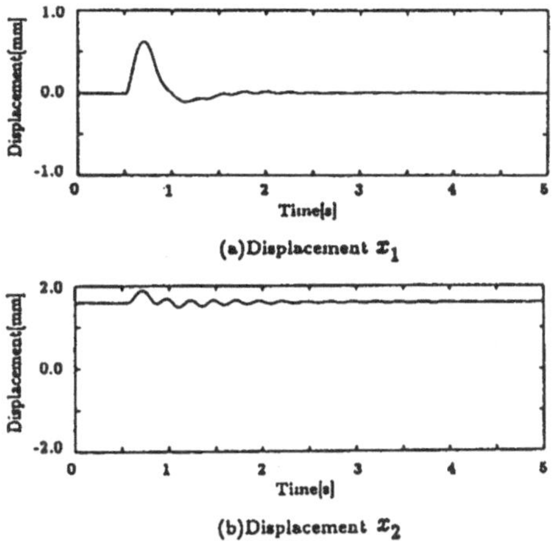

Fig. 9. Responses for step-type disturbance (M = 11.51 kg).

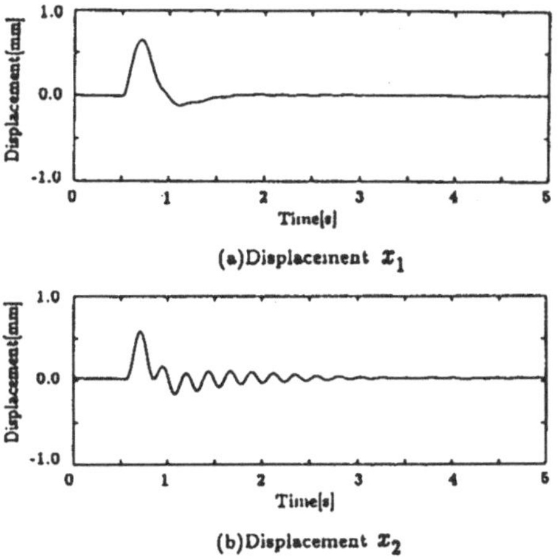

Fig. 10. Responses for step-type disturbance (m = 6.82 kg).

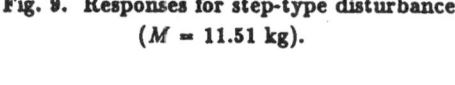

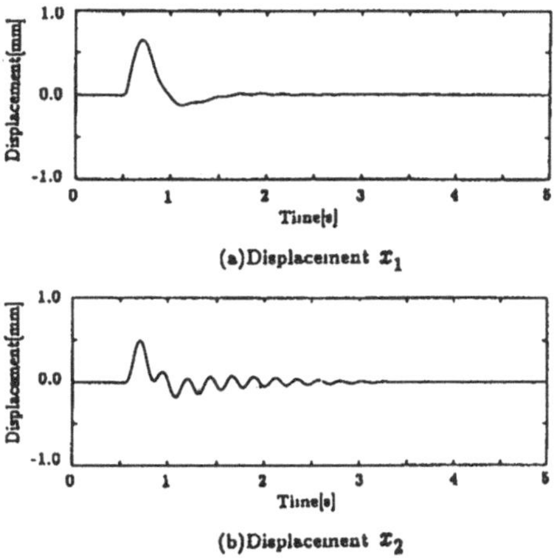

Fig. 11. Responses for step-type disturbance (R = 62.0 Ω).

Robust Control of Nonlinear Mechanical Systems
— Case Studies —

Koichi Osuka
Department of Mechanical Engineering,
University of Osaka Prefecture,
Sakai, Osaka 591, Japan.

1 Introduction

Since the 1960's, trajectory control problem of robot, especially manipulators, has been studied. In the early stage, since the dynamical equations of robot manipulators are highly nonlinear, many researchers were interested in how to attack such nonlinear properties[1]. However, in the early of 1980's, it was pointed out that the dynamical model of robot manipulators have some useful properties. As the results, main subject of the research has been shifted from 'How to attack the nonlinearlity' to 'How to treat the modelling errors of the dynamical model of robots[2]' . Here, it can be said that the following uncertainties cause the modelling errors. (i)*Structured Uncertainty; Uncertainty caused by the uncertainties of dynamical parameters such as mass or moment of inertia of links.* (ii)*Unstructured Uncertainty; Uncertainty caused by the uncertainties of parastical unmodelled dynamics of actuater system.*

From the above discussions, it is clear that to design a controller which makes the resulting control system robust to both uncertainties is important and necessary. However, it seems that no design method for such an ideal controller has been proposed.

In this paper, dividing the uncertainties into two cases, that is (i) and (ii), and then we introduce control strategy which suited for each case through some experiments.

2 Dynamical Model of Robot

In general, robots are consisted of n-rigid links and actuators installed in each joint. Often, we divide the dynamical model of robot into two parts. One is the dynamical model of rigid links and the other is that of the actuator systems.

Suppose that the links which construct the link system are all rigid and the inputs to this system are torque produced by the actuator systems. Then, letting $\theta = [\theta_1, \theta_2, \cdots, \theta_n]^T$ be joint angle vector and letting $\tau = [\tau_1, \tau_2, \cdots, \tau_n]^T$ be input torque vector, the dynamical equation of the link system can described as

$$< S > \qquad J(\theta)\ddot{\theta} + c(\theta, \dot{\theta}) = \tau. \tag{1}$$

Where, $J(\theta)\ddot{\theta}$ denotes n dimensional vector which denotes inertia torque term, $c(\theta,\dot{\theta})$ is n dimensional vector which represents torque due to centrifugal force, Coriolis force, gravity and viscosity friction. Notice that, once a robot given, the structure of the nonlinear functions in these vectors are decided. Furthermore, it is well known that the dynamical equation Eq.(1) has the following useful properties. $< P1 >$ *The matrix $J(\theta)$ is positive definite for all θ .* $< P2 >$ *All elements of $J(\theta)$ is bounded for all θ, and all elements of the vector $c(\theta,\dot{\theta})$ are bounded for all θ and all bounded $\dot{\theta}$.* $< P3 >$ *The equation satisfies the 'Matching Conditions'.*

By the way, most of actuator systems of robots are constructed from an actuator itself, a transmission box and an electrical amplifier. The input of this system is an electric signal u to the amplifier and its output is the torque τ form the actuator. Then the dynamical model of the actuator system can be written as follows:

$$< A > \begin{cases} \dot{x} = f(x,u) \\ \tau = h(x). \end{cases} \tag{2}$$

Where, $f(x,u)$ denotes a vector whose elements are nonlinear functions of x and u, and $h(x)$ is a vector whose elements are nonlinear functions of x. It should be noticed that it is difficult for us to decide the structures of the nonlinear functions in $f(x,u)$ and $h(x)$, and also difficult to specify the dimensions of x and u. Therefore, we should take account of the unstructured uncertainties in controlling the actuator system.

3 Robust Control Under Structured Uncertainty

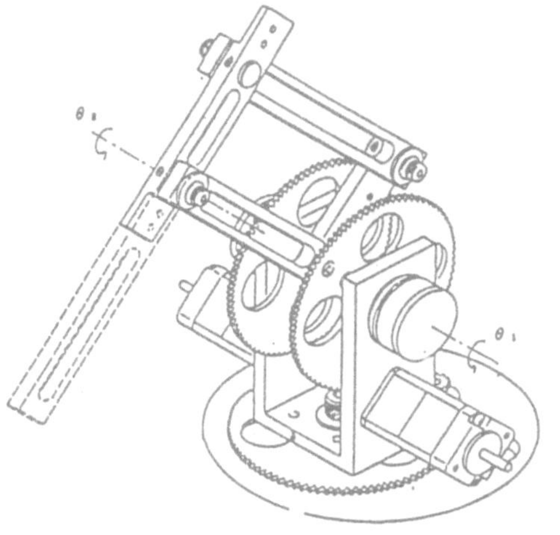

Fig.1 Miniature Robot

In this section, we introduce a small manipulator named Miniature Robot. Then using the robot, we show that the philosophy of high-gin feedback control is effective if the uncertainty is mainly structured one. We designed PD-Type Two Stage Robust Tracking Controller which proposed by the author and controlled the robot by using the controller[3]. Fig.1 shows the Miniature Robot. This robot has two degrees of freedom as shown in figure. The dynamical model of this robot can be described as $< S >$. We can measure $[\theta^T \quad \dot{\theta}^T]^T$ and the initial condition is $[\theta(0)^T \quad 0^T]^T$. Here, we define the following regions:

$$
\begin{aligned}
H_s(\epsilon_s) &= \{\theta \mid \theta_i^{min} + \epsilon_s < \theta_i < \theta_i^{max} - \epsilon_s \quad (i = 1, 2)\} \\
H_v(\epsilon_v) &= \{\dot{\theta} \mid \dot{\theta}_i^{min} + \epsilon_s < \dot{\theta}_i < \dot{\theta}_i^{max} - \epsilon_s \quad (i = 1, 2)\} \\
H_a(\epsilon_a) &= \{\ddot{\theta} \mid \|\ddot{\theta}\| < \epsilon_a\}
\end{aligned} \tag{3}
$$

where θ_i^{min} and θ_i^{max} are decided from the admissible configulation of the robot. $\dot{\theta}_i^{min}$ and $\dot{\theta}_i^{max}$ are decided from the admissible velocity of the robot. The scalars of ϵ_s, ϵ_a, ϵ_v are of the robot. ϵ_s is a constant.

It is important that, in the dynamical model of the miniature robot, the dynamics of the link system is dominant. Therefore, since $< P3 >$ can be applicable, a concept of high gain feedback control is effective. PD-Type Two Stage Robust Tracking Controller is a controller by which we can specify the following specifications. (a) *The desired trajectory θ_R can be specified.* (b) *The control accuracy ϵ can be specified.* (c) *If the initial condition error between desired trajectory and actual trajectory exist, we can specify the characteristics of convergence of initial error.* (d) *Robustness against structured uncertainties can be assured.* (e) *The structure of the resultant control system is simple.*

Notice that we can specify (c) explicitly.

The idea of the design method is the following. First, construct an intermediate linear model $< M >$ whose initial condition coincides with that of the robot $< S >$. Then, design a PD-Type robust controller $< C_R >$ so that $< S >$ may follow $< M >$ with the specified control accuracy, that is,

$$
\|[e(t)^T \;\; \dot{e}(t)^T]\| := \|[\theta(t)^T - \theta_M(t)^T \;\; \dot{\theta}(t)^T - \dot{\theta}_M(t)^T]\| < \epsilon \; (t \geq 0), \tag{4}
$$

where $\theta_M(t)$ denotes the output of $< M >$. Then, regard $< M >$ as a controlled object and construct a two-degree-of-freedom controller $< C_L >$ for $< M >$ so as to $< M >$ follows $\theta_d(t) = \theta_R(t) + \mathcal{L}^{-1}(G_e(s)e_0)$.

The following procedure shows the design method of the above controller.

Step0) *Formulation of design specifications*

(a) Specify the control accuracy ϵ ; $\| \; [\theta^T - \theta_d^T, \;\; \dot{\theta}^T - \dot{\theta}_d^T] \; \| < \epsilon$.

(b) Specify the admissible initial error bound E ; $\|e(0)\| < E$.

(c) Specify the convergence characteristics of $e(0)$; $G_e(s) = \frac{s+2\lambda}{(s+\lambda)^2}$.

(d) Plan the desired trajectory θ_R ; $\theta_R \in H_s(\epsilon_s)$, $\dot{\theta}_R \in H_v(\epsilon_v)$, $\ddot{\theta}_R \in H_a(\epsilon_a)$.
where, $\epsilon_s = E + \epsilon$, $\epsilon_v = 0.4\lambda E + \epsilon$.

Step1) *Design of PD-Type Two Stage Controller.*

Construct the intermediate linear model

$$
< M > \quad \ddot{\theta}_M = u_M \quad (\theta_M(0) = \theta(0), \; \dot{\theta}_M(0) = 0) \tag{5}
$$

and design the control law:

$$
\begin{array}{ll}
< C_L > & u_M = \ddot{\theta}_R - 2\lambda(\dot{\theta}_M - \dot{\theta}_R) - \lambda^2(\theta_M - \theta_R) \\
< C_R > & \left\{ \begin{array}{l} v = -kh_2(\dot{\theta} - \dot{\theta}_M) - kh_1(\theta - \theta_M) \\ \tau = \hat{J}(\theta)(u_M + v) + \hat{c}(\theta, \dot{\theta}) \end{array} \right.
\end{array} \tag{6}
$$

where, $\hat{J}(\theta)$ and $\hat{c}(\theta,\dot{\theta})$ denote nominal matrices of $J(\theta)$ and $c(\theta,\dot{\theta})$. $h_i(i = 1,2)$ are arbitrary positive scalar and k is the design parameter which is designed by using the specified control accuracy and the modelling errors. See Appendix and Fig.2.

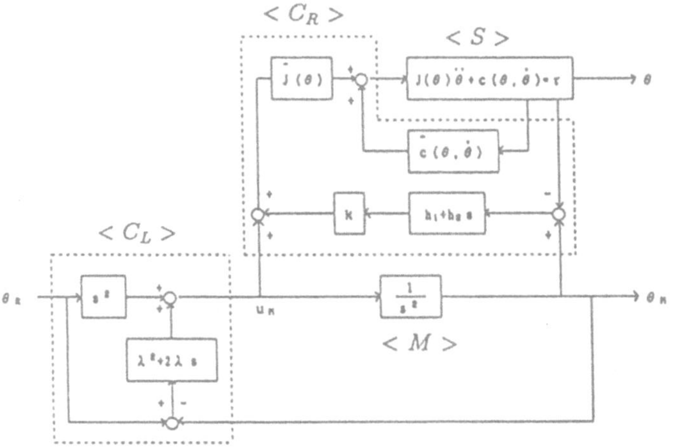

Fig.2 PD-Type Two Stage Robust Tracking Control System

To show the effectiveness of the control law (5)(6), we applied the law to the miniture robot[4]. Especialy, we show the experimental results by which the ability of specification of convergence property of e(0) can be recognized. At first, we desided a nominal model of the robot as follows.

$$
\begin{array}{l}
\hat{J}(\theta) = diag[1.63 \times 10^{-6} \quad 1.63 \times 10^{-6}] \quad (kgm^2) \\
\hat{c}(\theta, \dot{\theta}) = [0.325 \cos\theta_1 \quad 0.244 \cos\theta_2]^T \quad (Nm)
\end{array} \tag{7}
$$

Then, we chose the initial condition $[\theta(0)^T \ \dot{\theta}(0)^T]^T$ and the desired trajectory $[\theta_R^T \ \dot{\theta}_R^T]^T$ as follows.

$$
\begin{array}{l}
\theta(0) = \begin{bmatrix} 1.89 \\ 0.45 \end{bmatrix} \ (rad), \quad \dot{\theta}(0) = \begin{bmatrix} 0.0 \\ 0.0 \end{bmatrix} \ (rad/sec) \\
\theta_R = \begin{bmatrix} \theta_{R1} \\ \theta_{R2} \end{bmatrix} = \begin{bmatrix} 1.571\{-\cos(\frac{2\pi t}{T}) + 1.0\} + 1.571 \\ 0.785\{-\cos(\frac{2\pi t}{T}) + 1.0\} \end{bmatrix} (0 \le t \le T = 0.77 sec)
\end{array} \tag{8}
$$

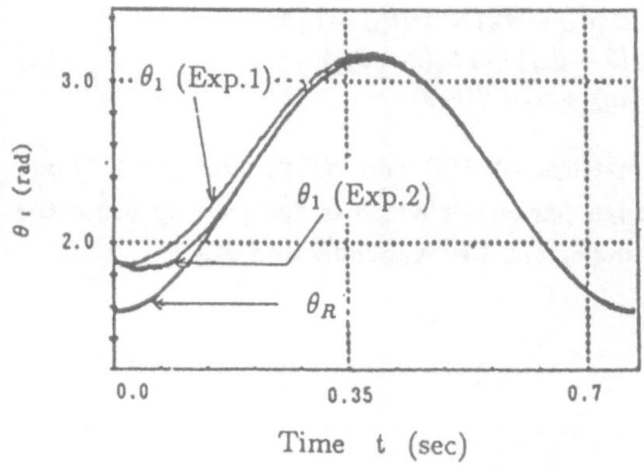

Fig.3 Experimental Results

Fig.3 shows the experimental results. Parameters are setted as $kh_1 = 50000, kh_2 = 100$. For experiment 1, we chose $\lambda = 10$ and for experiment 2, we chose $\lambda = 20$. From the figure, we can see that the convergence characteristics of initial error can be specified easily.

4 Robust Control Under Unstructured Uncertainty

In this section, we introduce a certain actuator named Rubbertuator.

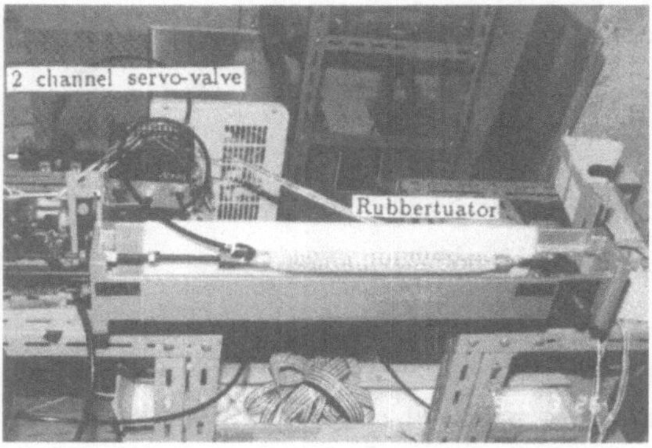

Fig.4 Rubbertuator

Using this actuator, we show that the philosophy of H_∞ control is effective if the uncertainty is mainly unstructured one.

Rubbertuator is a new actuator developed by BRIDGESTONE in Japan (see Fig.4). As shown in Fig.4, we use a pair of Rubbertuator like muscles of animals, that is, they are connected each other to rotate the load in push-pull mode. The sequence for operating the system is as follows.

Step0) Supply an electric current $I = 6.25(mA)$ to channel 1 and 2 of the servo-valve.
Step1) Repeat the following procedures every sampling period.
1) Detect the angle of the rotor, θ (pulse),by using a rotary encoder.
2) Normalize θ (pulse) as $\theta = \theta\ /147.67(rel)$ and compute the input signal,$u(rel)$.
3) Input $i_1 = I + 0.2u\ \ (mA)$ and $i_2 = I - 0.2u\ \ (mA)$ to the channel 1 and 2 of servo-valve respectively, and returen to 1).

In the following, we regard $u(rel)$ and $\theta(rel)$ as input and output of the system.

Since, the dynamical model of Rubbertuator is nonlinear system[5], the model should be described as a nonlinear state space equation $< A >$ shown in 2. However, it should be noticed that the structure of the nonlinear function in < A > and the dimension of the equation are not clear, but the nonlinearlity of the function is not so strong. This implies that we should not apply a high gain feedback control as descrived in the previous section. Therefore, we regard the system as linear system approximately and, obtaining Bode plots experimentally, design a linear robust controller. The specification of the controller which we are going to design is as follows: (a)*Ensure robust stability against modelling errors.* (b)*Have a good tracking property.* To do so, we design a controller according to the following procedure. First, fix the structure of control system as shown in Fig.5.

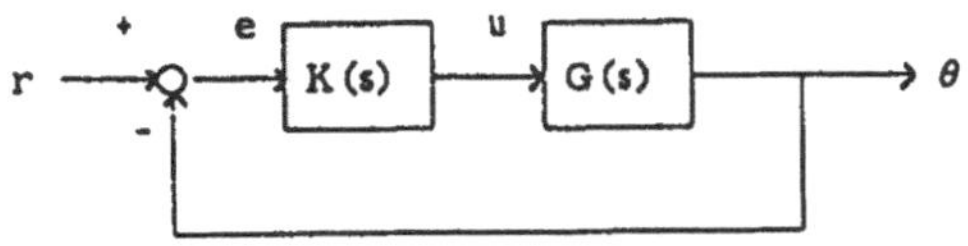

Fig.5 Control System

Where, $G(s)$ denotes transfer function of Rubbertuator which is regarded as a linear system, $K(s)$ denotes a controller which will be designed. Next, using the information about the frequency region in which large modelling errors may exist, decide a weighting function $W_2(s)$ for complementary sensitivity function $T(s)$.

Using the infrmation about the specification for control accuracy, decide a weighting function $W_1(s)$ for sensitivity function $S(s)$. Finaly, solve the Mixed-Sensitivity Problem, that is, design a controller to satisfy

$$\left\| \begin{bmatrix} \delta W_1(s)S(s) \\ W_2T(s) \end{bmatrix} \right\|_\infty < 1. \tag{9}$$

Where $\|\cdot\|_\infty$ denotes H_∞ norm[6]. This problem can be solved by setting a generalized plant in the H_∞ theory as

$$P(s) = \begin{bmatrix} \delta W_1(s) & -\delta W_1(s)G_0(s) \\ 0 & W_2(s)G_0(s) \\ 1 & -G_0(s) \end{bmatrix}. \tag{10}$$

To construct an H_∞ controller, we adopt the following design procedure.

Step0) Input various kinds of sine-waves to Rubbertuator, we got Bode diagram. See Fig.6. It should be noticed that, since Rubbertuator is a nonlinear system, we can see many lines in the figure.

Step1) From the Bode diagram, we decided a nominal plant model $G_0(s)$. The result is $G_0(s) = \frac{889.27}{s^2+23.74s+889.27}$, and we also show $|G_0(j\omega)|$ and $\angle G_0(j\omega)$ in Fig.6.

Step2) From Fig.6, calculate a multiplicative uncertainty $\Delta(j\omega)$ by $\Delta(j\omega) = \frac{G(j\omega)}{G_0(j\omega)} - 1$. for each $G(j\omega)$. Then, we obtained $|r(j\omega)|$as a pointwise maximal of $\Delta(j\omega)$.

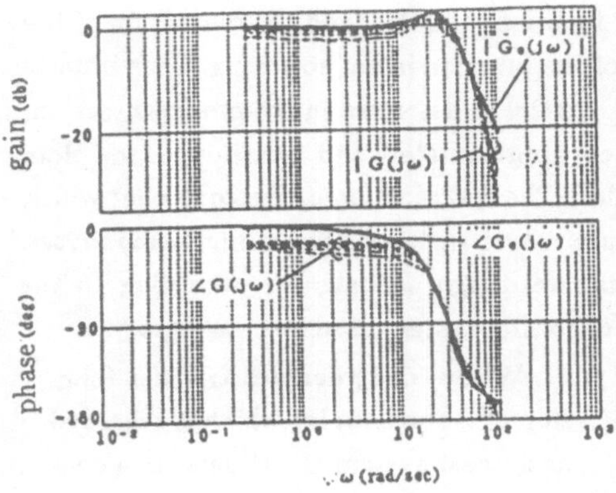

Fig.6 Bode Diagram

Step3) From a viewpoint of robust stability, we decided a weighting function $W_2(s)$ for complementary sensitivity function $T(s)$ so as to $|W_2(j\omega)| > |r(j\omega)|$. As the result, we chose $W_2(s) = \frac{s^2+20s+100}{290}$.

Step4) Decide $W_1(s)$: From a view point of control accuracy, we decided a weighting function $W_1(s)$ for sensitivity function $S(s)$ as $\delta W_1(s) = \frac{\delta}{(s+1)^2}$.

Step5) Constructing a generalized plant $P(s)$ as Eq.(10), designd an H_∞ controller $K(s)$ by using PC-MATLAB[6].

We could solve the problem when $\delta = 50$. The resulting controller can be written as

$$K(s) = \frac{43.669s^3 + 125.87s^2 + 4439.7s + 211640}{s^4 + 163.96s^3 + 4375.20s^2 + 8262.4s + 4050.2} \tag{11}$$

Fig.7 shows the experimental result. The desired trajectory is $r = u(t)$ $(0 \leq t \leq 4(sec))$. In the figure, the simulated result is also shown. From these results, the effectiveness of H_∞ can be recognized.

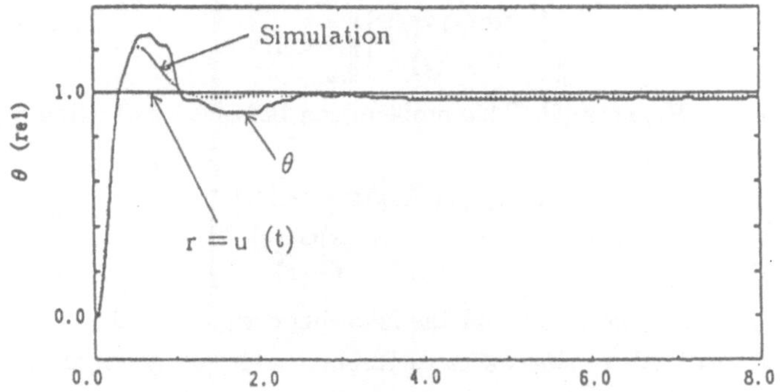

Fig.7 Experimental Result

5 Conclusion

In this paper, we showed some case studies of robust control of mechanical systems. In this paper, dividing the uncertainties in the dynamical model of robot systems into two cases, that is *(i) Structured Uncertainty* and *(ii) Unstructured Uncertainty*, and then we showed control strategy which suited for each case through some experiments.

We can say that the next step in the field of robot motion control is to consider both uncertainties at once.

Appendix

In this appnedix, we show how to design $\hat{J}(\theta)$, $\hat{c}(\theta,\dot{\theta})$, k[3].

(i)$\hat{J}(\theta)$ must be designed so as to boounded for arbitrary θ and to satisfy $D^* := (D^T + D) > 0$. Where, $D = J(\theta)^{-1}\hat{J}(\theta)$.

(ii) $\hat{c}(\theta,\dot{\theta})$ must be chosen so as to bounded for arbitrary θ and for arbitrary bounded $\dot{\theta}$.

(iii) k should be chosen so as to satisfy $k > k_0 := \frac{(h_1\epsilon + h_2 f_g)(h_1^2+h_2^2)^{1/2}}{h_1 h_2^2 J_d}$. Where, $J_d = \min_{\theta\in H_\theta(0)} \lambda_m(D^*)$, $f_g = \max_{\substack{\theta\in H_\theta(0)\\ \dot{\theta}\in H_v(\epsilon_v)\\ \|u_M\|<R}} = [J(\theta)^{-1}\hat{J}(\theta) - I]u_M + J(\theta)^{-1}[\hat{c}(\theta,\dot{\theta}) - c(\theta,\dot{\theta})]$, R$=\epsilon_a + 2\lambda\epsilon_v + \lambda^2\epsilon$, and $\lambda_m(\cdot)$ denotes minimum eigenvalue of matrix.

References

[1]J.J.Uicker: Dynamic Force Analysis of Spatial Linkages,ASME J. of Applied Mechanics 34, 418/424 (1967).

[2]K.Osuka:Adaptive Control of Nonlinear Mechanical Systems, Trans. of SICE, 22-2, 126/133 (1986)(in Japanese).

[3]K.Osuka,T.Sugie and T.Ono: PD-Type Two-Stage Robust Tracking Control for Robot Manipulators, Proc. U.S.A.-Japan Symposium on Flexible Automation, 153/160 (1988).

[4]N.Yoshida,K.Osuka and T.Ono: PD-Type Two-Stage Robust Tracking Control for Robot Manipulators-An Application to DAIKIN Miniature Robot-, Proc. of 20th ISIR, 1051/1058 (1989).

[5]K.Osuka, T.Kimura and T.Ono: H ∞ Control of a Certain Nonlinear Actuator, Proc. of The 29th IEEE CDC, 370/371 (1990).

[6]B.Francis: A Course in H_∞ Control Theory, Springer-Verlag (1987).

[7]R.Y.Chiang and M.G.Safonov: Robust-Control Toolbox, The Math Works, Inc. (1988)

High-Order Parameter Tuners for the Adaptive Control of Nonlinear Systems*

A. S. Morse
Department of Electrical Engineering
Yale University
P. O. Box 1968
New Haven, Ct., 06520-1968, USA

Abstract

A new method of parameter tuning is introduced which generates as outputs not only tuned parameters, but also the first $\bar{n}$ time derivatives of each parameter, $\bar{n}$ being a prespecified positive integer. It is shown that the algorithm can be used together with a suitably defined identifier-based parameterized controller to adaptively stabilize any member of a specially structured family of nonlinear systems.

Introduction

In the past few years there have been a number of attempts to extend existing adaptive control techniques to processes modelled by restricted classes of nonlinear dynamical systems {e.g., [1]-[3]}. Such efforts usually consist of showing that a particular algorithm is capable of stabilizing a specially structured linearly parameterized, nonlinear design model $\Sigma_D(p)$ for each fixed but unknown value of p. Techniques from nonlinear system theory are then used to characterize, in coordinate independent terms, those dynamical process models whose input-output descriptions match or are close to that of Σ_D for particular values of p in Σ_D's parameter space $\mathcal{P}$. This then is the class of admissible process models C_P to which the particular algorithm can be successfully applied.

With the preceding in mind it is natural to wonder just how far existing adaptive techniques {e.g., [4]} can be extended into the world of nonlinear systems without having to introduce significant modifications. For example, for those cases in which parameters enter the model in "just the right way" and nonlinearities are such that with knowledge of the parameters nonadaptive stabilization is possible, one might expect that without knowledge of the parameters and notwithstanding the nonlinear dynamics, adaptive stabilization could be achieved using familiar methods. Interestingly, this has proved *not* to be the case, even for very simply structured examples, when the nonlinearities are not globally Lipschitz. Within the very recent past, several new adaptive algorithms have been discovered {e.g., see [5]-[9]} which overcome some of these difficulties. One of the key features of these algorithms which sharply distinguishes them from more familiar algorithms described in [4] and elsewhere, is the absence of tuning error normalization. The purpose of this paper is to introduce a new method of parameter tuning, also employing an unnormalized tuning error, which can be used to obtain adaptive controllers with the same capabilities as those reported in [6].

The new tuner is of "high-order". By a tuner of *order* $\bar{n}$, is meant a tuning algorithm capable of generating as outputs not only tuned parameters, but also the parameter's first $\bar{n}$ time-derivatives. Most tuning algorithms are of order one. The only previous studies of high-order tuners we know of can be found in [10] and [11]. What distinguishes the algorithm introduced here from those considered in [10] and [11], is the absence of tuning error normalization.

The new high-order tuner Σ_T is described in §1 . In §2, we define one possible class C_P of process models Σ_P to which high-order tuning might be applied. In §3 we explain how to construct an "identifier-based parameterized controller" Σ_C {cf. [12]}, which together with Σ_T , is capable of adaptively stabilizing any process admitting a model in C_P. A brief stability analysis is carried out in §4.

*This research was supported by the National Science Foundation under grant number ECS-9012551

1. High-Order Tuner Σ_T

Consider the "error equation"

$$\dot{e} = -\lambda e + q_0(k - q_P)'w + \epsilon \tag{1}$$

where λ is a positive constant, q_0 is a nonzero, scalar, constant of unknown magnitude but known sign, q_P is an unknown constant m-vector, w is an m-vector of known time functions, e is a known scalar tuning error, ϵ is an exponentially decaying time function, and k is a vector of m parameters to be tuned. Equations of this form arise in connection with certain types of adaptive control problems, as will be shown in the sequel. Our objective here is to explain how to tune k in such a way as to cause both k and e to be bounded wherever they exist and also to cause e to have a finite $\mathcal{L}^2$ norm. The algorithm we propose will generate as outputs not only k, but also the first $\bar{n}$ derivatives of k, $\bar{n}$ being a prespecified positive integer. These derivatives can thus be used elsewhere in the construction of an overall adaptive controller. In order for the algorithm to have this property we shall have to make the following assumption.

Assumption 1: The first $\bar{n}-1$ derivatives of w are available signals.

High-Order Tuning Algorithm:

- If $\bar{n} = 1$, set

$$\dot{k} = -\text{sign}(q_0)we \tag{2}$$

- If $\bar{n} > 1$ pick a monic, stable polynomial $\bar{\alpha}(s)$ of degree $\bar{n}-1$ and let $(\bar{c}, \bar{A}, \bar{b})$ be a minimal realization of $\bar{\alpha}(0)/\bar{\alpha}(s)$; set

$$\begin{aligned} \bar{e} &= e - v & (3)\\ \dot{v} &= -\lambda v + h_0(k-h)'w & (4)\\ \dot{h}_0 &= (k-h)'w\bar{e} & (5)\\ \dot{h} &= -\text{sign}(q_0)w\bar{e} & (6)\\ \dot{X} &= \bar{A}X(I+D) + \bar{b}h'(I+D) & (7)\\ k' &= \bar{c}X & (8) \end{aligned}$$

where D is the $m \times m$ diagonal matrix whose ith diagonal entry is w_i^2, w_i being the ith component of w.

Remark 1: Note that for $\bar{n} > 1$, the definition of $(\bar{c}, \bar{A}, \bar{b})$ together with (6) to (8) ensure that the first $\bar{n}$ derivatives of k can be expressed as functions of $X, h, w, \bar{e}$ and the first $\bar{n}-1$ derivatives of w. Therefore, in view of Assumption 1, the first $\bar{n}$ derivatives of k are realizable from available signals.

Remark 2: The significant properties of the preceding algorithm will remain unchanged if D is replaced with any positive definite matrix $\bar{D}$ satisfying $\bar{D} \geq D$. One such matrix is $\bar{D} = w'wI$.

The main result of this paper is as follows.

High-Order Tuning Theorem: *Let $[0,T)$ be any subinterval of $[0,\infty)$ on which w is defined and continuous.*

- *For $\bar{n} = 1$, each solution $\{e,k\}$ to (1) and (2) is bounded on $[0,T)$ and e has a finite $\mathcal{L}^2$ norm.*
- *For $\bar{n} > 1$, each solution $\{e, v, h_0, h, X\}$ to (1), (3) - (8) is bounded on $[0,T)$ and e and v have finite $\mathcal{L}^2$ norms.*

Idea of proof: For $\bar{n} = 1$ the result is well known and can easily be deduced by evaluating derivative of the "partial Lyapunov function" $e^2 + |q_0|\|k - q_P\|^2$ along a solution to (1) and (2). For $\bar{n} > 1$ the theorem can be proved without much more effort using the partial Lyapunov function $\bar{e}^2 + (h_0 - q_0)^2 + |q_0|\|h - q_P\|^2 + \delta\sum_{i=1}^{\bar{n}} z_i'Pz_i$ where $z_i = x_i + \bar{A}^{-1}\bar{b}h_i$, x_i is the ith column of X, h_i is the ith component of h, P is the unique positive definite solution to the Lyapunov equation $P\bar{A} + \bar{A}'P = -I$, and $\delta = \frac{\lambda}{2\bar{n}\|P\bar{A}^{-1}\bar{b}\|^2}$. For a complete proof, see [13].

Remark 3: If an upper bound q^* for $|q_0|$ is known, the preceding algorithm can be simplified by setting $\bar{e}$ equal to e, eliminating (4) and (5), and replacing D with μD where μ is a positive scale factor of sufficiently large value; e.g., $\mu \geq \frac{2\bar{n}q^*\|\bar{c}'\|\|P\bar{A}^{-1}\bar{b}\|}{\lambda}$ where P is as above. Using this particular algorithm it is possible to construct an adaptive controller along the lines of [14], which is capable of stabilizing any siso process admitting a linear, minimum phase model whose dimension does not exceed a prespecified integer n and whose relative degree is either $\bar{n}$ or $\bar{n}+1$ [13].

2. Process Model Σ_P

Let n and n^* be fixed {i.e., known} integers satisfying $n \geq n^* > 0$ and let $N : \Re \to \Re$ be a memoryless, locally Lipschitz, nonlinear function satisfying $N(0) = 0$. The class C_P of process models Σ_P to be considered consists of siso systems of the form

$$\begin{aligned} \dot{x}_P &= A_P x_P + b_P u + b_N N(y) \\ y &= c_P x_P \end{aligned} \tag{9}$$

where (A_P, b_P, c_P) is a n_P–dimensional, stabilizable, detectable linear subsystem. Each $\Sigma_P \in C_P$ admits a block diagram representation of the form shown in Figure 1 where $t_P(s)$ and $t_N(s)$ are the strictly proper transfer functions $g_P\frac{\beta_P(s)}{\alpha_P(s)}$ and $\frac{\beta_N(s)}{\alpha_P(s)}$ respectively; here $\alpha_P(s)$ is the characteristic polynomial of A_P, $\beta_P(s)$ is monic and g_P is a nonzero constant called the *high frequency gain*.

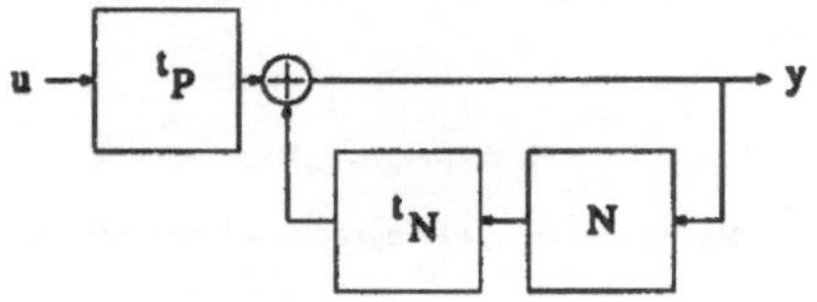

Figure 1: Process Model Σ_P

We shall make the following

Process Model Assumptions: For each $\Sigma_P \in C_P$

1. $n_P \leq n$
2. relative degree $t_P(s) = n^*$
3. $g_P > 0$
4. $\beta_P(s)$ is stable
5. N is known
6. relative degree $t_P(s) \leq$ relative degree $t_N(s)$

For the case when $t_N(s) = 0$ the preceding reduce to the process model assumptions made in the well-known classical model reference adaptive control problem originally solved in [15]. A more general version of the nonlinear problem considered here, with multiple nonlinear feedback loops, has recently been solved in [6], but without the benefit of high-order tuning.

Assumption 3 is not crucial and can easily be avoided using a Nussbaum gain [16]. Assumption 6 appears also not to be crucial and can probably be replaced, as in [7, 8, 9], with the much milder requirement that N have a continuous μth derivative, μ being either relative degree $\frac{t_P(s)}{t_N(s)}$ or zero, whichever is larger . The additional techniques needed for handling this case, which involve the introduction of "nonlinear damping" can be found in [9].

3. Parameterized Controller Σ_C

Our aim here is to construct an "identifier-based parameterized controller" Σ_C suitable for high-order tuning. In general, identifier-based parameterized controllers consist of two subsystems, one an "identifier" Σ_I whose function is to generate an identification error e and the other an "internal regulator" or "certainty equivalence controller" Σ_R whose output serves as the closed-loop control to the process {cf. [12]}.

As the first step in the construction of Σ_C , we select an appropriate "parameterized design model" Σ_D upon which the structure of Σ_I and the synthesis of Σ_R are to be based. For this, choose a positive number λ, a monic stable polynomial $\alpha_*(s)$ of degree $n^* - 1$, a controllable, observable realization (A_*, b_*, c_*, d_*) of $\frac{1}{\alpha_*(s)}$, and a n-dimensional, single-output, observable pair (c, A) with A stable. These selections determine the "direct control, design model" {cf. [12]} $\Sigma_D(p)$, parameterized by the vector $p = [\ p_0 \quad p_1' \quad p_2' \quad p_3' \quad p_4\ ]' \in \mathcal{P} = \Re \oplus \Re^n \oplus \Re^n \oplus \Re^n \oplus \Re$ and described by the equations

$$\begin{aligned}
\dot{x}_{D1} &= Ax_{D1} + p_1 y_D + p_2 u_D + p_3 N(y_D) \\
\dot{x}_{D2} &= A_* x_{D2} + b_*(u_D + cx_{D1} + p_4 N(y_D)) \\
\sigma_D &= c_* x_{D2} + d_*(u_D + cx_{D1} + p_4 N(y_D)) \\
\dot{y}_D &= -\lambda y_D + p_0 \sigma_D
\end{aligned} \tag{10}$$

Σ_D admits the following block diagram representation where $\frac{\pi(s,p_i)}{\alpha(s)}$ is the transfer function of (A, p_i, c).

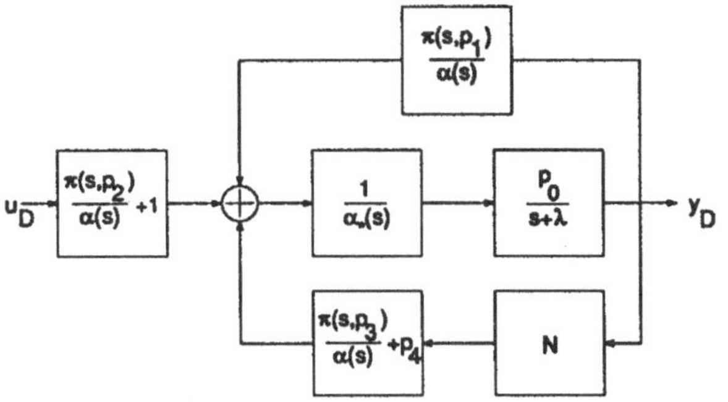

Figure 2: Direct Control Design Model

Σ_D has three properties which make it especially useful for the problem at hand.

Properties of Σ_D :

1. *Detectability:* For each fixed $p \in \mathcal{P}$, each solution $[\ x_{D1}' \quad x_{D2}'\ ]'$ to (10), along which $y_D(t) = u_D(t) = 0$, goes to zero exponentially fast.

2. *Output Stabilizability:* For each fixed $p \in \mathcal{P}$, the control $u_D = -cx_{D1} - p_4 N(y_D)$ "output stabilizes" Σ_D in that $y_D(t) = e^{-\lambda t} y_D(0)$.

3. *Matchability:* For each $\Sigma_P \in \mathcal{C}_P$, there exists a point[1] $q \in \mathcal{P}$ at which Σ_D "matches" Σ_P in the sense that $\Sigma_D(q)$ admits the same block diagram representation as Σ_P .

Because of matchability and the structure of Σ_D in (10), it is possible to write

$$\dot{y} = -\lambda y + q_0(q_P' w + c_* x_* + d_* u) + \epsilon_P \tag{11}$$

where w is a weighting vector generated by the identifier equations

$$\begin{aligned}
w' &= c_* H + d_* [\ z' \quad N(y)\ ] \\
\dot{H} &= A_* H + b_* [\ z' \quad N(y)\ ] \\
\dot{x}_* &= A_* x_* + b_* u \\
\dot{z} &= \begin{bmatrix} A' & 0 & 0 \\ 0 & A' & 0 \\ 0 & 0 & A' \end{bmatrix} z + \begin{bmatrix} c' & 0 & 0 \\ 0 & c' & 0 \\ 0 & 0 & c' \end{bmatrix} \begin{bmatrix} y \\ u \\ N(y) \end{bmatrix}
\end{aligned} \tag{12}$$

[1]To calculate such a value $q = [\ q_0 \quad q_1' \quad q_2' \quad q_3' \quad q_4\ ]'$, write $\alpha(s)$ for the characteristic polynomial of A, and let $\gamma(s)$ and $\rho(s)$ denote the unique quotient and remainder respectively of $(s+\lambda)\alpha(s)\alpha_*(s)$ divided by $\alpha_P(s)$. It is easy to verify that with $q_0 = g_P$ and q_1, q_2, q_3, and q_4 defined so the (A, b, q_1'), (A, b, q_2'), and (A, b, q_3', q_4) realize $\frac{\beta_P(s)\gamma(s) - \alpha(s)}{\alpha(s)}$, $\frac{\rho(s)}{g_P \alpha(s)}$ and $\frac{\beta_N(s)\gamma(s)}{\alpha(s)}$ respectively, q will have what's required.

Here ϵ_P is a linear combination of decaying exponentials depending on the eigenvalues of A and A_*, and $q_P = [\ q_1' \quad q_2' \quad q_3' \quad q_4\ ]'$ {cf, [15, 12]}. In view of (11) , it makes sense to use as an identification error

$$e = \hat{y}_D - y \tag{13}$$

where

$$\begin{aligned} \dot{\hat{y}}_D &= -\lambda \hat{y}_D + k_0 \hat{\sigma}_D & (14) \\ \hat{\sigma}_D &= k'w + c_* x_* + d_* u & (15) \end{aligned}$$

since this leads to the familiar error equation

$$\dot{e} = -\lambda e + (k_0 - q_0)\hat{\sigma}_D + q_0(k - q_P)'w - \epsilon_P \tag{16}$$

Here $k_0 \in \Re$, and $k = [\ k_1' \quad k_2' \quad k_3' \quad k_4\ ]' \in \Re^{3n+1}$ are parameters to be tuned. Σ_I is thus completely described by (12) to (15).

To define certainty equivalence control Σ_R , consider first the "surrogate" design model substate

$$\hat{x}_{D1} = [\ S(k_1) \quad S(k_2) \quad s(k_3)\ ]\, z \tag{17}$$

where $S(\zeta) = [\ (M^{-1}M_1)'\zeta \quad (M^{-1}M_2)'\zeta \quad \cdots \quad (M^{-1}M_n)'\zeta\ ]'\ \forall \zeta \in \Re^n$, M is the observability matrix of (c, A), M_i is the observability matrix of (e_i', A) and e_i is the ith unit vector in $\Re^n$ [12]. If N were linear, then in accordance with the Certainty Equivalence Output Stabilization Theorem [12], one would set[2]

$$u = -c\hat{x}_{D1} - k_4 N(y) \tag{18}$$

since $u_D = -cx_{D1} - p_4 N(y_D)$ output stabilizes Σ_D at p. This control would then define Σ_R .

Unfortunately, problems arise as soon as one tries to use (18) when N is nonlinear. The problem appears to stem from the use of certainty equivalence itself , since this is, in essence, a "static" or "frozen" parameter approach to the design of Σ_R [12] . The difficulty is that certainty equivalence promotes control laws such as (18), and does so without directly taking into account the actual time dependence of tuned parameters. Because of this, for certainty equivalence control to work, parameters must be tuned "slowly," at least on the average. This is typically accomplished using some form of tuning error normalization. Simple examples suggest that it is the slow tuning necessitated by the certainty equivalence approach which enables strongly nonlinear functions {i.e., functions which are locally but not globally Lipschitz} to foil adaptive controllers configured in this way.

What recent research has shown [5]-[9], is that there are alternative methods, not using tuning error normalization and not of the "static" certainty equivalence type which can be used with success, even if N is not globally Lipschitz. What we now propose is a method along these general lines.

To motivate what we have in mind, let us note first that the output stabilizing control $u_D = -cx_{D1} - p_4 N(y_D)$ upon which the static certainty equivalence control (18) is based, also exponentially stabilizes σ_D as defined in (10), since with this control

$$\alpha_*(\delta)\sigma_D = 0$$

where δ is the differential operator. From this, (12) and (15), it is easy to see that if k were held fixed, then (18) would exponentially stabilize $\hat{\sigma}_D$ since this control would cause $\hat{\sigma}_D$ to satisfy

$$\alpha_*(\delta)\hat{\sigma}_D = 0 \tag{19}$$

This suggests that to avoid the need for slow tuning, u should be chosen to make (19) really hold, even though k is not a constant. Since (15) and (12) imply that

$$\alpha_*(\delta)\hat{\sigma}_D = u + \alpha_*(\delta)(k'w) \tag{20}$$

[2]Using (17) and the easily derived identity $cS(\zeta) = \zeta'$, (18) can be rewritten in the more familiar form $u = -[\ k_1' \quad k_2' \quad k_3'\ ]\, z - k_4 N(y)$.

it is obvious that (19) can, in fact, be achieved with the *dynamic* certainty equivalence output stabilization law

$$u = -\alpha_*(\delta)(k'w) \tag{21}$$

Recalling that $\alpha_*(s)$ is of degree n^*-1, we see that control (21) can be realized without time function differentiation, provided the first n^*-1 derivatives of both k and w are available signals. In view of (12), w clearly has this property. Therefore, if things can be arranged so that error equation (16) takes on the form of (1), then a tuner of order n^*-1 can be constructed to endow k with the required properties. To get (16) into this form, its enough to simply set $k_0 = 0$, since if this is done (16) becomes

$$\dot{e} = -\lambda e + q_0(k - q_P)'w + \epsilon \tag{22}$$

where

$$\epsilon = -q_0\widehat{\sigma}_D - \epsilon_P \tag{23}$$

In view of (19), ϵ must go to zero exponentially fast. Henceforth, therefore we shall take (21) as the definition of Σ_R and we shall assume that Σ_T is a tuner of order $\bar{n} = n^*-1$ defined by (2) to (7)[3]. In addition, since $\widehat{y}_D$ is not needed, we shall take it to be zero in which case

$$e = -y \tag{24}$$

4. Stability Analysis

The overall adaptive control system to be examined consists of the high-order tuner Σ_T described by (2) to (8), process model Σ_P defined by (9), identifier Σ_I modeled by (12), (15), (24), and dynamic certainty equivalence control Σ_R given by (21). The steps involved in establishing global boundedness are as follows:

1. Using (12) it is easy to see that the component subvectors w_1, w_2, w_3, and w_4 of w satisfy the differential equations

$$\begin{aligned}
\alpha_*(\delta)(\delta I - A')w_1 &= c'y &(25)\\
g_P\beta_P(\delta)\alpha_*(\delta)(\delta I - A')w_2 &= c'\alpha_P(\delta)y - \beta_N(\delta)\alpha_*(\delta)(\delta I - A')w_3 &(26)\\
\alpha_*(\delta)(\delta I - A')w_3 &= c'N(y) &(27)\\
\alpha_*(\delta)w_4 &= N(y) &(28)
\end{aligned}$$

Since the process model assumptions imply that the transfer matrices

$$\frac{1}{\alpha_*(s)}(sI - A')^{-1}c', \quad \frac{\alpha_P(s)}{g_P\beta_P(s)\alpha_*(s)}(sI - A')^{-1}c', \quad \frac{\beta_N(s)}{g_P\beta_P(s)}$$

are each proper and stable, in the light of (25) and (26), w can be viewed as the output of a stable linear system Σ_w with state x_w and inputs y, w_3, and w_4. The interconnection of Σ_w with Σ_T determined by (22) and (24), is thus a dynamical system Σ of the form

$$\begin{aligned}
\dot{x}_\Sigma &= f(x_\Sigma, w_3, w_4, \epsilon)\\
w &= g(x_\Sigma, w_3, w_4, \epsilon)
\end{aligned} \tag{29}$$

where $x_\Sigma = \{y, v, h_0, h, X, x_w\}$ and f and g are analytic functions[4].

2. Now invoke the High-Order Tuning Theorem and (24), thereby establishing the boundedness of all components of x_Σ , except for x_w.

3. From (19) and the definition of ϵ in (23), it follows that ϵ is in $\mathcal{D}^{n^*-1}$, the space of all vector-valued, bounded functions with bounded time derivatives up to and including those of degree n^*-1. From the boundedness of y, (27) and (28) it is clear that w_3 and w_4 are also in $\mathcal{D}^{n^*-1}$. The definition of Σ_w together with the boundedness of its inputs imply that x_w is bounded as well.

[3] If $n^* = 1$, set $\bar{n} = 1$.
[4] If $\bar{n} = 1$, $x_\Sigma = \{y, k, x_w\}$

4. The preceding characterizes Σ as an analytic dynamical system with a bounded state and an input $\{\epsilon, w_3, w_4\}$ in $\mathcal{D}^{n^*-1}$. From this it follows that w must be in $\mathcal{D}^{n^*-1}$. The equations defining Σ_T show that k must be in $\mathcal{D}^{n^*-1}$ as well. From these observations, and (21) it follows that u is bounded.

5. The boundedness of u and y and the assumption that (c_P, A_P) is detectable, imply that x_P is bounded. The boundedness of u and y also implies that the identifier state components H, x_* and z defined by (12) are bounded.

6. The preceding establishes the boundedness of complete adaptive system state $\{v, h_0, h, X, x_P, H, x_*, z\}$ (or $\{k, x_P, H, x_*, z\}$ for the case $\bar{n} = 1$), wherever it exists. From this and the smoothness of the overall adaptive system it follows that its state must exist and be bounded on $[0, \infty)$. It also follows from the boundedness of $\dot{y}$ and the finiteness of the $\mathcal{L}^2$-norm of y established by the High-Order tuning Theorem and (24), that y goes to zero as $t \to \infty$ [17].

Concluding Remarks

The aim of this paper has been to introduce a new high-order tuning algorithm and to discuss its possible use in the synthesis of adaptive controllers for special families of nonlinear systems. While the resulting controllers, like their predecessors in [5]-[9], are a good deal more complicated than the standard certainty equivalence controllers discussed in [4], the former can adaptively stabilize many types of nonlinear systems which the latter cannot.

It is interesting to note that because the overall adaptive system under consideration here has been developed without the presumption of slowly tuned parameters, the stability analysis differs somewhat from that used in the study of more familiar adaptive systems {eg., see [12, 4]}. Whereas the latter typically make use of small signal perturbation analysis, the Bellman Gronwall Lemma, and the like, the reasoning used here does not and consequently is somewhat less technical.

The algorithms developed in this paper can of course also be applied to linear process models. Whether or not the departure from static certainty equivalence control this paper suggests can improve performance in such applications, remains to be seen.

References

[1] S. Sastry and A. Isidori, "Adaptive Control of Linearizable Systems," *IEEE Trans. Auto. Control*, AC-34, Nov. 1989, pp 1123-1131.

[2] R. Marino, I. Kanellakopoulos and P. V. Kokotovic, "Adaptive Tracking for Feedback Linearizable Systems," *Proc. 28th CDC*, Tampa, 1989, pp. 1002-1007.

[3] L. Praly, G. Bastin, J.-B Pomet and Z. P. Jiang, "Adaptive Stabilization of Nonlinear Systems", Ecole Nationale Superieure des Mines de Paris Technical Report No. A236, October, 1990

[4] S. Sastry and M. Bodson, *Adaptive Control: Stability, Convergence and Robustness*, Prentice-Hall, Inc., Englewood Cliffs, NJ, 1989.

[5] I. Kanellakopoulos, P. V. Kokotovic and A. S. Morse, "Systematic Design of Adaptive Controllers for Feedback Linearizable Systems," *IEEE Trans. Auto. Control*, to appear.

[6] I. Kanellakopoulos, P. V. Kokotovic and A. S. Morse, "Adaptive Output-Feedback Control of Systems with Output Nonlinearities," *Foundations of Adaptive Control*, Springer-Verlag, Berlin, 1991.

[7] R. Marino and P. Tomei, "Global Adaptive Observers and Output-feedback Stabilization for a Class of Nonlinear Systems, *Foundations of Adaptive Control*, Springer-Verlag, Berlin, 1991.

[8] Z. P. Jiang and L. Praly, "Iterative Designs of Adaptive Controllers for Systems with Nonlinear Integrators," *Proc. 1991 IEEE CDC*, submitted.

[9] I. Kanellakopoulos, P. V. Kokotovic and A. S. Morse, "Adaptive Output-Feedback Control of a Class of Nonlinear Systems," Coordinated Science Laboratory Technical Report DC-131, submitted for publication.

[10] D. R. Mudgett, ***Problems in Parameter Adaptive Control***, Yale University Doctoral Dissertation, 1988.

[11] D. R. Mudgett and A. S. Morse, "High-Order Parameter Adjustment Laws for Adaptive Stabilization," *Proc. 1987 Conference on Information Sciences and Systems*, March, 1987.

[12] A. S. Morse, "Towards a Unified Theory of Parameter Adaptive Control - Part 2: Certainty Equivalence and Implicit Tuning," *IEEE Trans. Auto. Control*, to appear.

[13] A. S. Morse, "High-Order Parameter Tuners for the Adaptive Control of Linear and Nonlinear Systems" *Proc. U. S.-Italy Joint Seminar on Systems, Models, and Feedback: Theory and Applications,"* Capri,1992.

[14] A. S. Morse, "A $4(n+1)$-Dimensional Model Reference Adaptive Stabilizer for any Relative Degree One or Two Minimum Phase System of Dimension n or Less," *Automatica*, v. 23, 1987, pp.123-125.

[15] A. Feuer and A. S. Morse, "Adaptive Control of Single-Input, Single-Output Linear Systems," *IEEE Trans. Auto. Control*, AC-23, August, 1978, pp. 557-569.

[16] D. R. Mudgett and A. S. Morse, "Adaptive Stabilization of Linear Systems with Unknown High-Frequency Gains," *IEEE Trans. Auto Control*, AC-30, June, 1985, pp.549-554.

[17] M. A. Aizerman and F. G. Gantmacher, *Absolute Stability of Regulator Systems*, Holden-Day, 1964,p.58.

DESIGN OF ROBUST ADAPTIVE CONTROL SYSTEM WITH A FIXED COMPENSATOR

H. OHMORI and A. SANO

Department of Electrical Engineering, Keio University,
3-14-1 Hiyoshi, Kohoku-ku, Yokohama 223, Japan

Abstract. This paper presents a design method of model reference robust adaptive control systems (MRRACS) with fixed compensator for a single-input single-output(SISO) linear time-invariant continuous-time plant in the presence of bounded unknown deterministic disturbances and unmodeled dynamics. The fixed compensator is designed based on $\mathcal{H}^\infty$-optimal theory. The aim of the proposed control structure is realization of harmonizing the adaptive control to the robust control.

1. INTRODUCTION

The model reference adaptive control systems(MRACS) are designed under the assumption that the plant dynamics are exactly presented by one member of a specified class of models. Unfortunately, it has been shown (Rohrs and co-workers, 1982) that bounded disturbances and/or unmodeled dynamics can make the system unstable.

Several modified adaptive control laws have been proposed(surveyed by Ortega and Tang, 1989) to achieve the robust stability and performance in the face of the presence of unmodeled dynamics and/or disturbances.

In this paper we consider the model reference adaptive control of a linear time-invariant plant in the presence of unknown deterministic disturbances and unmodeled dynamics. The key idea of the proposed adaptive control system is that the system is consists of two feedback loops: one performs the model-matching adaptively and the other includes an error feedback controller we call "the fixed compensator", which can compensate the disturbances and model uncertainty.

The main objective of this paper is to propose the $\mathcal{H}^\infty$ design method of the fixed compensator that can achieve the global stability of the proposed adaptive control system and that can reduce the effect of disturbances and robust performance. The fixed compensator is obtained as the solution for the $\mathcal{H}^\infty$-optimal design problem.

2. SYSTEM DESCRIPTION

Consider the following single-input, single-output(SISO), linear-time-invariant(LTI) system:

$$\begin{cases} A(s)x(t) &= u(t) + d_u(t) \\ y(t) &= B(s)(1 + L(s))x(t) + d_y(t) \end{cases} \tag{1}$$

where

$$A(s) \equiv s^n + \sum_{i=0}^{n-1} a_i s^i, \ B(s) \equiv \sum_{i=0}^{m} b_i s^i \tag{2}$$

where $y(t)$ and $u(t)$ are the measured output and input respectively, $d_u(t)$ and $d_y(t)$ represent the disturbances at input and output respectively, and $x(t)$ is the partial state. $L(s)$ denotes unmodeled dynamics which represents an unstructured uncertainty, that is dominant at high frequency domain.

We will make the following assumptions: *(A1)* The order n and $m(\leq n-1)$ are known. Furthermore $A(s)$ and $B(s)$ are relatively prime; *(A2)* The unknown deterministic disturbances $d_u(t)$ and $d_y(t)$ are bounded; *(A3)* The zeros of the polynomial $B(s)$ lie strictly inside the left half-plane, i.e. the plant is nonminimum phase system; *(A4)* $b_m > 0$. *(A5)* There exists a bounded scalar function $\ell(j\omega)$ such that

$$|L(j\omega)| < \ell(j\omega), \quad \forall \omega \tag{3}$$

The desired output $\{y_M(t)\}$ satisfies the following reference model:

$$A_M(s) y_M(t) \ = \ B_M(s) r(t) \tag{4}$$

where

$$A_M(s) \equiv s^{n_M} + \sum_{i=0}^{n_M-1} a_{Mi} s^i, B_M(s) \equiv \sum_{i=0}^{m_M} b_{Mi} s^i \tag{5}$$

We also impose the following additional constraints: *(A5)* $A_M(s)$ is asymptotically stable polynomial; *(A6)* $\{r(t)\}$ is uniformly bounded; *(A7)* $n - m \leq n_M - m_M$.

The design problem to be considered here is to synthesize a MRRACS for the plant in the presence of disturbances and unmodeled dynamics which can cause the tracking error $e(t)(\equiv y(t) - y_M(t))$ to approach zero as closely as possible.

3. CONTROLLERS STRUCTURE

In this section we propose the configuration including the fixed compensator which can not only reduce the effect of disturbances and unmodeled dynamics but also maintain the exact model-matching. We choose any asymptotically stable polynomial $T(s)$ which is described by

$$T(s) \ \equiv \ s^{n-m-1} + \sum_{i=0}^{n-m-2} t_i s^i. \tag{6}$$

There exist uniquely polynomials $R(s)$ and $S(s)$ which can satisfy the following Diophantine equation in polynomials:

$$T(s) A_M(s) \ = \ A(s) R(s) + S(s) \tag{7}$$

where

$$R(s) \equiv s^{n-m-1} + \sum_{i=0}^{n-m-2} r_i s^i, \; S(s) \equiv \sum_{i=0}^{n-1} s_i s^i \tag{8}$$

Furthermore by defining

$$B_R(s) \equiv R(s)B(s) - b_m H(s) \equiv \sum_{i=1}^{n-1} b_{Ri} s^i \tag{9}$$

where $H(s)$ is an asymptotically stable monic polynomial of the order $n-1$, the plant Eq.(1) can be rewritten as

$$T(s)A_M(s)y(t) = H(s)\left[\theta_*^T \xi(t)\right] + R(s)d(t) \tag{10}$$

where

$$\begin{aligned} \theta_* &\equiv \Big[b_m, b_{R(n-2)}, \cdots, b_{R0}, \\ &\qquad s_{n-1}, s_{n-2}, \cdots, s_0\Big]^T \in \mathcal{R}^{2n} & (11) \\ \xi(t) &\equiv \Big[u(t), \frac{s^{n-2}}{H(s)}u(t), \cdots, \frac{1}{H(s)}u(t), \\ &\qquad \frac{s^{n-1}}{H(s)}y(t), \cdots, \frac{1}{H(s)}y(t)\Big]^T \in \mathcal{R}^{2n} & (12) \\ d(t) &\equiv B(s)L(s)u(t) + \bar{d}(t) & (13) \\ \bar{d}(t) &\equiv A(s)d_y(t) + B(s)(1 + L(s))d_u(t) & (14) \end{aligned}$$

Now we propose a MRRACS(see Fig.1) with the fixed compensator $F(s)$ which is given by

$$u(t) = \frac{1}{\hat{\theta}_1(t)}\left[-\sum_{i=2}^{2n} \hat{\theta}_i(t)\xi_i(t) + \frac{T(s)B_M(s)}{H(s)}v'(t)\right] \tag{15}$$

where $\hat{\theta}_i(t)$ is an adjustable parameter and

$$\begin{aligned} & v'(t) \equiv r(t) - v(t) & (16) \\ & M(s)v(t) = N(s)e(t); \; F(s) \equiv \frac{N(s)}{M(s)} & (17) \\ & \hat{\theta}(t) \equiv [\hat{\theta}_1(t), \hat{\theta}_2(t), \cdots, \hat{\theta}_{2n}(t)]^T & (18) \end{aligned}$$

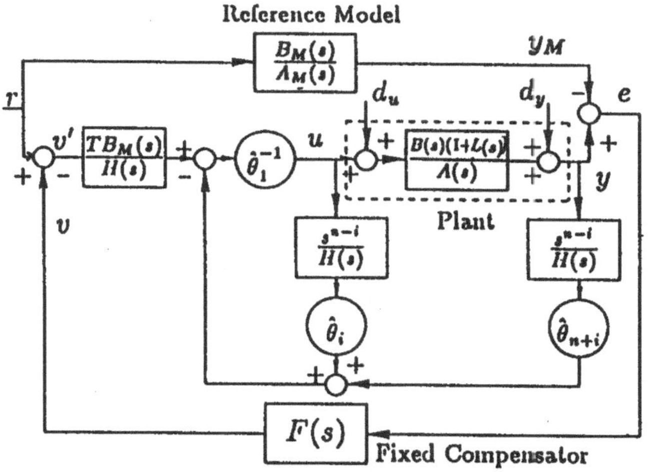

Fig.1. The proposed adaptive control system using the fixed compensator.

The proposed MRRACS with $F(s) = 0$ reduces to the ordinary MRACS structure. Using Eq.(4), the adaptive control law Eq.(15) can be rewritten as

$$T(s)A_M(s)y_M(t) = H(s)\left[\hat{\theta}^T(t)\xi(t)\right] + T(s)B_M(s)v(t) \tag{19}$$

From Eqs. (10) and (19), we have

$$\begin{aligned} e(t) &= \frac{H(s)}{T(s)A_M(s)}\left[\psi^T(t)\xi(t) - \frac{T(s)B_M(s)}{H(s)}\frac{N(s)}{M(s)}e(t)\right] \\ &\quad + \frac{R(s)}{T(s)A_M(s)}d(t) \end{aligned} \tag{20}$$

where

$$\psi(t) \equiv \theta_* - \hat{\theta}(t) \tag{21}$$

We can rewritten Eq.(20) as

$$\begin{aligned} e(t) &= \frac{H(s)}{T(s)A_M(s)}(1 + L_1(s))\left[\psi^T(t)\xi(t) - \frac{T(s)B_M(s)}{H(s)}\frac{N(s)}{M(s)}e(t)\right] \\ &\quad + \frac{R(s)}{T(s)A_M(s) + S(s)L(s)}d_1(t) \end{aligned} \tag{22}$$

where

$$L_1(s) \equiv \frac{A(s)R(s)}{T(s)A_M(s) + S(s)L(s)}L(s) \tag{23}$$

$$d_1(t) \equiv A(s)L(s)y_M(t) + \bar{d}(t) \tag{24}$$

It is worth noting that $d_1(t)$ becomes bounded disturbances. We can obtain the following the theorem.

Theorem 1 : *Let the non-minimal expression of the plant Eq.(10) be given together with the adaptive control law Eq.(15) with the fixed compensator $F(s)$. Then we can get the tracking error $e(t)$ which described by*

$$e(t) = W_1(s)\left[\psi^T(t)\xi(t)\right] + W_d(s)d_1(t) \tag{25}$$

where

$$W_1(s) = \frac{H(s)M(s)}{T(s)V(s)}(1 + L_2(s)), \tag{26}$$

$$W_d(s) = \frac{R(s)M(s)}{T(s)V(s) + S_M(s)L(s)}, \tag{27}$$

$$V(s) \equiv A_M(s)M(s) + B_M(s)N(s), \tag{28}$$

$$S_M(s) \equiv M(s)S(s) + T(s)B_M(s)N(s), \tag{29}$$

$$L_2(s) \equiv \frac{M(s)A(s)R(s)}{T(s)V(s) + S_M(s)L(s)}L(s) \tag{30}$$

Remark 1 : *If the fixed compensator $F(s)$ and the unmodeled dynamics $L(s)$ are absence, the error transfer function $W_1(s)$, the disturbance transfer function $W_d(s)$ and $d_1(t)$ are obtained as follows:*

$$W_1(s) = \frac{H(s)}{T(s)A_M(s)}, \; W_d(s) = \frac{R(s)}{T(s)A_M(s)}, \tag{31}$$

$$d_1(t) = A(s)d_y(t) + B(s)d_u(t) \tag{32}$$

Comparing the above disturbance transfer function $W_d(s)$ with one in Eq.(26) and Eq.(27), it is obvious that the presence of the fixed compensator can achieve the disturbance rejection if it is designed suitably.

Remark 2 : *Note that the proposed system including the fixed compensator can be implemented if $V(s)$ is a stable polynomial, and that it has absolutely no relation to the controller parameters which accomplish the exact model matching. Hence introducing the fixed compensator is not violate the exact model matching condition.*

Remark 3 : *In the case of the ordinary exact model matching, since the characteristic polynomial of the disturbances is $T(s)A_M(s)$, if $A_M(s)$ must be chosen as the polynomial having slow convergence mode then the convergence mode of disturbances is also slow. While in the proposed exact model matching with the fixed compensator, since the characteristic polynomial of the system becomes $T(s)V(s)$, we can set the convergence property of disturbances independed of $A_M(s)$.*

4. ADAPTIVE SCHEME AND STABILITY ANALYSIS

From this section we assume that the unmodeled dynamics $L(s)$ is absence. If the plant model has the unmodeled dynamics, we can also prove the stability of the proposed adaptive system in the presence of the unmodeled dynamics by using the normalizationof adaptive law (Ioannou and Tsakalis, 1986).

The adjustable controller parameter $\hat{\theta}(t)$ is updated by the adaptation algorithm, we adopt the σ-modification approach:

$$\hat{\theta}(t) = \hat{\theta}_I(t) + \hat{\theta}_P(t), \tag{33}$$

$$\dot{\hat{\theta}}_I(t) = -\sigma\hat{\theta}_I(t) + \Gamma_I z(t)\varepsilon(t), \tag{34}$$

$$\hat{\theta}_P(t) = \Gamma_P z(t)\varepsilon(t), \tag{35}$$

$$\sigma > 0, \Gamma_I = \Gamma_I^T > 0, \quad \Gamma_P = \Gamma_P^T > 0$$

where $\varepsilon(t)$ is augmented error and $z(t)$ is the filtered signal as follows:

$$\varepsilon(t) = e(t) - e_a(t) \tag{36}$$

$$e_a(t) \equiv W(s)(\hat{\theta}(t)W_2^{-1}(s) - W_2^{-1}(s)\hat{\theta}(t))^T\xi(t) \tag{37}$$

$$z(t) \equiv \frac{1}{W_2(s)}\xi(t) \tag{38}$$

where $W_2(s)$ is chosen such that $W(s) \equiv W_1(s)W_2(s)$ is strictly positive realness.

It will be shown that the adaptive control system using the above adaptive scheme is globally stable, i.e. all signal in this system are bounded. We can easily get the following equation:

$$\varepsilon(t) = W(s)\left[\psi(t)^T z(t)\right] + d'(t) \tag{39}$$

$$d'(t) \equiv W_d(s)d(t) \tag{40}$$

Main theorem of the stability of proposed adaptive system including the fixed compensator is below:

Theorem 2 : *If $W(s)$ is strictly positive real and the disturbances term $d'(t)$ is bounded, the proposed adaptive control law Eq.(15) and Eqs. (33)-(35) accomplish that for all initial condition tracking error $e(t)$ convergences a certain closed region Q, and all signal in the adaptive control system are bounded.*

The performance of the control system can be envaluted from the region above. So if the region Q can be smaller, we can realize the high performance. Next section we will propose the design method of fixed compensator which can make this region smaller.

5. DESIGN SCHEME OF FIXED COMPENSATOR

In this section, it is shown that the design problem of the fixed compensator in the direct adaptive control system becomes the design problem of a stable controller which internally stabilizes the reference model, which and performs low sensitivity property.

We use the following notation. $\mathcal{R}$ is a field of real numbers and C_{+e} denotes the extended right half-plane, i.e. $\{s : \mathcal{R}_e(s) \geq 0\} \cup \{\infty\}$. $\mathcal{RH}^\infty$ is the set of all proper real-rational functions which have no poles in C_{+e}. $\mathcal{U}$ is the class of a unit function, whose element belongs to $\mathcal{RH}^\infty$ and its inverse is also in $\mathcal{RH}^\infty$. Furthermore, $G(\mathcal{E})$ denotes the set $\{G(s) :^\forall s \in \mathcal{E}\}$.

The fixed compensator $F(s)$ should be designed to satisfy the following two requests for the closed loop system: *(R1)* there exists $W_2(s)$ such that $W(s)$ in Eq.(39) is strictly positive realness; *(R2)* $\mathcal{Q}$ is as small as possible. The first requirement is essential to achieve the global stability of the proposed adaptive system via theorem 2. The second requirement is necessary to reduce the region $\mathcal{Q}$ for the enhanced control performance.

If both the denominator and numerator polynomials of $W_1(s)$ in Eq.(26) are asymptotically stable polynomials then the proper $W_2^{-1}(s)$ exists such that $W(s) = W_1(s)W_2(s)$ is strictly positive realness. Hence if both the denominator $M(s)$ in Eq.(17) and the characteristic polynomial $V(s)$ in Eq.(28) are asymptotically stable polynomials then first request *(R1)* can be satisfied. It is obvious that the design problem of $F(s)$ such that $M(s)$ and $F(s)$ are asymptotically stable polynomials becomes the design problem of a stable stabilizing controller for $G_M(s)$. Since $G_M(s)$ is stable, $G_M(s)$ automatically satisfies the Parity-Interlacing-Property, i.e. it is strongly stabilizable. It is well-known that the class of all stable stabilizing compensator $F(s)$ for $G_M(s)$ can be represented as follows:

$$\begin{aligned} \mathcal{S}_-(G_M) &= \mathcal{S}(G_M) \cap \mathcal{RH}^\infty \\ &= \left\{ F(s) = \frac{U(s)-1}{G_M(s)} : \ U(s) \in \mathcal{U},\ 1 = U(\mathcal{Z}_{C_{+e}}) \right\} \end{aligned} \tag{41}$$

where $\mathcal{Z}_{C_{+e}} \equiv \{s : G_M(s) = 0, s \in C_{+e}\}$ and $U(s)$ is a unit free parameter.

From Eqs.(28) and (41), by using $U(s)$, $W_d(s)$ in Eq.(27) is represented as

$$\begin{aligned} W_d(s) &= \frac{R(s)}{T(s)A_M(s)} \cdot \frac{1}{1 + G_M(s)F(s)} \\ &= \frac{R(s)}{T(s)A_M(s)} \cdot \frac{1}{U(s)} \end{aligned} \tag{42}$$

On the other hand, since $F(s)$ should also satisfy the request *(R2)*, we should desire that $F(s)$ minimizes the following cost function:

$$J(U) \equiv \|X\|_\infty, \quad X(s) \equiv T_D(s) \cdot \frac{1}{U(s)} \tag{43}$$

where $T_D(s) \in \mathcal{U}$ is a weighting function.

Combining the requests in Eqs.(41) and (43), we can formulate the design problem of the fixed compensator $F(s)$ as the following $\mathcal{H}^\infty$ sub-optimal problem:

Find $X(s) \in \mathcal{U}$ that satisfies

$$\begin{aligned} &\|X\|_\infty < \gamma \in \mathcal{R}, \\ &\text{w.r.t. } X(\mathcal{Z}_{C_{+e}}) = T_D(\mathcal{Z}_{C_{+e}}),\ X \in \mathcal{U} \end{aligned} \tag{44}$$

Several approaches for finding the solution of the problem mentioned above have been considered. Hara and Vidyasagar(1989) have formulated the sensitivity minimization and robust stabilization with stable controllers as two interpolation-minimization problems for unit functions. Ito, Ohmori and Sanó(1991) have shown an uncomplicated algorithm for attaining low sensitivity property by a stable controller is able to be presented based on a certain Nevanlinna-Pick interpolation problem. Using either approach, we can get the sub-optimal solution $X^{sub}(s)$. Then the resulting fixed compensator is obtain by

$$F(s) = \frac{\frac{T_{D(s)}}{X^{sub}(s)} - 1}{G_M(s)} \tag{45}$$

Remark 4 : *The proposed design of the fixed compensator can practically realize the high performance of the disturbance reduction, because the structure of the disturbances is not assumed a prior.*

Remark 5 : *It is note that if the plant model has the unmodeled dynamics $L(s)$, the above-mentioned design problem of the fixed compensator becomes the design problem of a stable robust stabilizing compensator.*

6. CONCLUSIONS

We have presented a MRRACS for the plant in the presence of the deterministic disturbances and unmodeled dynamics. The fixed compenstor in the MRRACS can designed based on the $\mathcal{H}^\infty$ optimal control theory in order to realize high performance of the control. The same configuration should be applicable in the discrete time case, furthermore it can be briefly extend to the multivariable systems.

REFERENCES

Hara, S., and M. Vidyasagar (1989). Sensitivity minimization and robust stabilization by stable controller. *Proc. Int. Symp. on Mathmatical Theory of Networks and Systems.*

Ioannou, P., and K. Tsakalis (1986). A robust direct adaptive controller. *IEEE Trans. Automat. Contr., AC-31*, No.11, 1033-1043.

Ito, H., H. Ohmori and A. Sano (1991). A design of stable controller attaining low sensitivity property. submitted for the *IEEE Transaction s on Automatic Control.*

Ortega, R., and Yu Tang (1989). Robustness of Adaptive Controllers– a Survey. *Automatica, 25*, No.5, 651-677.

Rohrs, C. E., L. Valavani, M. Athand, and G, Stein (1982). Robustness of adaptive control algorithms in the presence of unmodeled dynamics. *Proceedings of the 21st IEEE Conference on Decision and Control*, Orlando, FL, 3-11.

Synthesis of Servo mechanism Problem via H_∞ Control

Shigeyuki. Hosoe
Nagoya University, Japan

Feifei. Zhang Michio. Kono
Tokyo University of Mercantile Marine, Japan

1. Introduction

The robust servodesign is a classical but important problem in control engineering and has attracted considerable attention heretofore(, see [1]- [3] and the references cited therein). The central notion employed in the work was the internal model principle. To assure the robust tracking property of a closed loop system irrespective of plant uncertainty to a class of desired trajectories (which is assumed to be generated by a signal generator), it is known that the loop necessarily contains the dynamics(, called internal model) representing the one for the signal generator in the feedforword path. Also if it is the case, servodesign can be converted to a regulator-design and thus various design methods(LQG, pole-assignment, etc.) developed for the latter can be used.

In this paper, the servodesign problem is considered in the framework of H_∞ control. It is formulated as a kind of mixed sensitivity problem, aiming at constructing a feedback system which has the low sensitive characteristics and stability robustness with respect to plant variations. It will be shown that if we put modes corresponding to the dynamics of the signal generator into the weighting function then the resulting controller must eventually possesses an internal model.

Considered are both the output feedback case and the case where some of the states or all the states of the plant are directly measurable without measurement noise. In the latter case, it is revealed that under some mild conditions the order of the controller is reduced by the number of independent observations.

Notation:

RH_∞ : the set of stable and rational proper transfer function matrices
BH_∞ : the subset of contractive matrices in RH_∞

$$\left[\begin{array}{c|c} A & B \\ \hline C & D \end{array}\right] = C(sI-A)^{-1}B + D$$

$$DHM(A,S) = (A_{11} + SA_{21})^{-1}(A_{12} + SA_{22})$$

$$F_l(A,S) = A_{11} + A_{12}S(I - A_{22}S)^{-1}A_{21}$$

$$X = Ric\begin{bmatrix} A & -P \\ -Q & -A^T \end{bmatrix} : XA + A^T - XPX + Q = 0\ ,\ A - PX\text{: stable}$$

2. Problem Formulation

Consider the system in *Fig*.1, where

R(s): the transfer function matrix of the reference signal generator which is assumed, without loss of generality, to be antistable.

P(s)= $\left[\begin{array}{c|c} A_p & B_p \\ \hline C_p & 0 \end{array}\right]$: the transfer function matrix of the plant which is minimal.

$W_1(s)$, $W_2(s)$: the weighting functions.

K(s): the transfer function matrix of the controller.

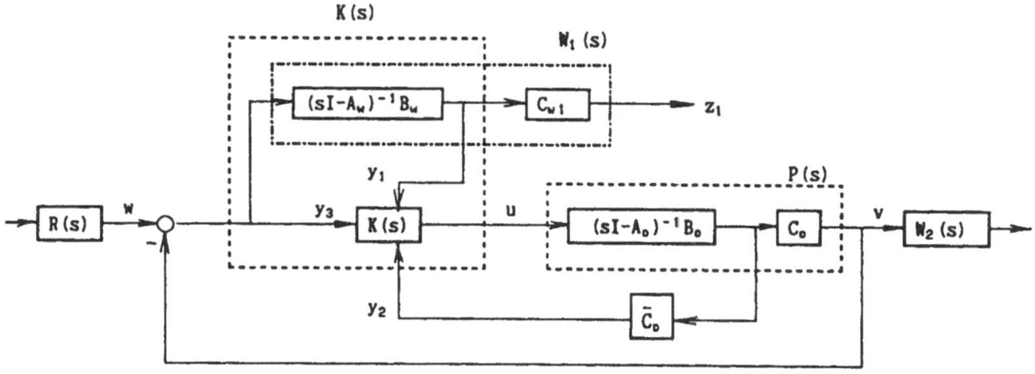

Fig.1 Weighted feedback system

Define vectors x_{w0}, w, y_1, y_2, y_3, u, v and z_1, z_2 as indicated in the figure. The vector y_2 represents the measurement variable which can be observed without measument noise. Also, denote by Φ_0 the open loop transfer function matrix from y_3 to v when the loop is opened at the point marked by ×. Then the sensitivity function S(s) and the co-sensitivity function T(s) evaluated at the point × are given respectively by

$$S(s) = (I + \Phi_0)^{-1} \tag{2.1}$$

$$T(s) = \Phi_0(I + \Phi_0)^{-1} \tag{2.2}$$

Now, the servodesign problem to be considered here is stated as follows:

Find a controller K(s) which ensures the following design objectives.

1) The closed system is internally stable.
2) The system has low sensitive and robustly stable characteristics.
3) The system achieves output regulations to reference inputs generated by R(s) and this is preserved under small perturbation of the plant.

It is well known that 1) and 2) can be realized by appropriately choosing $W_1(s)$ and $W_2(s)$ and solving the following kind of H_∞ control problems:

Problem
Find all stabilizing K(s) such that

$$\left\| \begin{matrix} W_1(s)S(s) \\ W_2(s)T(s) \end{matrix} \right\|_\infty < 1 \tag{2.3}$$

Observe that this is a mixed sensitivity problem but it differs from the usual formulation in the following points. First let us see that the controller's transfer function is given by K(s), not by $\bar{K}(s)$ (; see Fig.1) and that $(sI-A_w)^{-1}$ is shared with $W_1(s)$ and K(s). This implies that in our formulation the states of the realization of $W_1(s)$ is assumed to be measurable and can be used for control purposes. Next, notice that $\Phi_0 = P(s)K(s)$ if $\bar{C}_p = 0$. Therefore in this case the problem coincides with the usual mixed sensitivity problem[4] and the present result provides an alternative way for deriving the solution.

Now, to treat the third objective in an unified framework, we choose

$$W_1(S) = \frac{1}{\alpha(s)} W_S(s) \tag{2.4}$$

where $\alpha(s)$ is the least common denominator of R(s) and $W_S(s)$ is a minimal phase transfer function matrix such that $W_1(\infty) = 0$. Looking at Fig.1, it can be seen that this choice of $W_1(s)$ enforces that the controller has $\frac{1}{\alpha(s)}$ in the feedforword path and the modes of $\alpha(s)$ are not canceled since the closed loop is stable. Therefore if the problem has a solution, then the internal model principle is automatically satisfied.

Finally we assume that $W_2(s)$ is chosen so that

$$W_2(s)P(s) = \left[\begin{array}{c|c} A_p & B_p \\ \hline C_{w2} & I \end{array}\right] \tag{2.5}$$

Before proceeding to the next section, let us rewrite Fig.1 as in Fig.2 (standard form). The generalized plant G(s) is given by

$$G(s) = \left[\begin{array}{cc|c:c} A_w & -B_w C_p & B_w & 0 \\ 0 & A_p & 0 & B_p \\ \hline C_{w1} & 0 & 0 & 0 \\ 0 & C_{w2} & 0 & I \\ \hdashline I & 0 & 0 & 0 \\ 0 & \bar{C}_p & 0 & 0 \\ 0 & -C_p & I & 0 \end{array}\right] \stackrel{s}{=} \left[\begin{array}{c|cc} A & B_1 & B_2 \\ \hline C_1 & 0 & D_{12} \\ C_2 & D_{21} & 0 \end{array}\right] \tag{2.6}$$

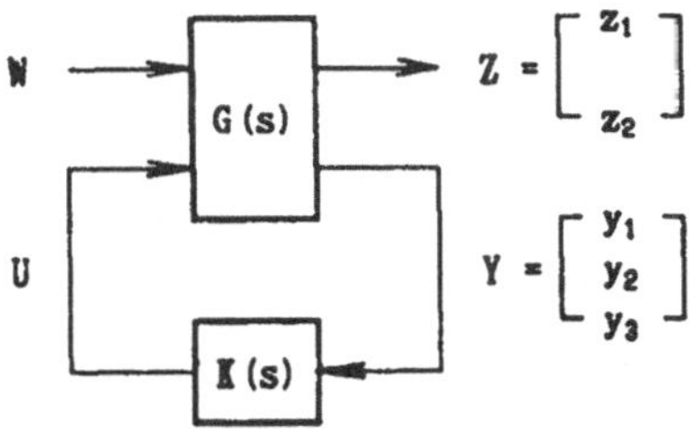

Fig.2 Feedback system in standard form

3. Solution

In this section, the solution will be given to the servodesign problem just formulated. First, we state it in somewhat generalized setting. Write

$$G(s) = \left[\begin{array}{c|cc} A & B_1 & B_2 \\ \hline C_1 & 0 & D_{12} \\ C_2 & D_{21} & 0 \end{array}\right] \tag{3.1}$$

to which, the following assumptions are made.

A1) $D_{12} = \begin{bmatrix} 0 \\ I_{m2} \end{bmatrix} \in R^{p1 \times m2}$

A2) $D_{21} = \begin{bmatrix} 0 \\ I_{m1} \end{bmatrix} \in R^{p2 \times m1}$

A3) $C_2 = \begin{bmatrix} C_{21} \\ C_{22} \end{bmatrix}$, $C_{21} \in R^{(p2-m1) \times n}$:row full rank, $C_{22} \in R^{m1 \times n}$

A4) (A, B_2, C_2): stabilizable and detectable

A5) $\begin{bmatrix} A - j\omega & B_1 \\ C_2 & D_{21} \end{bmatrix}$: column full rank for $0 \leq \omega < \infty$

A6) $\begin{bmatrix} A - j\omega & B_2 \\ C_1 & D_{12} \end{bmatrix}$: column full rank for $0 \leq \omega < \infty$

From A3), C_{21} can be arranged as

$$C_{21} = [\, I_{p2-m1} \quad 0 \,] \tag{3.2}$$

Corresponding to this partitioning of C_{21}, we denote $C_1 = [\, C_{1l} \quad C_{1r} \,]$, $A - B_1 C_{22} = \begin{bmatrix} A_1 & A_2 \\ A_3 & A_4 \end{bmatrix}$, and $C_{22} = [\, C_{22l} \quad C_{22r} \,]$, respectively. Further, for the simplicity of notation, we assume that A_2 and A_4 are arranged as

$$\begin{bmatrix} A_2 \\ A_4 \end{bmatrix} = \begin{bmatrix} A_{21} & 0 \\ A_{411} & 0 \\ A_{421} & A_{422} \end{bmatrix}, \quad A_{411} : \text{stable}, \quad (A_{21}, A_{411}) \text{: observable} \tag{3.3}$$

i.e., A_2 and A_4 has the observable canonical form. Notice that the assumption " A_{411} is stable " does not loss any generality, since there exist F such that $F_1 A_{21} + A_{411}$ is stable, and this stabilization of A_{411} can be realized by an appropriate coordinate transformation on $A - B_1 C_{22}$.

Now let

$$X = Ric \begin{bmatrix} A - B_2 D_{12}^T C_1 & B_1 B_1^T - B_2 B_2^T \\ -C_{11}^T C_{11} & -(A - B_2 D_{12}^T C_1)^T \end{bmatrix} \tag{3.4}$$

$$\hat{Y}_4 = Ric \begin{bmatrix} A_{422}^T & \hat{C}_{22} \\ 0 & -A_{422} \end{bmatrix} \tag{3.5}$$

where $\hat{C}_{22}$ is the (2-2) block of $-C_{1r}^T C_{1r} + C_{22r}^T C_{22r}$. Write

$$Y_4 = \begin{bmatrix} 0 & 0 \\ 0 & \hat{Y}_4 \end{bmatrix} \tag{3.6}$$

$$Y = \begin{bmatrix} 0 & 0 \\ 0 & Y_4 \end{bmatrix} \tag{3.7}$$

Notice that Y_4 stabilizes

$$\bar{A}_4 = A_4 + Y_4 (C_{1r}^T C1r - C_{22r}^T C_{22r}) \tag{3.8}$$

and Y satisfies

$$Y(A - B_1 C_{22})^T + (A - B_1 C_{22}) Y + Y(C_1^T C_1 - C_2^T C_2))Y = 0 \tag{3.9}$$

Theorem

For the generalized plant described by (3.1) and satisfying the assumptions A1)-A6): There exists a internally stabilizing controller $\bar{K}(s)$ such that

$$\| G_{11} + G_{12} \bar{K} (I - G_{22} \bar{K})^{-1} G_{21} \|_\infty < 1 \tag{3.10}$$

if and only if there exist $X \geq 0$ and $\hat{Y}_4 \geq 0$ ($Y \geq 0$) satisfying

$$\begin{bmatrix} X & XY \\ YX & X \end{bmatrix} > 0, \qquad det \begin{bmatrix} I & Y \\ X & I \end{bmatrix} \neq 0 \tag{3.11}$$

In this case all the controllers $\bar{K}(s)$ are given by

$$\bar{K}(s) = DHM(K_D, [Q, V]) \tag{3.12}$$

$$K_D = \left[\begin{array}{c|c:cc} \bar{A}_b + B_{1e}C_{21}Z & \bar{B}_2 & \bar{B}_{1e} & -\bar{B}_1 \\ \hline \bar{C}_1 Z & I_{m2} & 0 & 0 \\ \hdashline \bar{C}_{22}Z & 0 & 0 & I_{m1} \\ C_{21}Z & 0 & I_{p2-m1} & 0 \end{array}\right] \tag{3.13}$$

or

$$\bar{K}(s) = F_l(K_F; [Q, V]) \tag{3.14}$$

$$K_F = \left[\begin{array}{c|cc:c} \bar{A}_b + B_{1e}C_{21}Z - \bar{B}_2\bar{C}_1 Z & -\bar{B}_{1e} & \bar{B}_1 & \bar{B}_2 \\ \hline -\bar{C}_1 Z & 0 & 0 & I \\ \hdashline -\bar{C}_{22}Z & 0 & I & 0 \\ -C_{21}Z & I & 0 & 0 \end{array}\right] \tag{3.15}$$

for arbitrary $Q \in BH_\infty$ and $V \in RH_\infty$, where $\bar{B}_{1e} \in R^{n\times(p2-m1)}$ is chosen so that it stabilizes $\bar{A}_b + \bar{B}_{1e}C_{21}Z$, and

$Z = (I - YX)^{-1}$
$\bar{A}_b = A - B_1C_{22} + Y(C_1^T C_1 - C_2^T C_2)$
$\bar{B}_1 = B_1 + YC_{22}^T$
$\bar{B}_2 = B_2 + YC_1^T D_{12}$
$\bar{C}_1 = D_{12}^T C_1 + B_1^T X$
$\bar{C}_{22} = C_{22} + B_1^T X$

Notice that the order of K_D or K_F is equal to n, the order of the generalized plant. Below, we shall show that if V is appropriately chosen, this can be reduced.

From the definitions,

$$\begin{aligned} \bar{A}_b &= \begin{bmatrix} A_1 & A_2 \\ A_3 + Y_4(C_{1r}^T C_{1l} - C_{22r}^T C_{22l}) & A_4 + Y_4(C_{1r}^T C_{1r} - C_{22r}^T C_{22r}) \end{bmatrix} \\ &= \begin{bmatrix} A_1 & A_2 \\ A_3 + Y_4(C_{1r}^T C_{1l} - C_{22r}^T C_{22l}) & \bar{A}_4 \end{bmatrix} \end{aligned}$$

$$Z = \begin{bmatrix} I & 0 \\ (I - Y_4X_4)^{-1}Y_4X_3 & (I - Y_4X_4)^{-1} \end{bmatrix} \stackrel{s}{=} \begin{bmatrix} I & 0 \\ Z_3 & Z_4 \end{bmatrix}$$

where, X_4 is the (2-2) block of X. Therefore, if $\bar{B}_{1e}$ is selected as

$$\bar{B}_{1e} = -\begin{bmatrix} A_1 \\ A_3 + Y_4(C_{1r}^T C_{1l} - C_{22r}^T C_{22l}) \end{bmatrix} - \begin{bmatrix} \alpha I \\ 0 \end{bmatrix}$$

the A-matrix of K_D becomes as (, notice $C_{21}Z = C_{21}$)

$$\bar{A}_b + \bar{B}_{1e} C_{21} Z = \left[\begin{array}{c|c} -\alpha I & A_2 \\ \hline 0 & \bar{A}_4 \end{array}\right]$$

Next, using the definition of DHM, $\bar{K}(s)$ is represented as

$$\bar{K}(s) = D(s)^{-1} N(s)$$

$$\begin{aligned}
[D(s) \quad N(s)] &= \left[\begin{array}{c|c:cc} \bar{A}_b + B_{1e} C_{21} Z & \bar{B}_2 & \bar{B}_{1e} & -\bar{B}_1 \\ \hline (\bar{C}_1 + Q\bar{C}_{22} + V C_{21}) Z & I & V & Q \end{array}\right] \\
&= \left[\begin{array}{cc|c:cc} -\alpha I & A_2 & \bar{B}_{2u} & \bar{B}_{1eu} & -\bar{B}_{1u} \\ 0 & \bar{A}_4 & \bar{B}_{2d} & \bar{B}_{1ed} & -\bar{B}_{1d} \\ \hline \times & \times & I & V & Q \end{array}\right]
\end{aligned}$$

in which the C-matrix is

$$[\bar{C}_{1l} + Q\bar{C}_{22l} + (\bar{C}_{1r} + Q\bar{C}_{22r}) Z_3 + V \quad (\bar{C}_{1r} + Q\bar{C}_{22r}) Z_4]$$

Therefore if we set $V = -(\bar{C}_{1l} + Q\bar{C}_{22l}) - (\bar{C}_{1r} + Q\bar{C}_{22r}) Z_3$, one gets

$$[D(s) \quad N(s)] = \left[\begin{array}{c|c:cc} \bar{A}_4 & \bar{B}_{2d} & \bar{B}_{1ed} & -\bar{B}_{1d} \\ \hline (\bar{C}_{1r} + Q\bar{C}_{22r}) Z_4 & I & V & Q \end{array}\right]$$

Hence the reduced order controller K(s) is obtained as

$$\bar{K}(s) = DHM(K_{Dr}, Q) \tag{3.16}$$

$$K_{Dr} = \left[\begin{array}{c|c:cc} \bar{A}_4 & \bar{B}_{2d} & \bar{B}_{1ed} & -\bar{B}_{1d} \\ \hline \bar{C}_{1r} Z_4 & I & -\bar{C}_{1l} - \bar{C}_{1r} Z_3 & 0 \\ \bar{C}_{22r} Z_4 & 0 & -\bar{C}_{22l} - \bar{C}_{22r} Z_3 & I \end{array}\right] \tag{3.17}$$

or in the form of linear feedback transformation,

$$\bar{K}(s) = F_l(K_{Fr}; Q) \tag{3.18}$$

$$K_{Fr} = \left[\begin{array}{c|cc:c} \bar{A}_4 - \bar{B}_{2d} \bar{C}_{1r} Z_4 & -\bar{B}_{1ed} - \bar{B}_{2d}(\bar{C}_{1l} + \bar{C}_{1r} Z_3) & \bar{B}_{1d} & \bar{B}_{2d} \\ \hline -\bar{C}_{1r} Z_4 & -\bar{C}_{1l} - \bar{C}_{1r} Z_3 & 0 & I \\ -\bar{C}_{22r} Z_4 & -\bar{C}_{22l} - \bar{C}_{22r} Z_3 & I & 0 \end{array}\right] \tag{3.19}$$

Observe that the order of the controller is reduced by the number of the independent rows of C_{21}. It would be of interest if this result is compared with the minimal order observers.

Now, let us apply the above result to the servodesign problem. The correspondences between the expressions of (2.6) and (3.1) are given by

$$A = \begin{bmatrix} A_w & -B_w C_p \\ 0 & A_p \end{bmatrix}, B_1 = \begin{bmatrix} B_w \\ 0 \end{bmatrix}, B_2 = \begin{bmatrix} 0 \\ B_p \end{bmatrix}, D_{12} = \begin{bmatrix} 0 \\ I \end{bmatrix}, D_{21} = \begin{bmatrix} 0 \\ 0 \\ I \end{bmatrix}$$

$$C_1 = \begin{bmatrix} C_{w1} & 0 \\ 0 & C_{w2} \end{bmatrix}, C_{21} = \begin{bmatrix} I & 0 \\ 0 & \bar{C}_p \end{bmatrix}, C_{22} = [0 \quad -C_p]$$

where, $\bar{C}_p$ and A_p are arranged as $\begin{bmatrix} \bar{C}_p \\ A_p \end{bmatrix} = \begin{bmatrix} I & 0 \\ A_{p1} & A_{p2} \\ A_{p3} & A_{p4} \end{bmatrix}$, and thus

$$C_{21} = \begin{bmatrix} I & 0 & 0 \\ 0 & I & 0 \end{bmatrix}, \qquad A - B_1 C_{22} = \left[\begin{array}{cc|c} A_w & 0 & 0 \\ 0 & A_{p1} & A_{p2} \\ \hline 0 & A_{p3} & A_{p4} \end{array}\right] = \begin{bmatrix} A_1 & A_2 \\ A_3 & A_4 \end{bmatrix}$$

Looking at these expressions, let us see that assumptions A1), A2), A3) are trivially satisfied. Using the assumption that (A_p, B_p, C_p) is minimal and the definition of $W_2(s)$, it can be seen that A4) is equivalent to

A4') $\begin{bmatrix} A_p - sI & B_p \\ C_p & 0 \end{bmatrix}$: row full rank for $\forall s$ satisfying $\alpha(s) = 0$.

Also, it would be easy to verify that A5) is equivalent to

$$\text{rank} \begin{bmatrix} A_p - j\omega I \\ \bar{C}_p \end{bmatrix} = n_p.$$

There is no simple expression for A6), and it must be directly checked.

The controller K(s) is obtained by (3.14) and the structure indicated in Fig.1, The result is given by

$$K(s) = F_l(K_{op}(s), Q) \tag{3.20}$$

$$K_{op}(s) = K_{F_r}(s) \begin{bmatrix} 0 & (sI - A_w)^{-1} B_w & 0 \\ I & 0 & 0 \\ 0 & I_{m1} & 0 \\ 0 & 0 & I_{m2} \end{bmatrix} \tag{3.21}$$

4. References

1) W.M. Wonham: Linear Multivariable Control - A Geometric Approach, Berlin : Springer-Verlag (1979).
2) B.A. Francis and M. Vidyasagar: Algebraic and Topological Aspects of the Regulator Problem for Lumped Linear Systems, Automatica, Vol. 19, No.1, pp.87- 90(1983).
3) M. Ikeda and N. Suda: Synthesis of Optimal Servosystems, Transitions of the SICE, Vol.24, No.1, pp40-46 (1989).
4) F. Zhang, S. Hosoe and M. Kono: Synthesis of Robust Output Regulators via H^∞-Control, Preprints of First IFAC Symposium on Design Method of Control Systems, pp.366-371 (1991).

H_∞-suboptimal Controller Design of Robust Tracking Systems

Toshiharu SUGIE*, Masayuki FUJITA** and Shinji HARA***

* Div. of Applied Systems Science, Kyoto University, Uji, Kyoto 611, Japan
** Dept. of Electrical and Computer Engineering, Kanazawa University, Kodatsuno, Kanazawa 920, Japan
*** Dept. of Control Engineering, Tokyo Institute of Technology Meguro-ku, Tokyo 152, Japan

1 Introduction

In the control system design, one of the fundamental issue is the robust tracking problem [1, 2], whose purpose is to find a controller satisfying (a) internal stability and (b) zero steady state tracking error in the presence of plant perturbations. However, in addition to (a) and (b), control systems are generally required to have both (c) good command response and (d) satisfactory feedback properties such as enough stability margin and low sensitivity. Concerning to (d), it is widely accepted that H_∞ control framework is quite powerful for robust stabilization and the loop shaping.

The purpose of this paper is to characterize all controllers which satisfy (a) internal stability, (b) robust tracking, (c) the exact model matching and (d) the specified H_∞ control performance related to the above feedback properties. First, we derive a necessary and sufficient condition for such a controller to exist, based on the authors' former result on (i) the H_∞ control problem with $j\omega$-axis constraints [6] and (ii) the two-degree-of-freedom robust tracking systems [7]. Then, we parametrize all such controllers, where the free parameter of the H_∞ suboptimal controller is fully exploited.

Notation: $\mathbf{C}_+$ denotes the set of all $s \in \mathbf{C}$ satisfying $\mathrm{Re}[s] \geq 0$. The sets of all real rational and stable proper real rational matrices of size $m \times n$ are denoted by $\mathbf{R}(s)^{m\times n}$ and $\mathbf{RH}_\infty^{m\times n}$, respectively. For $G(s) \in \mathbf{RH}_\infty^{m\times n}$, we define $\|G\|_\infty := \sup_\omega \|G(j\omega)\|$, where $\|\cdot\|$ denotes the largest singular value. A subset of $\mathbf{RH}_\infty^{m\times n}$ consisting of all $S(s)$ satisfying $\|S\|_\infty < 1$ is denoted by $\mathbf{BH}_\infty^{m\times n}$. If no confusion arises we drop the size "$m \times n$".

2 Problem formulation

We consider a linear time-invariant system $\Sigma(C,\ P)$ described by

$$y = Pu \tag{2.1}$$

$$u = C\begin{bmatrix} r \\ y \end{bmatrix} := [C_1,\ -C_2]\begin{bmatrix} r \\ y \end{bmatrix} \tag{2.2}$$

where $P \in \mathbf{R}(s)^{m\times q}$ $C \in \mathbf{R}(s)^{q\times 2m}$ denote the plant to be controlled and the compensator to be designed, respectively. The vectors u, y, and r are q-dimensional control input, m-dimensional controlled output and m-dimensional command signal, respectively.

Suppose that r is described by

$$r = Rr_0, \qquad R = \frac{1}{\prod_{i=1}^{k}(s - j\omega_i)} I_m \tag{2.3}$$

where R is a command signal generator whose poles are distinct and r_0 is unknown constant vector.

The system $\Sigma(C,P)$ is said to be *internally stable* if all the transfer matrices from $\{d_1, d_2, d_3\}$ to $\{u, y\}$ are in $\mathbf{RH}_\infty$ when we replace $\{r, u, y\}$ by $\{r + d_1, u + d_2, y + d_3\}$, where $\{d_1, d_2, d_3\}$ are virtual external inputs inserted at the input channels of P and C.

Concerning to the internal stability, we have the following lemma [7].

Lemma 1: For given P, all compensators $C = [C_1,\ -C_2]$ such that $\Sigma(C,\ P)$ is internally stable are given by

$$C_1 = (D + C_2 N)K, \qquad \forall K \in \mathbf{RH}_\infty \tag{2.4}$$

$$C_2 \in \Omega(P) \tag{2.5}$$

where $P = ND^{-1}$ is a right coprime factorization (rcf) of P over $\mathbf{RH}_\infty$, and $\Omega(P)$ denotes the set of all closed loop stabilizer of P.

Remark 2.1: Strictly speaking, $C_2 \in \Omega(P)$ means that $\tilde{D}_C D + \tilde{N}_C N$ is unimodular over $\mathbf{RH}_\infty$, where $C_2 = \tilde{D}_C^{-1}\tilde{N}_C$ is a left coprime factorization of C_2.

Since the design of C that achieves internal stability is equivalent to the choice of $\{K, C_2\}$ subject to (2.4) and (2.5), C is referred to $C = C(K, C_2)$ hereafter.

Note that the transfer matrix G_{yr} from r to y of $\Sigma(C,\ P)$ is given by

$$\begin{aligned} G_{yr} &= P(I+C_2P)^{-1}C_1 \\ &= NK \end{aligned} \tag{2.6}$$

from (2.4), which implies that the role of K is to specify the command response of $\Sigma(C, P)$. Also, it is easy to see that the role of C_2 is to make the control system robust against the modeling errors and disturbances. The recognition of these roles is quite important to solve the problem stated below.

Now we consider the following four requirements:

(s1): (*Stability*) $\Sigma(C,\ P)$ is internally stable.

(s2) (*H_∞ suboptimality*) The feedback compensator C_2 satisfies the following H_∞ norm performance, i.e.,

$$\|\Phi\|_\infty := \left\| \begin{bmatrix} W_s S \\ W_t T \end{bmatrix} \right\|_\infty < 1 \tag{2.7}$$

$$S := (I+PC_2)^{-1}, \quad T := I - S \tag{2.8}$$

where W_s and W_t are given weighting matrices.

(s3): (*Robust tracking*) There exits an open neighborhood of P, say $\mathcal{D}$, such that

$$\Sigma(C,\ P') \text{ is internally stable} \tag{2.9}$$

$$(I_m - G'_{yr})R \in \mathrm{RH}_\infty \tag{2.10}$$

hold for every $P' \in \mathcal{D}$, where G'_{yr} denote the transfer matrix from r to y of $\Sigma(C,\ P')$.

(s4): (*Exact model matching*) Let $G_M \in \mathrm{RH}_\infty^{m\times m}$ be the given desired model, then the transfer matrix G_{yr} of $\Sigma(C,P)$ satisfies

$$G_{yr} = G_M \tag{2.11}$$

Remark 2.2: The problem of finding a C_2 which satisfies (s2) is the so called mixed sensitivity problem, which often appears in the practical H_∞ applications(e.g, [5, 3]). The requirement (s3) means that $y(t)$ tracks $r(t)$ without steady-state error against the plant perturbations. On the other hand, (s4) is one of the most direct ways to get the desired command response.

Now we are in the position to state the problem:

Problem: For given P, W_s, W_t, R and G_M, find all compensators C which satisfy the requirements (s1) $\sim$ (s4) if they exist.

In order to utilize the existing H_∞ control results, we make the following assumptions:

(A1) P does not have any pole on the $j\omega$-axis, and rank $P(j\omega) = m$ for any ω.

(A2) W_s is in $\mathbf{RH}_\infty$. W_t has no poles in $\mathbf{C}_+$ and W_tP is biproper. Both of them has no zeros in the $j\omega$-axis.

3 Solvability condition

In this section, we derive the solvability condition of our problem.

Let γ_* be the infimum of $\|\Phi\|_\infty$ achieved by stabilizing compensators, i.e.,

$$\gamma_* = \inf_{C_2 \in \Omega(P)} \|\Phi\|_\infty \tag{3.1}$$

Then, we obtain the following result.

Theorem 1: Under the assumptions (A1) and (A2), for given P, W_s, W_t, R and G_M, there exists an compensator $C = C(K, C_2)$ satisfying all of the four requirements (s1) $\sim$ (s4) if and only if the following four conditions hold:

(i)
$$\gamma_* < 1 \tag{3.2}$$

(ii)
$$\|W_t(j\omega_i)\| < 1 \quad \text{for} \quad i = 1 \sim k \tag{3.3}$$

(iii)
$$G_M(j\omega_i) = I_m \quad \text{for} \quad i = 1 \sim k \tag{3.4}$$

(iv) There exits a $K_M \in \mathbf{RH}_\infty$ such that the following equation holds.

$$G_M = N K_M \tag{3.5}$$

where $P = ND^{-1}$ is an rcf over $\mathbf{RH}_\infty$.

In order to prove the theorem, we need several lemmas.

The following two lemmas are directly obtained from [7].

Lemma 2: For given P and G_M, there exits a compensator C satisfying (s1) and (s4) if and only if the condition (iv) holds. In addition, a stabilizing compensator $C = C(K, C_2)$ satisfies (s4) if and only if $G_M = NK$ holds.

Lemma 3: For given P and R, there exits a compensator C satisfying (s1) and (s3) if and only if

$$\operatorname{rank} N(j\omega_i) = m \quad \text{for} \quad i = 1 \sim k \tag{3.6}$$

holds. In addition, a stabilizing compensator $C = C(K, C_2)$ satisfies (s3) if and only if

$$[NK](j\omega_i) = I_m, \quad \text{for} \quad i = 1 \sim k \tag{3.7}$$

$$D_C(j\omega_i) = 0, \quad \text{for} \quad i = 1 \sim k \tag{3.8}$$

where $P = ND^{-1}$ and $C_2 = N_C D_C^{-1}$ are rcf's over $\mathbf{RH}_\infty$.

The following result is easily obtained from [6].

Lemma 4: Under the assumptions (A1) and (A2), suppose $\gamma_* < 1$ holds for given P, W_s and W_t, and there exists an internally stabilizing compensator satisfying $\Phi(j\omega_i) = \Phi_i$ ($i = 1 \sim k$) for given constant matrices Φ_i. Then there exits a $C_2 \in \Omega(P)$ satisfying both $\|\Phi\|_\infty < 1$ and

$$\Phi(j\omega_i) = \Phi_i \quad \text{for} \quad i = 1 \sim k \tag{3.9}$$

if and only if the following inequality holds.

$$\|\Phi_i\| < 1 \quad \text{for} \quad i = 1 \sim k \tag{3.10}$$

In the above lemma, the sufficiency is important.

(Proof of Theorem 1) *Necessity:* Suppose (s1) $\sim$ (s4) hold. It is obvious that (i) holds from (s1) and (s2). Also Lemmas 2 and 3 yields (iii) and (iv). So we will show that (ii) holds. Concerning to C_2, the requirement (s3) is equivalent to (3.8), which also equals to $S(j\omega_i) = 0$ and $T(j\omega_i) = I$. Therefore, under the assumptions (A1) and (A2), this implies that the stabilizing compensator C_2 satisfies

$$\Phi(j\omega_i) = \Phi_i := \begin{bmatrix} 0 \\ W_t(j\omega_i) \end{bmatrix} \quad \text{for} \quad i = 1 \sim k \tag{3.11}$$

From Lemma 4 with $\|\Phi_i\| = \|W_t(j\omega_i)\|$, we obtain (ii).

Sufficiency: Suppose (i) $\sim$ (iv) hold. Then there exists a stabilizing C_2^* satisfying $\|\Phi\|_\infty < 1$ and (3.11). Since (3.11) means (3.8), it is enough for us to choose $C = C(K_M, C_2^*)$. This completes the proof. (QED)

Remark 3.1: We can calculate the infimum of $\|\Phi\|_\infty$ subject to (s1) and (s3), which is given by

$$\max\{\gamma_*,\ \ \|W_t(j\omega_i)\|\ \ (i = 1 \sim k)\} \tag{3.12}$$

Wu and Mansour [9] discuss the similar problem in the special case where P is stable and $\Phi := C_2(I + PC_2)^{-1}$. So our result is a generalization of [9].

4 Characterization of all solutions

In this section, we characterize all the solutions of our problem.

The next lemma is well known (e.g., [4]).

Lemma 5: Under the assumptions (A1) and (A2), suppose $\gamma_* < 1$ holds for given P, W_t and W_s. Then all stabilizing C_2 which satisfy $\|\Phi\|_\infty < 1$ are described by

$$C_2 = (\Pi_{11}S + \Pi_{12})(\Pi_{21}S + \Pi_{22})^{-1} \qquad \forall S \in \mathbf{BH}_\infty^{q\times m} \tag{4.1}$$

where the matrices $\Pi_{ij} \in \mathbf{RH}_\infty$ are determined from the data $\{P, W_s, W_t\}$ via two Riccati equations, and $\Pi := \begin{bmatrix} \Pi_{11} & \Pi_{12} \\ \Pi_{21} & \Pi_{22} \end{bmatrix}$ is unimodular over $\mathbf{RH}_\infty$. (see [4] for the explicit form of Π)

Based on Theorem 1 and the above lemma, we obtain the following result.

Theorem 2: Under the assumptions (A1) and (A2), suppose the conditions (i) $\sim$ (iv) of Theorem 1 hold for given P, W_s, W_t, R and G_M. Then all the compensators $C = C(K, C_2)$ satisfying (s1) $\sim$ (s4) are given by

$$K = K_M + WQ_K \qquad \forall Q_K \in \mathbf{RH}_\infty^{(q-m)\times m} \tag{4.2}$$

$$C_2 = (\Pi_{11}S + \Pi_{12})(\Pi_{21}S + \Pi_{22})^{-1} \qquad \forall S \in \mathbf{BH}_\infty^{q\times m} \tag{4.3}$$

subject to

$$[\Pi_{21}S + \Pi_{22}](j\omega_i) = 0 \ \text{ for } \ i = 1 \sim k \tag{4.4}$$

where $W \in \mathbf{RH}_\infty^{q\times(q-m)}$ is a matrix satisfying

$$NW = 0, \quad \mathrm{rank}W = q - m \tag{4.5}$$

for the rcf $P = ND^{-1}$, and $K_M \in \mathbf{RH}_\infty$ is a matrix satisfying (3.5).

(Proof) As for K, noting that $G_M = NK_M$, it is easy to see that the general solution of $G_M = NK$ is given by (4.2).

Concerning to C_2, note that the right hand side of (4.1) is an rcf of C_2, because

$$\text{rank}\begin{bmatrix}\Pi_{11}S+\Pi_{12}\\ \Pi_{21}S+\Pi_{22}\end{bmatrix} = \text{rank}\begin{bmatrix}\Pi_{11} & \Pi_{12}\\ \Pi_{21} & \Pi_{22}\end{bmatrix}\begin{bmatrix}S\\ I_m\end{bmatrix} = m, \quad \text{for} \quad \forall s \in \mathbf{C}_+ \cup \{\infty\} \tag{4.6}$$

holds. So (4.4) is equivalent to (3.8), which means the requirement (s3) is satisfied. (s1) and (s2) are satisfied automatically as long as $S \in \mathbf{BH}_\infty$. This completes the proof. (QED)

Remark 4.1: Concerning to the parametrization of the H_∞ controllers, (4.1) is often given by the linear fractional transformation form

$$C_2 = H_{11} + H_{12}S(I - H_{22}S)^{-1}H_{21} \qquad \forall S \in \mathbf{BH}_\infty^{q\times m} \tag{4.7}$$

Since the following relation holds

$$\begin{bmatrix}\Pi_{11} & \Pi_{12}\\ \Pi_{11} & \Pi_{11}\end{bmatrix} = \begin{bmatrix}H_{11} & H_{12}\\ I & 0\end{bmatrix}\begin{bmatrix}0 & I\\ H_{21} & H_{22}\end{bmatrix}^{-1} \tag{4.8}$$

between H_{ij} and Π_{ij} (see [4]), (4.4) is equivalent to

$$[H_{22}S](j\omega_i) = I \ \text{ for } \ i = 1 \sim k \tag{4.9}$$

Remark 4.2: In this paper, we have treated the plant P given by (2.1). However, by using the robust tracking result of [8], it is easy to extend our result to the case where the plant is described by

$$\begin{bmatrix}y\\ z\end{bmatrix} = \begin{bmatrix}P_1\\ P_2\end{bmatrix}u \tag{4.10}$$

where the measurement z may not be equal to the output y.

5 Conclusion

In this paper, we have discussed the problem to find a controller which satisfies (a) internal stability, (b) robust tracking, (c) the exact model matching and (d) the specified H_∞

control performance related to the feedback properties. First, we have obtained a necessary and sufficient condition for such a controller to exist. Then, we have parametrized all such controllers, where (i) the two-degree-of-freedom compensator configuration and (ii) the free parameter of the H_∞ suboptimal controller are fully exploited.

References

[1] E.J.Davison, The robust control of a servomechanism problem for linear time-invariant multivariable systems, *IEEE Trans. Autom. Contr.*, 21, 25-34 (1976)

[2] B.A.Francis and M.Vidyasagar, Algebraic and topological aspects of the regulator problem for lumped linear systems, *Automatica*, 19, 87-90 (1983)

[3] M. Fujita, F. Matsumura and M. Shimizu : H^∞ robust control design for a magnetic suspension system ; *Proc. of 2nd Int. Sympo. on Magnetic Bearing*, pp. 349 ~ 355 (1990)

[4] R.Kondo and S.Hara, On cancellation in H_∞ optimal controllers, *Systems & Control Letters*, 13, 205-210 (1989).

[5] H. Kuraoka et al. : Application of H_∞ optimal Design to Automotive Fuel Control ; *Proc. 1989 American Control Conf.*, pp. 1957 ~ 1962 (1989)

[6] T. Sugie and S. Hara, H_∞-suboptimal control problem with boundary constraints, *Systems & Control Letters*, 13, 93-99 (1989).

[7] T. Sugie and T. Yoshikawa, General solution of robust tracking problem in two-degree-of-freedom control systems, *IEEE Trans. Autom. Contr.*, 31, 552-554 (1986)

[8] T. Sugie and M. Vidyasagar, Further results on the robust tracking problem in two d.o.f. control systems, *Systems & Control Letters*, 13, 101-108 (1989).

[9] Q.H. Wu and M. Mansour, Robust output regulation for a class of linear multivariable systems, *Systems & Control Letters*, 13, 227-232 (1989)

Stability and Performance Robustness of ℓ^1 Systems with Structured Norm-Bounded Uncertainty

M. Khammash
Iowa State University
Ames, Iowa 50011

Abstract

An overview of recent advances in the robustness techniques in the ℓ^1 setting are discussed. In particular, results on the stability and performance robustness of multivariable systems in the presence of norm-bounded structured perturbations are presented. For robustness analysis, the computation of nonconservative conditions will be highlighted as one of the most attractive features of this approach. In addition, a useful iteration technique for robustness synthesis is presented. Finally, extension of the robustness techniques to the time-varying case is outlined and then used to provide necessary and sufficient conditions for the robustness of sampled-data systems in the presence of structured uncertainty.

1 Introduction

When designing or analyzing systems, a single system model rarely provides complete information about the true behavior of the physical system at hand. For this reason, robustness considerations must be taken into account in the analysis and design procedures. Robustness can be divided into two, not unrelated, types: robustness to *signal* uncertainty, and robustness to *system* uncertainty. When $\mathcal{L}^2$ signals are present, the $\mathcal{H}^\infty$ theory provides a systematic means for designing controllers which have the effect of minimizing the influence of uncertain signals on the desired outputs. In the $\mathcal{H}^\infty$ setting, the robustness to system uncertainty can be treated using the μ function of Doyle [5] which is particularly suited for handling structured uncertainty. But when the uncertain signals at hand are assumed to be bounded in magnitude and thus belong to $\mathcal{L}^\infty$ (or ℓ^∞ in the discrete-time case), the $\mathcal{L}^1$ (ℓ^1) theory provides the means for designing controllers which have optimal or near optimal properties in terms of reducing the effect of the disturbances on the desired outputs. This paper is concerned with the stability and performance robustness of ℓ^1 systems in the presence of structured uncertainty. The stability robustness of systems with unstructured uncertainty has been solved by Dahleh and Ohta [2] in the discrete-time case. These results are extended to the case of structured uncertainty. In addition, the performance robustness problem is solved by showing it can be equivalent to a stability robustness problem having one additional fictitious uncertainty block. These results apply equally well to continuous-time and discrete-time systems. However, modern control systems often implement a digital controller to a continuous time plant via sample and hold devices. Since the

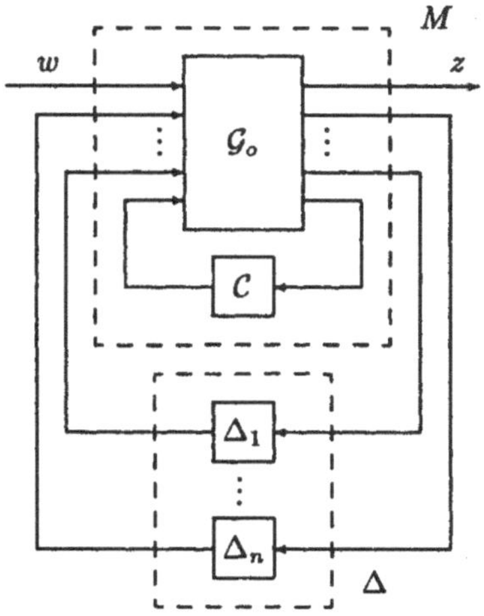

Figure 1: System with Structured Uncertainty.

resulting sampled-data system is time-varying, the robustness of such a system cannot be treated using the robustness techniques for time-invariant nominal systems. This paper also presents necessary and sufficient conditions for the robustness of sampled-data systems in the presence of structured uncertainty. The results are stated in terms of specral radii of certain matrices obtained using the state-space description of the continuous-time plant and the discrete-time controller. Finally, the performance robustness of sampled-data systems is solved whereby a certain equivalence between the stability and performance problems is shown to hold.

The remaining part of this paper is divided into two main parts: the robustness of time-invariant systems and the robustness of sampled-data systems. In the first part, motivation is given and the general robustness problem for time-invariant systems is setup. Then, some results regarding the performance robustness are presented, followed by the solution for the robustness analysis problem. Finally, the robustness synthesis problem is addressed and an iterative scheme for robustness synthesis is presented. In the second part of the paper, the robustness of sampled-data systems in the presence of structured uncertainty is addressed. The problem is formally set up. Next, the performance robustness of sampled-data systems is discussed, and its relation to stability robustness is outlined. Finally, necessary and sufficient conditions for the robustness of sampled-data systems are presented.

2 Robustness of Time-Invariant Systems

2.1 Setup

We start by setting up the robustness problem for time-invariant systems. Consider the system in figure 1. M represents the nominal part of the systems composed of the nominal linear time-invariant plant and the linear stabilizing controller. M can be either continuous-time or discrete-time. We will assume it is discrete-time,

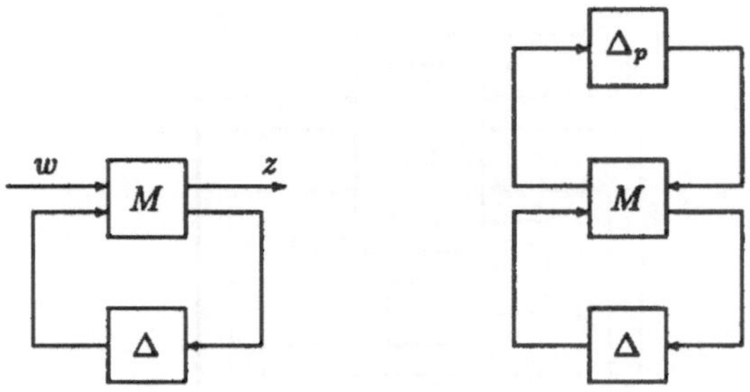

Figure 2: Stability Robustness vs. Performance Robustness

although the results carry over with obvious modifications to the continuous-time case. w in the figure represents the unknown disturbances. The only assumption on w is that it is bounded in magnitude, i.e. it belongs to the space ℓ^∞, and we assume that $\|w\|_\infty \leq 1$. z, on the other hand, is the regulated output of interest. $\Delta_1,\ldots,\Delta_n$ denotes system perturbations modelling the uncertainty. Each perturbation block Δ_i is causal and has an induced ℓ^∞-norm less than or equal to 1. Therefore, each Δ_i belongs to the set $\underline{\Delta} = \{\Delta \ : \ \Delta \text{ is causal, and } \|\Delta\| \ := \sup_{u\neq 0} \frac{\|\Delta u\|_\infty}{\|u\|_\infty} \leq 1\}$ and can be time-varying and/or nonlinear. The Δ_i's are otherwise independent. Consequently, we define the class of admissible perturbations as follows:

$$\mathcal{D}(n) := \{\Delta = diag(\Delta_1,\cdots,\Delta_n) \ : \ \Delta_i \in \underline{\Delta}\}. \tag{1}$$

The system in the figure achieves robust stability if it is stable for all admissible perturbations. It is said to achieve robust performance if it achieves robust stability and at the same time $\mathcal{T}_{zw} < 1$ for all admissible perturbations, where $\mathcal{T}_{zw}$ is the map between the input w and the output z. Note that when the perturbation Δ is zero, $\|\mathcal{T}_{zw}\|$ is the induced ℓ^∞ norm of the nominal system and is equal to the ℓ^1 norm of the impulse response of the map $\mathcal{T}_{zw}$.

2.2 Stability Robustness vs. Performance Robustness

In this subsection, we discuss the unique relationship that holds between stability robustness and performance robustness. When M is time-invariant, a certain interesting equivalence between stability robustness and performance robustness holds. More specifically, a performance robustness problem can be treated as a stability robustness problem. Consider the two systems in Fig. 2. The first system in the figure, System I, is that corresponding to the robust performance problem. By adding a fictitious perturbation block, Δ_P, where $\Delta_P \in \underline{\Delta}$, we get System II. System II, therefore, corresponds to a stability robustness problem with $n+1$ perturbation blocks. The following theorem establishes the relation between the two systems:

Theorem 1 *System I achieves robust performance if and only if System II achieves robust stability.*

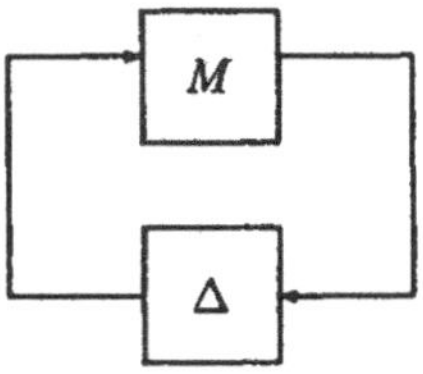

Figure 3: Stability Robustness Problem

The proof of this theorem will be omitted and may be found in [9]. A similar relationship holds when the perturbations are time-invariant and when $w \in \ell^2$ as has been shown in [6]. As a result of this theorem, we can focus our efforts on finding stability robustness conditions. Conditions for performance robustness will automatically follow from the stability conditions.

2.3 Robustness Conditions in the Presence of SISO Perturbations

In this subsection, we provide necessary and sufficient conditions for stability robustness in the presence of structured SISO perturbations. These conditions are shown to be easily computable even for systems with a large number of perturbation blocks. We will do so for a nominal system M with n perturbation blocks as appears in Fig. 3. As a result, M has n inputs and n outputs. Since M is linear time-invariant and stable, each component M_{ij} has an impulse response belonging to ℓ^1. The ℓ^1 norm of the impulse response of M_{ij} is equal to the induced ℓ^∞ norm of M_{ij}. We refer to this quantity as $\|M_{ij}\|_{\mathcal{A}}$. Given state-space representation of M, $\|M_{ij}\|_{\mathcal{A}}$ can be computed arbitrarily accurately. As a result, we can define the following nonnegative $n \times n$ matrix:

$$\widehat{M} = \begin{bmatrix} \|M_{11}\|_{\mathcal{A}} & \cdots & \|M_{1n}\|_{\mathcal{A}} \\ \vdots & & \vdots \\ \|M_{n1}\|_{\mathcal{A}} & \cdots & \|M_{nn}\|_{\mathcal{A}} \end{bmatrix}. \tag{2}$$

With this definition, we can state the following necessary and sufficient conditions for robust stability of the system in Fig. 3:

Theorem 2 *The following are equivalent:*

1. *The system in Fig. 3 is stable for all $\Delta \in \mathcal{D}(n)$, i.e. robustly stable.*
2. *$(I - M\Delta)^{-1}$ is ℓ^∞- stable for all $\Delta \in \mathcal{D}(n)$.*
3. *$\rho(\widehat{M}) < 1$, where $\rho(.)$ denotes the spectral radius.*
4. *$\inf_{R\in\mathcal{R}} \|R^{-1}MR\|_{\mathcal{A}} < 1$, where $\mathcal{R} := \{diag(r_1, \ldots, r_n) : r_i > 0\}$.*

The proof of this theorem can be found in [9] and in [8]. Item *3.* in this theorem provides a very effective means for robustness analysis. A number of algorithms exist for the fast computation of the spectral radius of a nonnegative matrix. These algorithms take advantage of the Perron-Frobenius theory of nonnegative matrices. (see [1]). Among the interesting properties of nonnegative matrices is that the spectral radius of a nonnegative matrix is itself an eigenvalue of the matrix, and is often referred to as the Perron root of that matrix. Corresponding to that root, there exists an eigenvector which is itself nonnegative. One way of computing $\rho(\widehat{M})$ appears in [13]. It is summarized as follows: Suppose $\widehat{M}$ is primitive ($\widehat{M}^m > 0$ for some positive integer m). Let $x > 0$ be any n vector. Then

$$\min_i \frac{(\widehat{M}^{m+1}x)_i}{(\widehat{M}^m x)_i} \leq \rho(\widehat{M}) \leq \max_i \frac{(\widehat{M}^{m+1}x)_i}{(\widehat{M}^m x)_i}.$$

Furthermore, the upper and lower bounds both converge to $\rho(\widehat{M})$ as $m \to \infty$. If the $\widehat{M}$ were not primitive, it can be perturbed by adding $\epsilon > 0$ to each element making each entry of $\widehat{M}$ positive and thus making $\widehat{M}$ primitive. ϵ can be chosen so that its effect on the spectral radius is arbitrarily small. An explicit bound for the size of ϵ required has been derived in [10].

As a consequence of *3* of the previous theorem, the robustness analysis problem is completely solved. Item *4.* of the same theorem provides a useful method for the synthesis of robust controllers. This is the next subject of discussion. The dependence of M on the controller can be reflected through its dependence on the Youla parameter Q. More specifically, M can be written as:

$$M(Q) = T_1 - T_2 Q T_3, \tag{3}$$

where T_1, T_2, and T_3 are stable and depend only on the nominal plant $\mathcal{G}_o$, while Q is a stable rational function which determines the stabilizing controller. See [7] for more details. The robustness synthesis problem can therefore be stated as follows:

$$\inf_{Q \text{ stable}} \rho(\widehat{M}(Q)). \tag{4}$$

However, it is not difficult to see that this problem is a nonconvex optimization problem. Instead, we can use *4.* of Theorem 2 to rewrite the robustness synthesis problem as follows:

$$\inf_{Q \text{ stable}} \inf_{R \in \mathcal{R}} \|R^{-1}\widehat{M}(Q)R\|_{\mathcal{A}}. \tag{5}$$

The optimization involves two parameters, R and Q. With R fixed, the optimization problem with respect to Q is a standard ℓ^1-norm minimization problem which can be readily solved using linear programming as in [3] and [11] for example. With Q fixed, the optimization problem with respect to R is nonconvex. However, the global minimizer (if one exists) or an R yielding arbitrarily close values to the global minimum (if no minimizer exists) can be easily found, thanks to the theory of nonnegative matrices. It turns out that if $\widehat{M}(Q)$ is primitive, the infimum is in fact a minimum and it is achieved by $R = diag(r_1, \ldots, r_n)$ where $(r_1, \ldots, r_n)^T$ is an eigenvector corresponding to the Perron root $\rho(\widehat{M}(Q))$. Moreover, this eigenvector can be computed easily using power methods. In particular, let $x^{(0)} > 0$ be any n vector. Given any $x^{(i-1)}$ we can define $x^{(i)} := \frac{\widehat{M}x^{(i-1)}}{\|\widehat{M}x^{(i-1)}\|}$. It can be shown [13]

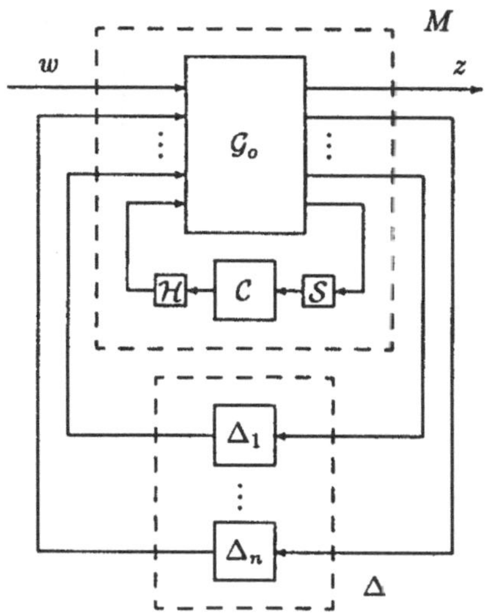

Figure 4: Sampled-Data System with Structured Uncertainty.

that $\lim_{i\to\infty} x^{(i)}$ exists and is equal to the desired eigenvector corresponding to $\widehat{M}$. Thus the optimum R, when Q is fixed, can be easily computed. As a result, we can setup the following iteration scheme which provides a sequence of Q's, and hence controllers, yielding successively smaller values for $\rho(\widehat{M})$. The iteration which is somewhat similar to the D-K iteration in the μ synthesis goes as follows:

1. Set $i := 0$, and $R_o := I$.
2. Set $Q_i := arg\ \inf_{Q\ stable} \|R^{-1}(T_1 - T_2QT_3)R\|_{\mathcal{A}}$.
3. Set $R_i := arg\ \inf_{R\in\mathcal{R}} \|R^{-1}(T_1 - T_2Q_iT_3)R\|_{\mathcal{A}}$.
4. Set $i := i + 1$. Go to step 2.

Finally, it should be mentioned that even though the iteration scheme above does not guarantee the limiting controller achieves the global minimum for the spectral radius, it does provide a useful way of obtaining controllers having good robustness properties.

3 Robustness of Sampled-Data Systems

In Fig. 4 the robustness problem of sampled-data systems is shown. In the figure, $\mathcal{G}_o$ is a continuous-time, linear, time-invariant nominal plant. C, on the other hand, is a discrete-time controller. $\mathcal{S}$ is a sampling device with sampling period T_s, while $\mathcal{H}$ is a first order hold with the same period. The interconnection of $\mathcal{G}_o$ and C and the sample and hold devices comprises M which is now a linear periodically time-varying system with period T_s. The definition for stability robustness and performance robustness is the same as before except that now the input w and the output z are $\mathcal{L}^\infty$ signals, and the Δ_i's are continuous-time norm-bounded operators.

3.1 Stability Robustness vs. Performance Robustness

For sampled-data systems, M is time-varying. In [10] it was shown that the unique relationship between stability robustness and performance robustness which holds for time-invariant M ceases to hold when M is time-varying. It was shown that while performance robustness still implies stability robustness, stability robustness *does not* in general imply performance robustness. Examples demonstrating this fact can be easily constructed. However, in the case of sampled-data systems, M has a special structure. It is *periodically* time-varying. This fact is significant enough to make the equivalence between the two notions of robustness hold in the same way it does in the time-invariant case! We state this fact in the next theorem whose proof can be found in [10].

Theorem 3 *Let M and Δ in System I and System II of Fig. 2 be the same as the M and Δ in Fig. 4, i.e. those corresponding to a sampled-data system. Then System I achieves robust performance if and only if System II achieves robust stability.*

So, here again, we need only get conditions for robust stability in the presence of structured perturbations. Robust performance can be treated in the same way as robust stability. The required necessary and sufficient conditions for robust stability of sampled-data systems are discussed next.

3.2 Conditions for Robustness of Sampled-Data Systems

In this section, it will be shown that the spectral radius still plays a significant role in the robustness of sampled-data systems. The conditions for robustness are not derived in terms of the spectral radius of a single nonnegative matrix. Instead, it is obtained in terms of the spectral radii of nonnegative matrices in a certain parametrized family.

Even though conditions can be derived for a general linear time-invariant plant $\mathcal{G}_o$ and controller C, we will restrict our treatment for a plant and a controller having a finite dimensional state-space representation. So let $\mathcal{G}_o$ and C have the following state-space representation:

$$\mathcal{G}_o = \left(\begin{array}{c|ccc} A & B_1 & \dots & B_{n+1} \\ \hline C_1 & D_{11} & \dots & D_{1,n+1} \\ \vdots & \vdots & & \vdots \\ C_n & D_{n1} & \dots & D_{n,n+1} \\ C_{n+1} & 0 & \dots & 0 \end{array}\right), \qquad C = \left(\begin{array}{c|c} A_c & B_c \\ \hline C_c & D_c \end{array}\right) \tag{6}$$

where the last row in the D matrix of $\mathcal{G}_o$ is zero so that the sample and hold operations make sense. Each M_{ij} has a certain kernel representation, say $M_{ij}(t,\tau)$ for $0 \le t, \tau < \infty$, so that for any $u \in \mathcal{L}^\infty$

$$(M_{ij}u)(t) = \int_0^\infty M_{ij}(t,\tau)u(\tau)d\tau \tag{7}$$

where $M_{ij}(t,\tau) = \hat{M}_{ij}(t,\tau) + \sum_{k=0}^{\infty} \hat{m}_{ij}^k(t)\delta(t - t_k - \tau)$ (see [4] for more details). Because M_{ij} is periodic, it follows that $M_{ij}(t,\tau) = M_{ij}(t + kT_s, \tau + kT_s)$ for $k = 0, 1, \ldots$.

To completely characterize the system, we only need to know

$$M_{ij}^k(t,\tau) := M_{ij}(t + kT_S, \tau)$$

where $M_{ij}^{(k)}(t,\tau)$ takes value in $[0,T_s]$ x $[0,T_s]$. We can now state the following theorem:

Theorem 4 *Let M in Fig. 3 correspond to that of the sampled-data system in Fig. 4. Then the given sampled-data system is robustly stable if and only if*

$$\max_{t_i < [0,T_s]} \rho \left(\begin{bmatrix} \|\beta_{11}(t_1)\|_1 & \cdots & \|\beta_{1n}(t_1)\|_1 \\ \vdots & & \vdots \\ \|\beta_{n1}(t_n)\|_1 & \cdots & \|\beta_{nn}(t_n)\|_1 \end{bmatrix} \right) < 1 \tag{8}$$

where $\beta_{ij}(t) = \{\|M_{ij}^o(t)\|_{\mathcal{A}}, \|M_{ij}^1(t)\|_{\mathcal{A}}, \cdots\} \in \ell^1$, and $\|M_{ij}^k(t)\|_{\mathcal{A}} := \int_0^{T_s} |M_{ij}(t,\tau)| d\tau$.

The proof of this theorem can be found in [10]. From the theorem statement it is clear that unless we can compute $\|\beta_{ij}(t_i)\|_1$, the theorem will not be of much practical use. To compute this norm, it is should be enough to find $\|M_{ij}^k(t)\|_{\mathcal{A}}$ where $k = 0, 1, \ldots, N$, for N sufficiently large to yield an arbitrarily accurate estimate of $\|\beta_{ij}\|_1$. The size of N can be determined a priori from information from $\mathcal{G}_o$ and C.

We can indeed compute $M_{ij}^k(t,\tau)$ of a sampled-data system using state space representation of $\mathcal{G}_o$ and C. It can be shown that $M_{ij}^o(t,\tau) = D_{ij}(t,\tau)$, and $M_{ij}^k(t,\tau) = \hat{C}_i(t)\hat{A}^{k-1}\hat{B}_j(\tau)$ where

$$\left(\begin{array}{c|c} \hat{A} & \hat{B}_j(\tau) \\ \hline \hat{C}_i(t) & \hat{D}_{ij}(t,\tau) \end{array} \right) =$$

$$\left(\begin{array}{cc|c} e^{AT_s} + \psi(T_s)B_{n+1}D_cC_{n+1} & \psi(T_s)B_{n+1}C_c & e^{A(T_s-\tau)B_j} \\ B_cC_{n+1} & A_c & 0 \\ \hline C_ie^{At} + \Gamma_i(t)D_cC_{n+1} & \Gamma(t)C_c & C_ie^{A(t-\tau)}1(t-\tau) \\ & & +D_{ij}\delta(t-\tau) \end{array} \right) \tag{9}$$

where $\psi(t) = \int_0^t e^{A\tau} d\tau$ and $\Gamma_i(t) = C_i\psi(t)B_{n+1} + D_{i,n+1}$.

References

[1] A. Berman and R. Blemmons, *Nonnegative Matrices in the Mathematical Sciences*, Academic Press, New York, 1979.

[2] M. A. Dahleh and Y. Ohta, "A Necessary and Sufficient Condition for Robust BIBO Stability," *Systems & Control letters* 11, 1988, pp. 271-275.

[3] M. A. Dahleh and J. B. Pearson, "ℓ^1 Optimal Feedback Controllers for MIMO Discrete Time Systems," *IEEE Transactions on Automatic Control*, Vol. AC-32, No. 4, April 1987, pp. 314-322.

[4] C. Desoer and M. Vidyasagar, *Feedback Systems: Input-Output Properties*, Academic Press, New York, 1975.

[5] J. C. Doyle, "Analysis of Feedback Systems with Structured Uncertainty," *IEEE Proceedings*, Vol. 129, PtD, No. 6, pp. 242-250, November, 1982.

[6] J. C. Doyle, J. E. Wall, and G. Stein, "Performance and Robustness Analysis for Structured Uncertainty," *Proceedings of the 20th IEEE Conference on Decision and Control*, 1982, pp. 629-636.

[7] B. Francis, *A Course in $\mathcal{H}^\infty$ Control Theory*, Springer Verlag, 1987.

[8] M. Khammash and J. B. Pearson, " Robustness Synthesis for Discrete-Time Systems with Structured Uncertainty", to be presented in 1991 ACC, Boston, Massachusetts.

[9] M. Khammash and J. B. Pearson, "Performance Robustness of Discrete-Time Systems with Structured Uncertainty", *IEEE Transactions on Automatic Control*, vol. AC-36, no. 4, pp. 398-412, 1991.

[10] M. Khammash, "Necessary and Sufficient Conditions for the Robustness of Time-Varying Systems with Applications to Sampled-Data Systems," *IEEE Transactions on Automatic Control*, submitted.

[11] J.S. McDonald and J.B. Pearson, "Constrained Optimal Control Using the ℓ^1 Norm," *Rice University Technical Report*, No. 8923, July 1989. (Submitted for publication)

[12] J. Shamma and M. Dahleh, "Time-Varying vs. Time-Invariant Compensation for Rejection of Persistent Bounded Disturbances and Robust Stabilization", *IEEE Transactions on Automatic Control*, to appear.

[13] R. Varga, *Matrix Iterative Analysis*, Prentice Hall, 1963.

A STUDY ON A VARIABLE ADAPTIVE LAW

Seiichi SHIN
Institute of information Sciences and Electronics, University of Tsukuba
1-1-1 Ten-nodai, Tsukuba, 305 JAPAN

Toshiyuki KITAMORI
Department of Mathematical Engineering and Information Physics
Faculty of Engineering, University of Tokyo
7-3-1 Hongo, Bunkyo-ku, Tokyo, 113 JAPAN

ABSTRACT

This paper presents a variable adaptive law, which is robust to bounded disturbances. Combining with the dead zone, the proposing adaptive law uses a sort of the σ-modified adaptive law in the small estimation error region. Viability of the adaptive law, in comparing with the dead zone type adaptive law, is shown by theoretical analysis and a simple numerical simulation.

1. INTRODUCTION

Robust adaptive control systems are now widely studied.[1] Adaptive control system subject to bounded noises is one of these robust adaptive control systems. The major adaptive laws used in a such adaptive control system are the dead zone type adaptive law[2] and the σ-modified adaptive law.[3] The former law needs an upper bound of a norm of the bounded disturbance. This bound is sometimes set to be too conservative, so that the estimation error remains large. Although the σ-modified adaptive law does not need such a bound, the leakage term in the law hinders convergence of the estimation error in a small region. Moreover, it causes a bursting phenomena.

This study proposes a combination of these two adaptive laws in order to achieve good performance in control. The basic deterministic adaptive law is used in the large error region similar to the dead zone type and the σ-modified adaptive law is used in the small error region, where the adaptation is stopped in the dead zone type. Therefore, the proposing adaptive law has a switching property from the deterministic to the σ-modified adaptive laws dependent on the magnitude of the estimation error. The next section describes precious descriptions of the adaptive control system considered here and the proposing adaptive law. Basic property of the adaptive law is analyzed theoretically in the section 3 and a simple numerical simulation is presented in section 4. In these sections, the viability of the proposing adaptive law is shown comparing with the dead zone type.

2. ADAPTIVE SYSTEM

According to 2), the controlled object considered here is a controllable n-th order linear system described by,

$$\dot{x} = A_p x + bu + w \tag{1}$$

where $x \in \mathrm{R}^n$, $u \in \mathrm{R}$, and $w \in \mathrm{R}^n$ respectively are the state, the input, and the disturbance. An upper bound w_0 of $\|w\|$ is known, where $\|x\|$ is the Euclid norm when x is a vector and an induced norm from the Euclid norm when x is a matrix. $A_p \in \mathrm{R}^{nxn}$ is unknown and $b \in \mathrm{R}^n$ is known here.

The objective is the model following. Therefore, the problem considered here is how to generate the control input u so as to make the state x follow x_m, which is the state of the reference model denoted by,

$$\dot{x}_m = A_m x_m + br \tag{2}$$

where, r is a bounded command input, $A_m \in \mathrm{R}^{nxn}$ is a stable matrix. Then, there exists a positive definite matrix $P \in \mathrm{R}^{nxn}$, which satisfies,

$$A_m^T P + P A_m = -Q \tag{3}$$

for arbitrary positive definite matrix $Q \in \mathrm{R}^{nxn}$, where ^{T} means the transpose of matrices or vectors. We further assume that there exists a finite $k^* \in \mathrm{R}^n$, which satisfies,

$$A_p + bk^{*T} = A_m \tag{4}$$

This is called as the model matching condition.

Now, the control input is set to be,

$$u = r + \hat{k}^T x \tag{5}$$

where, $\hat{k} \in \mathrm{R}^n$ is an adjustable gain, which is specified in the later. Let the estimation error $e \in \mathrm{R}^n$ and the parameter error $k \in \mathrm{R}^n$ be,

$$e = x - x_m \tag{6}$$

$$k = \hat{k} - k^* \tag{7}$$

Then, from (1)-(7), we get an error system as follows,

$$\dot{e} = A_m e + bk^T x + w \tag{8}$$

Peterson and Narendra[2)] proposed an adaptive control system combining (8) and the dead zone type adaptive law;

$$\dot{k} = -e^T P b x, \qquad if\ e^T P e > e_0 \tag{9}$$

$$\dot{k} = 0 \qquad if\ e^T P e \leq e_0 \tag{10}$$

where,

$$e_0 = 4\alpha^2 \lambda_M^3(P) \lambda_m^{-2}(Q) w_0^2 \tag{11}$$

and $\lambda_M(P)$ and $\lambda_m(Q)$, respectively are the maximum and minimum eigenvalues of the matrix Q. The positive real number α (> 1) is a safety margin. It should be set to be large if the upper bound w_0 is uncertain, when the threshold e_0 becomes large and the control quality is decreased. We call the error e is in the region S_1 if $e^T P e > e_0$, and in the region S_2 if $e^T P e \leq e_0$

With the adaptive law, (9)-(10), Peterson and Narendra[2)] showed convergence of the error in S_2 .However, when the safety margin is large, the error remains large since the adaptive law is terminated in S_2. Instead of (10), we use the σ-modeified adaptive law;

$$\dot{\hat{k}} = -\sigma\hat{k} - e^T P b x + \sigma k_j, \qquad if\ e^T P e \leq e_0 \tag{12}$$

where, σ is a positive real number, which is specified in the later. The vector $k_j \in R^n$ is an value of $\hat{k}$ at the instance when the error crosses the boundary from S_1 to S_2 and the suffix j means number of crossing the boundary from S_1 to S_2. Therefore, j is set to be zero initially when $t=0$, and is increased at $t=t_j$, when the error crosses the boundary from S_1 to S_2. The initial value k_0 is the same of the initial value of $\hat{k}$.

With the parameter error k, (12) can be rewritten as,

$$\dot{k} = -\sigma k - e^T P b x + \sigma(k_j - k^*), \qquad if\ e^T P e \leq e_0 \tag{12'}$$

After all, we use the adaptive law (9) when error of (8) is in S_1 and the adaptive law (12') when error in S_2. Equation (12') means parameter error k becomes very small when the term $e^T P b x$ is small and k_j is close to k^*. We analyze this property in the next section.

3. ANALYSIS

Let a non-negative definite function v be,

$$v = e^T P e + k^T k \tag{13}$$

From (9) and (11),

$$\|e\| > 2\alpha\lambda_M(P)\lambda_M^{-1}(Q)w_0 \tag{14}$$

in S_1. Then, the same analysis done in 2) leads to

$$\dot{v} = -e^T Q e + 2e^T P w \leq -\lambda_m(Q)\|e\|^2 + 2\lambda_M(P)\|e\|w_0$$
$$< -4\alpha(\alpha-1)\lambda_M^2(P)\lambda_M^{-1}(Q)w_0^2 < 0 \tag{15}$$

in S_1. Since v is monotonically decreased in S_1, The error e goes to the region S_2 in a finite time.

On the other hand, the definition of S_2 leads to,

$$\|e\| < 2\alpha\lambda_M^{2/3}(P)\lambda_m^{-1}(Q)w_0 \tag{16}$$

in S_2. Furthermore, from the definition of v and k_j, we get,

$$v_j = e_0 + \|k^* - k_j\|^2 \tag{17}$$

where v_j means the instance value of v when $t=t_j$, that is, e goes into S_2. Therefore,

$$\dot{v} = -e^T Q e + 2e^T P w - 2\sigma k^T k + 2\sigma k^T(k_j - k^*)$$
$$\leq -\lambda_m(Q)\|e\|^2 + 2\lambda_M(P)\|e\|w_0 - \sigma\|k\|^2 + \sigma\|k_j - k^*\|^2$$
$$\leq -\delta(\lambda_M(P)\|e\|^2 + \|k\|^2) + \sigma v_j - 4\alpha\lambda_M^{5/2}(P)\lambda_m^{1/2}(P)\lambda_m^{-2}(Q)w_0^2(\sigma\alpha\lambda_M^{1/2}(P) - \lambda_m(Q))$$
$$\leq -\delta v + \sigma v_j - 4\alpha\lambda_M^{5/2}(P)\lambda_m^{1/2}(P)\lambda_m^{-2}(Q)w_0^2(\sigma\alpha\lambda_M^{1/2}(P) - \lambda_m(Q)) \tag{18}$$

where,

$$\delta = min(\lambda_m(Q)\lambda_M^{-1}(P), \sigma) \tag{19}$$

From (18), if,

$$\delta = \sigma \tag{20}$$

and,

$$\sigma\alpha\lambda_M^{1/2}(P)-\lambda_m(Q)>0 \tag{21}$$

then v is decreased and be less than v_j after $t=t_j$, while e remains S_2.Therefore, a sufficient condition for decreasing v in S_2 is,

$$\lambda_m(Q)(\alpha\lambda_M^{1/2}(P)\lambda_m(P))^{-1}<\sigma\leq\lambda_m(Q)\lambda_M^{-1}(P) \tag{22}$$

In order to ensure existence of σ, which satisfy the inequality (22),

$$\alpha>\lambda_M^{1/2}(P)\lambda_m^{-1/2}(P) \tag{23}$$

Therefore, we can set σ satisfies (22) when safety margin α is larger than $\lambda_M^{1/2}(P)\lambda_m^{-1/2}(P)$. In this case, v is also decreased in S_2 while it remains constant value v_j in S_2 if the adaptive law (10) is used in the region S_2. This improvement is achieved by using error information in S_2 unlike the adaptive law (10), where no adaptation is performed in S_2.It is pointed out that the parameter σ depends only on P and Q as shown in (22) and does not depend on any unknown parameter of the controlled object (1). Namely, if σ is set to be its upper limit as follows,

$$\sigma=\lambda_m(Q)\lambda_M^{-1}(P) \tag{24}$$

from (18), v is decreased as,

$$v\leq v_j-4\alpha\lambda_M^{7/2}(P)\lambda_m^{-1/2}(P)\lambda_m^{-2}(Q)w_0^2(\alpha\lambda_M^{1/2}(P)\lambda_m^{1/2}(P)-\lambda_M(P)) \tag{25}$$

The analysis performed here shows that non-negative definite function v is decreased monotonically in S_1 and it becomes less than v_j in S_2. Therefore, v is bounded at all time, so that all variables remain bounded, that is, the proposed adaptive system is stable.

4. NUMERICAL EXAMPLE

We show here a simple numerical simulation result in order to illustrate the property analyzed in the former section. The controlled object and the reference model used here are,

$$A_p=\begin{bmatrix}0&2\\2&1\end{bmatrix}, A_m=\begin{bmatrix}-1&0\\0&-2\end{bmatrix}$$

$$b=[1\ \ 1]^T \tag{26}$$

Therefore,

$$k^*=[-1\ \ -2]^T \tag{27}$$

Matrices P and Q are,

$$P=\begin{bmatrix}1&0\\0&1\end{bmatrix}, Q=\begin{bmatrix}2&0\\0&4\end{bmatrix} \tag{28}$$

Then,

$$\lambda_M(P)=\lambda_m(P)=1,\ \lambda_M(Q)=4,\ \lambda_m(Q)=2 \tag{29}$$

The disturbance is,

$$w=[\sin(2\pi t/5)\ \ \sin(2\pi t/5)]^T \tag{30}$$

Therefore, w_0 becomes $\sqrt{2}$. The command signal is set to be,

$$r=10\sin(2\pi t/5) \tag{31}$$

From these settings ,the same as those of 2), we get,

$$e_0=4\alpha^2\lambda_M^3(P)\lambda_m^{-2}(Q)w_0^2=2\alpha^2 \tag{32}$$

and,

$$2/\alpha < \sigma \leq 2,\ \alpha \geq 1 \tag{33}$$

The equation (33) corresponds to (22) and (23). Here, we set σ as 2. This corresponds to (24). The initial values of the controlled object, the reference model, and estimate $\hat{k}$ are all set to be zero.

First, we use the adaptive law (9) and (10), which are proposed in 2). The results are Fig. 1 and Fig. 2. It is shown that the function v is decreased in the region S_1 and the adaptation is stopped in S_2. Figures 3 and 4 are the results with (9) and (12), which is the proposing adaptive law. The function v is decreased both in S_1 and in S_2 and goes to a small limit cycle. The control quality in Fig. 4 is improved comparing with Fig. 2.

5. CONCLUSION

We analyzed here that the novel robust adaptive law, which is a combination of the dead zone and σ-modification, improves the control quality.

REFERENCE

1) K. S. Narendra and A. M. Annaswamy, *Stable Adaptive Systems*, Prentice Hall, 1989
2) B. B. Peterson and K. S. Narendra: Bounded error adaptive control, *IEEE Trans. Automatic Control*, vol. AC-27, no. 6, pp. 1161-1168 (1982)
3) P. A. Ioannou and P. V. Kokotovic: Instability analysis and improvement of adaptive control, *Automatica*, vol. 20, no. 5, pp. 583-594 (1984)

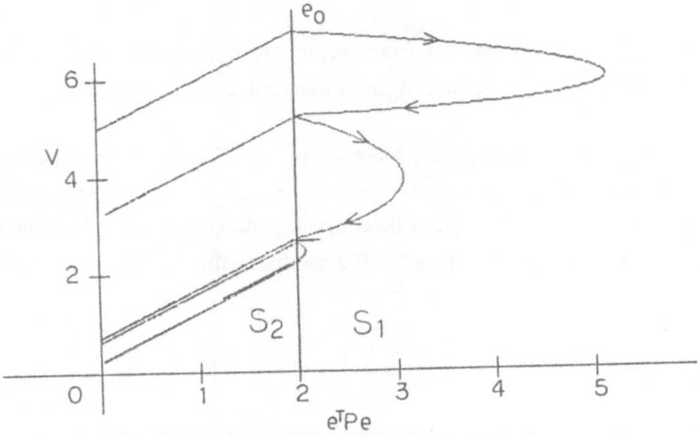

Figure 1. Lyapunov function *v* versus the control error e^TPe with the dead zone type adaptive law.

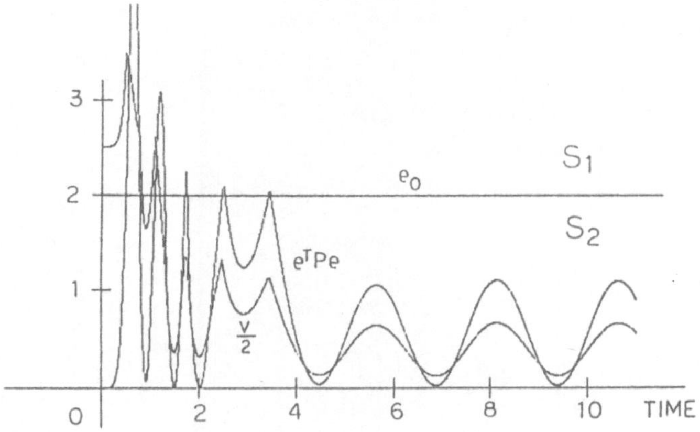

Figure 2. Time responses of Lyapunov function v and the control error e^TPe with the dead zone type adaptive law.

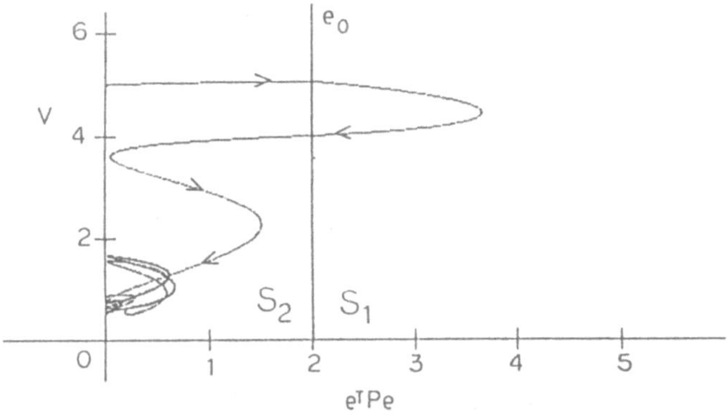

Figure 3. Lyapunov function v versus the control error e^TPe with the the proposing adaptive law.

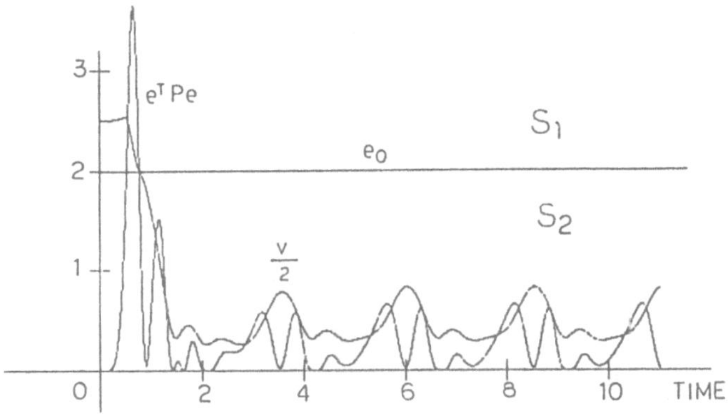

Figure 4. Time responses of Lyapunov function v and the control error e^TPe with the proposing adaptive law.

MODEL REFERENCE ROBUST CONTROL[1]

Minyue Fu
Department of Electrical and Computer Engineering
University of Newcastle, N.S.W. 2308 Australia

Abstract

Classical model reference adaptive control schemes require the following assumptions on the plant: A1) minimum phase; A2) known upper bound of the plant order; A3) known relative degree; and A4) known sign of high frequency gain. It is well-known that the robustness of the adaptive systems is a potential problem, and it requires many sophisticated techniques to fix it. In this paper, we consider the same model reference control problem via robust control. By further assuming that the boundedness of the parameter uncertainties of the plant (which is a very weak assumption), we show that a linear time-invariant dynamic output feedback controller can be constructed to give the following property: the closed-loop system is internally stable and its transfer function is arbitrarily close to the reference model. This method provides simple controllers and good robustness. It also has potential to cope with large size fast time-varying uncertainties.

1 Introduction

Both adaptive control theory and robust control theory have been developed to accommodate a wide range of uncertainties. Although experts have not agreed on the distinction between these two theories, one of their important differences is that adaptive controllers are nonlinear and time-varying (NLTV) while robust controllers are *usually* linear and time-invariant (LTI). Despite of the overwhelming progress made recently on the robust control theory, we find ourselves constantly "bothered" with the following fundamental question:

Q1. Can an LTI controller really compete with an NLTV controller?

[1]This work is supported by the Australian Research Council.

A positive answer to the above question has been given to the important case of quardratic stabilization of linear systems which are subject to a type of norm-bounded uncertainty. It is shown in this case that NLTV controllers offer no advantage over their LTI partners for the (see, e.g., [1]). However, since the adaptive control is often concerned with performance (such as model matching) and structured uncertainty in the parameter (coefficient) space, the above result does not apply and the question Q1 still deserves serious attention.

In this paper, we consider the problem of model reference control of uncertain linear systems and address the question Q1 from a different perspective:

Q2. Under what conditions can we find an LTI controller such that the closed-loop system is stable and "matches" a given reference model in certain sense for all admissible uncertainties?

Note that this is exactly the question asked by the adaptive control theory except that an NLTV controller is allowed there. It is well known that all the classical model reference adaptive control (MRAC) schemes (developed prior to the 1980's) require the following standard assumptions:

A1. The plant is of minimum phase;

A2. The upper bound of the plant order is known;

A3. The relative degree of the plant is known;

A4. The sign of the high-frequency gain is known;

B1. The reference model has the same relative degree as the plant; and

B2. The reference input is bounded and piecewise continuous.

With these assumptions, the MRAC schemes can guarantee that the closed-loop system becomes stable and converges to the reference model. However, the robustness of the classical adaptive schemes is known to be a serious problem, and it requires sophisticated techniques to fix it (see, e.g., [2]). Moreover, besides the complexity problem of the "robustified" adaptive controllers, the degree of robustness is often small, and additional information on the system (such as the size of the unmodeled dynamics and a bound on the input disturbance) is usually required.

The focal point of this paper is to answer the question Q2 by showing that a result similar to that given by MRAC can be achieved by a model reference robust control (MRRC) technique, provided there are some additional very mild assumptions. More precisely, suppose assumptions A1)-A4) and the following additional ones are satisfied:

A5. The set of admissible uncertain parameters is compact and known;

A6. The transfer function of the reference model is stable and strictly proper.

Then, we provide a technique for constructing a simple LTI controller which guarantees the robust stability of the closed-loop system and that the H_∞ norm of the difference in the transfer functions of the closed-loop system and the reference model can be made to be arbitrarily small. Several advantages of this MRRC method are obvious: an LTI controller allows simple analysis of the system performance and robust stability; it provides easy implementation and simulation; furthermore, robustness margins for input and output disturbances and additional unstructured perturbations can be computed by using the sensitivity and complementary sensitivity of the closed-loop system.

The result explained above is achieved by using feedback which possibly involves high gains. But we find in simulations that feedback gains are usually moderate when the plant uncertainties are not large and the requirement on model matching is not severe. The feedback gains need to be high when the plant is subject to large uncertainties and/or the plant is very different from the reference model.

Having established the MRRC method for time-invariant uncertainty, we look into the problem of time-varying uncertainty which the adaptive control theory has difficulty with. We find that the MRRC approach may also be suitable for accommodating a certain class of time-varying uncertainties. This point will be made via some discussions and a conjecture.

The endeavor of this paper should not be interpreted as an de-emphasis of adaptive control, it should rather be viewed as an attempt to have a better understanding of both the adaptive and robust control theories and as an exercise in our course of searching for better adaptive control schemes. More investigation on this subject is needed.

2 Related Work on Robust Control

There are three robust control design methods which are most pertinent to our MRRC technique.

The quantitative feedback theory (QFT) by Horowitz and his colleagues [3] provides a first systematic procedure for designing robust controllers. This method uses a two-degree-of-freedom controller to achieve desired frequency response and the stability margin of the closed-loop system. More specifically, a loop compensator is used to reduce uncertainty and assuring closed-loop stability while a prefilter is used to shape the closed-loop input-output transfer function. This design method can be regarded as a model reference control method. However, the reference model is not given in terms of a transfer function, and the phase of the closed-loop input-output transfer function is not paid attention to. The QFT method is somewhat empirical because trials and errors are needed for a proper design. There is also a concern about the effectiveness of this design method for multi-input-multi-output systems.

Barmish and Wei [4] employs a unity-feedback controller for robust stabilization of an uncertain linear time-invariant plant satisfying assumptions A1-A4 and some mild conditions. They show that an LTI stable and minimum-phase stabilizer can always be constructed for the uncertain plant. However, the model reference control problem is not treated in [4] because only one degree of freedom is used.

The problem of model reference robust control was recently studied by Sun, Olbrot and Polis [5]. They consider linear single-input-single-output plants satisfying Assumptions similar to A1-A6, and use the so-called "modeling error compensation" technique to show that a stabilizing controller can be constructed such that the closed-loop input-output transfer function is made to be arbitrarily close to the reference model over a finite bandwidth. However, the construction and analysis of the controller seems very complicated. Furthermore, model matching is done only on a finite bandwidth, this restriction, as we shall see, can be lifted.

3 A New Approach to MRRC

In this section, we use a two-degree-of-freedom controller to solve the MRRC problem for plants and reference models satisfying assumptions A1-A6. The

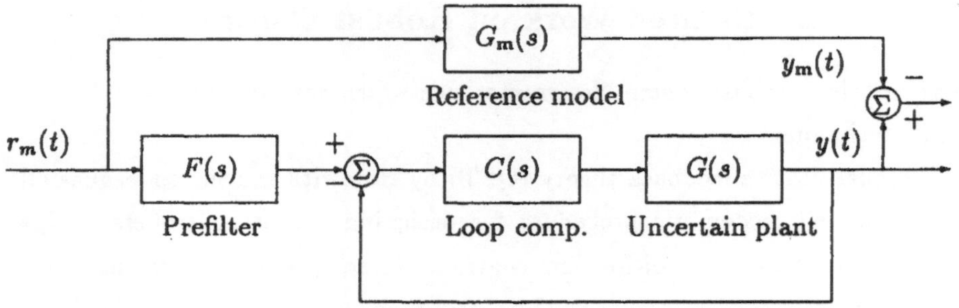

Figure 1: Model Reference Robust Control

schematic diagram of the closed-loop system is shown in Figure 1. The transfer function $G(s)$ of the plant takes the following form:

$$G(s) = \frac{N(s)}{D(s)} = \frac{\sum_{i=0}^{m} b_i s^i}{s^n + \sum_{i=0}^{n-1} a_i b_i}, \quad b_m \neq 0 \,. \tag{1}$$

The nominal model of the plant is denoted by $G_0(s)$, and it can be chosen arbitrarily although its choice might effect the controller. The closed-loop input-output transfer function will be denoted by $G_c(s)$, i.e.,

$$G_c(s) = \frac{F(s)C(s)G(s)}{1 + C(s)G(s)} \quad . \tag{2}$$

The main result is presented as follows:

Theorem 1. *Consider an SISO uncertain linear plant $G(s)$ and a linear reference model $G_m(s)$ satisfying assumptions A1-A6. Then, given any (arbitrarily small) $\varepsilon > 0$, there exist a stable transfer function $F(s)$ and a stable and minimum-phase transfer function $C(s)$ such that the closed-loop system given in Figure 1 is robustly stable and that the difference between the transfer functions of the closed-loop system and the reference model has H_∞ norm less than ε.*

To assist this theorem, we provide a simple algorithm for constructing the controller. The design of $C(s)$ is simplified from the robust stabilization technique in [4], and that of $F(s)$ is motivated by the QFT [3].

Construction of $C(s)$ and $F(s)$:

- Choose $N_c(s)$ to be any (n-m-1)th stable polynomial;

- Choose $1/D_c(s,\sigma)$ to be any parameterized stable unity gain low-pass filter (parameterized in $\sigma > 0$) with relative degree no less than n-m-1, and and the cutoff frequency $\to \infty$ as $\sigma \to 0$. For example, take

$$1/D_c(s) = \frac{1}{(\sigma s + 1)^{n-m-1}} \; ; \tag{3}$$

- Take

$$C(s) = K \frac{N_c(s)}{D_c(s,\sigma)} \tag{4}$$

$$F(s) = L(s,\rho) G_m(s) \left[\frac{C(s)G_0(s)}{1 + C(s)G_0(s)} \right]^{-1} \tag{5}$$

where $L(s,\rho)$ is either a unity gain or a parameterized unit-gain low-pass filter with cutoff frequency $\to \infty$ when $\rho \to 0$. K, σ and ρ are design parameters to be tuned;

- Choose K sufficiently large and σ and ρ sufficiently small (which are guaranteed to exist) such that

$$||G_c(s) - G_m(s)||_\infty < \varepsilon \tag{6}$$

for all admissible $G(s)$ satisfying assumptions A1)-A5).

Remark 1. The purpose of the stable polynomial $N_c(s)$ is to assure that $N_c(s)G(s)$ is a minimum phase transfer function with relative degree equal to one so that with $C(s) = KN_c(s)$ and sufficiently large K, the characteristic equation of the closed-loop system is robustly stable and its sensitivity function is sufficiently small. The reason for $1/D_c(s,\sigma)$ to be a low-pass filter of the required form is to guarantee the properness of $C(s)$ while preserving the above property of the closed-loop system when σ is sufficiently small. The function $F(s)$ is used to shape the closed-loop transfer function so that it approximates the reference model. The low-pass filter $L(s,\rho)$ is optional, only for reducing the unnecessary high-frequency gain of the closed-loop system. These points should be considered in tuning the controller. The proof of Theorem 1 is omitted, for the construction of the controller and this remark have explained the validity of the result.

4 Beyond Time-invariant Parameter Uncertainty

The purpose of this section is to look into the possibility of using the MRRC technique to deal with time-varying parameter uncertainty. In certain sense, the discussions in this section might be "the center of gravity" of the paper.

As we pointed out earlier, the adaptive control theory is applicable only to uncertainties which are time-invariant (or varying "sufficiently slowly"). The robust control methods mentioned in Section 2 and the MRRC technique in Section 3 are in general also restricted to time-invariant uncertainties due to their dependency on frequency domain analysis. However, the fact the MRRC technique requires only a simple LTI controller seems to suggest that it might be able to tolerate time-varying uncertainty better the adaptive schemes. This intuition is supported by a number of simulations carried out by the author. Motivated by these simulations, it is suspected by the author that a result similar to Theorem 1 in Section 3 also holds for plants with time-varying uncertainties.

To be more precise, we consider an uncertain plant described by the following input-output differential equation:

$$y^{(n)} + \sum_{i=0}^{n-1} a_i(t) y^{(i)} = \sum_{i=0}^{m} b_i(t) u^{(i)} \tag{7}$$

where $u(t)$ and $y(t)$ are the input and output of the plant, respectively. It is assumed that the assumptions A1-A6 hold. Here, the minimum-phase should be interpreted as that the zero dynamics of the plant:

$$\sum_{i=0}^{m} b_i(t) u^{(i)} = 0 \tag{8}$$

is "stable" in certain sense. Then, the following conjecture is put forward:

Conjecture: *Consider the uncertain plant in (7) and reference model $G_m(s)$ satisfying assumptions A1-A6. Then, under some additional mild assumptions and an appropriate definition for stability, there exists a linear time-invariant controller such that the closed-loop system is stable and the induced norm of the operator from $r_m(t)$ to $y(t) - y_m(t)$ can be made arbitrarily small.*

Possible additional assumptions might be that the coefficients $a_i(t)$ and $b_i(t)$ are differentiable and their derivatives are bounded. And investigation is needed for different types of stability such as Lyapunov, bounded-input-bounded-output, and exponential. Since time-varying systems are considered, frequency domain analysis is no longer valid. Therefore, new techniques need to be developed for proving or disproving the conjecture. A promising avenue is to apply the singular perturbation theory, this will be studied.

5 Conclusion

In this paper, we have demonstrated the potentials of an MRRC technique. This technique provides a non-H_∞ method for designing a low order robust LTI controller to solve the model reference problem, and it is able to handle large size parametric uncertainties. An important feature of this technique is its potential of handling large-size fast time-varying parametric uncertainties, this is a matter which deserves further research.

It is, however, noted that the MRRC technique potentially requires a high-gain feedback controller, when the size of uncertainties is large and/or the plant is vastly different from the reference model. The tradeoffs between the MRAC and MRRC techniques need to be further investigated.

REFERENCES

[1] M. A. Rotea and P. P. Khargonekar, "Stabilizability of linear time-varying and uncertain linear systems," *IEEE Trans. Auto. Contr.*, vol. 33, no. 9, pp. 884–887, 1988.

[2] R. Ortega and Y. Tang, "Robustness of adaptive controllers–a survey," *Automatica*, vol. 25, no. 5, pp. 651–677, 1989.

[3] I. Horowitz, "Quantitative feedback theory," *Proc. IEE, Part-D*, vol. 129, pp. 215–226, 1982.

[4] B. R. Barmish and K. H. Wei, "Simultaneous stabilizability of single input single output systems," in *Modelling, Identification and Robust Control*, (North-Holland, Amsterdam), 1986.

[5] J. Sun, A. W. Olbrot, and M. P. Polis, "Robust stabilization and robust performance using model reference control and modelling error compensation," in *Proc. International Workshop on Robust Control*, (San Antonio, Texas), March 1991.

PARAMETRIZATION OF H_∞ FI CONTROLLER AND H_∞/H_2 STATE FEEDBACK CONTROL

Tsutomu Mita and Kang-Zhi Liu
Department of Electrical and Electronics Engineering
Chiba University 1-33, Yayoi-cho, 260, Chiba

ABSTRACT

In this paper, we will derive complete parametrization of the H_∞ FI controller and that of the H_∞ state feedback (SF) controller. And we will determine the obtained dynamical free parameters so that they minimizes an H_2 control performance and will show the condition when the whole controller becomes a constant feedback gain other than the central solution F_∞.

1. Introduction

We have several approaches to solve the standard H_∞ control problem (Kimura et al.[1], Doyle et al.[2]) and many related results have been obtained, e.g., the H_2/H_∞ control was studied by Bernsteain et al.[3], Doyle et al.[4] and Rotea et al.[5] and special constant gains of the state feedback H_∞ control were obtained by Perterson [6] and Zhou et al. [7]. However little is known about the capability of the free parameters of the H_∞ controller. We can expect that these free parameters are used to improve the time response.

In this paper we first correct the parametrization of the H_∞ controller of the FI (Full Information) problem derived by Doyle et al. [2] since they have neglected the parametrization of a null space of $G_{21}(s)$ of the generalized plant. Then using the corrected result we will derive the parametrization of the H_∞ state feedback (SF) controller.

Using the obtained parametrization of SF controller, we will first characterize all SF controllers which are reduced to constant gains and we will secondly determine these free parameters so that they minimize an H_2 control performance.

We will omit the argument s whenever it is clear from the contexts and will adopt the following notations:

- $HM\{Z,Q\}=(Z_{11}Q+Z_{12})(Z_{21}Q+Z_{22})^{-1}$, $LF\{Z,Q\}=Z_{11}+Z_{12}Q(I-Z_{22}Q)^{-1}Z_{21}$ for $Z=\{Z_{ij}\}:i,j=1\sim2$.
- RH_2: a strictly proper RH_∞ function

  BH_2: an RH_2 function of which H_∞ norm is strictly less 1.

2. Parametrization of FI controller and SF controller

The generalized plant is given by

$$\dot{x}=Ax+B_1w+B_2u \tag{1a}$$

$$z=C_1x+D_{12}u \tag{1b}$$

$$y=x \tag{1c}$$

where $x\in R^n$, $u\in R^p$, $w\in R^r$ and $z\in R^m$ are the state variable, the control input, the disturbance input and the controlled output, respectively. We assume that both (A,B_2) and (A,B_1) are stabilizable, (C_1,A) is detectable and B_1 is of column full rank as well as the following orthogonal condition

$$C_1^TD_{12}=0,\quad D_{12}^TD_{12}=I_p \tag{2}$$

We define the closed loop system as

$$z=G(s)w \tag{3}$$

when a dynamical state feedback control

$$u=K(s)x \tag{4}$$

is applied to (1) where K(s) is a proper rational controller. Then the main problem is to parametrize K(s) which satisfy the internal stability as well as

$$\|G\|_\infty<1 \tag{5}$$

We will call this problem the SF (State Feedback) problem.

To derive the parametrization of K(s) of SF problem we introduce the FI (Full Information) problem [2] in which output equation (1c) and the control law (4) are replaced by

$$y_I=\begin{bmatrix}I_n\\0\end{bmatrix}x+\begin{bmatrix}0\\I_r\end{bmatrix}w \tag{6}$$

and

$$u=K_I(s)y_I=[K_x(s),K_w(s)]\begin{bmatrix}x\\w\end{bmatrix} \tag{7}$$

respectively. The problem is to find a proper rational $K_I(s)$ that assures internal stability and satisfies $\|G_I(s)\|_\infty<1$ where $G_I(s)$ is the closed loop transfer function between w and z when (7) is applied to (1).

We can parametrize all internally stabilizing controller using LF (linear fractional transformation) form.

Lemma 1. If we choose an F stabilizing $A+B_2F$ and define

$$\Phi(s)=sI-(A+B_2F) \tag{8}$$

then we will have the following results.

(I) All controllers K(s) which stabilize (1) are represented by

$$K(s)=LF\{\Sigma,Q\} \tag{9a}$$

where

$$\Sigma=\begin{bmatrix}F, & I_p\\-I_n, & \Phi^{-1}B_2\end{bmatrix} \tag{9b}$$

and Q is a free parameter matrix that satisfies $Q(s)\in RH_\infty$.

(<u>II</u>) All controllers $K_I(s)$ which stabilize (1) are represented by

$$K_I(s)=LF\{\Sigma_I,(Q_x,Q_w)\} \tag{10a}$$

where

$$\Sigma_I=\begin{bmatrix} (F,\ 0), & I_p \\ \begin{bmatrix} -I_n, & 0 \\ 0 & ,-I_r \end{bmatrix}, & \begin{bmatrix} \Phi^{-1}B_2 \\ 0 \end{bmatrix} \end{bmatrix} \tag{10b}$$

and Q_x and Q_w are free parameters that satisfy $Q_x(s),\ Q_w(s)\in RH\infty$.

<u>Proof</u>: We will set $H=B_2F$ in Youla's parametrization for SF case and $C_2=(I,0)^T, D_{21}=(0,I)^T$ and $H=(B_2F,0)$ for FI case [11].

Then every SF controller assuring internal stability is represented by [11]

$$K(s)=HM\{Z,R\} \quad : R(s)\in RH\infty \tag{11a}$$

where

$$Z=\begin{bmatrix} M_2, & -Y_2 \\ N_2, & -X_2 \end{bmatrix}=\begin{bmatrix} I+F\Phi^{-1}B_2, & F\Phi^{-1}B_2F \\ \Phi^{-1}B_2 & ,-I+\Phi^{-1}B_2F \end{bmatrix} \tag{11b}$$

Motivated by the fact that $K(s)=F$ when $R(s)=-F$, we define

$$R(s)=-F+Q(s) \quad :Q(s)\in RH\infty \tag{11c}$$

Then substituting this $R(s)$ into (11a) and transforming the result to LF form to get (9). For FI case, $K_I(s)$ can be represented by

$$K_I(s)=HM\{Z_I,(R_x,R_w)\} : R_x,\ R_w\in RH\infty \tag{12a}$$

where

$$Z_I=\begin{bmatrix} M_2 & ,\ (-Y_2,\ 0) \\ \begin{bmatrix} N_2 \\ 0 \end{bmatrix}, & \begin{bmatrix} -X_2, & 0 \\ 0 & ,-I \end{bmatrix} \end{bmatrix} \tag{12b}$$

Then defining

$$(R_x,R_w)=(-F,0)+(Q_x,\ R_w):Q_x,\ Q_w\in RH\infty \tag{12c}$$

to get (10) by the same reason. <u>Q.E.D</u>

Since we assume that the $H\infty$ control problem is solvable in this paper, we introduce

$$F\infty=-B_2{}^TX \tag{13a}$$

where X is a stabilizing positive semi-definite solution of

$$XA+A^TX+X(B_1B_1{}^T-B_2B_2{}^T)X+C_1{}^TC_1=0 \tag{13b}$$

and replace all the F in Lemma 1 by $F\infty$ so that

$$F=F\infty, \quad \Phi(s)=sI-(A+B_2F\infty) \tag{14}$$

Then all solutions of the FI problem can be derived as follows.

<u>Theorem 1</u> (<u>FI problem</u>): All $K_I(s)$ satisfying $\|G_I(s)\|\infty<1$ as well as the internal stability are represented by

$$K_I(s)=LF\{\Sigma_I,(Q_x,\ Q_w)\} \tag{15a}$$

In this expression, Q_x and Q_w are given by

$$Q_x(s)=[I+N(s)B_1{}^TX\Phi^{-1}(s)B_2]^{-1}[N(s)B_1{}^TX-W_x(s)] \tag{15b}$$

$$Q_w(s)=[I+N(s)B_1{}^TX\Phi^{-1}(s)B_2]^{-1}[-N(s)-W_w(s)] \tag{15c}$$

where N is a free parameter matrix satisfying $N(s)\in BH\infty$ and W_x, W_w also free parameters satisfying

$$W_x(s)\Phi^{-1}(s)B_1+W_w(s)=0 \;:\; W_x(s),\ W_w(s)\in RH_\infty \tag{15d}$$

Proof: Since $A+B_2F_\infty$ is stable, it follows from (12) and Lemma 1 that all controllers assuring the internal stability can be written as

$$K_I(s)=HM\{Z_I,(-F_\infty+Q_x,Q_w)\} \tag{16a}$$

$$=LF\{\Sigma_I,(Q_x,Q_w)\}=(F_\infty,0)-(I-Q_x\Phi^{-1}B_2)^{-1}(Q_x,Q_w) \tag{16b}$$

Using (16a) we can transform G_I into a model matching form [11]

$$G_I(s)=T_1-T_2QT_3 \tag{17a}$$

where

$$Q=[-F_\infty+Q_x,Q_w],\quad T_3=\begin{pmatrix}\Phi^{-1}(s)B_1\\ I_r\end{pmatrix} \tag{17b}$$

DGKF [2] derived a solution of the FI problem in their item FI.5 as follows

$$K_I(s)=[F_\infty-N(s)B_1{}^TX,\ N(s)] \quad :N(s)\in BH_\infty \tag{18}$$

Equating (16b) and (18) to get

$$Q_x=(I+NB_1{}^TX\Phi^{-1}B_2)^{-1}NB_1{}^TX\ ,\quad Q_w=-(I+NB_1{}^TX\Phi^{-1}B_2)^{-1}N \tag{19}$$

However (19) does not reflect Q(s) orthogonal to T_3 since $N\equiv0$ is obtained from

$$(Q_x,Q_w)T_3=(I+NB_1{}^TX\Phi^{-1}B_2)^{-1}N(I_n-B^T{}_1X\Phi^{-1}B_1)\equiv0 \tag{20}$$

Therefore we need additional free parameters satisfying

$$(Q_x,Q_w)T_3\equiv0\ :(Q_x,\ Q_w)\in RH_\infty \tag{21}$$

for a complete parametrization of $K_I(s)$. These free parameters are introduced as W_x, $W_w\in RH_\infty$ satisfying (15d). But since

$$(I+NB_1{}^TX\Phi^{-1}B_2)^{-1} \tag{22}$$

is unimodular for all $N\in BH_\infty$ (see APPENDIX 1) we can add

$$-(I+NB_1{}^TX\Phi^{-1}B_2)^{-1}W_x,\qquad -(I+NB_1{}^TX\Phi^{-1}B_2)^{-1}W_w \tag{23}$$

to Q_x and Q_w in (19), respectively, to get the complete parametrization (15).

Q.E.D.

We will parametrize the SF controller using Theorem 1. At first we define Π as an orthogonal complement of B_1 satisfying

$$\begin{pmatrix}\Pi\\ B_1{}^+\end{pmatrix}(\Pi^T,B_1)=I_n:\qquad B_1{}^+=(B_1{}^TB_1)^{-1}B_1{}^T \tag{24}$$

Since (1a) is written as $sx=Ax+B_1w+B_2u$, we can extract w, denoted $\hat{w}$, from x such that

$$\hat{w}=B_1{}^+(sx-Ax-B_2u)=B_1{}^+\{(sI-A-B_2F_\infty)x+B_2F_\infty x-B_2u\}=B_1{}^+\{\Phi(s)x-B_2v\} \tag{25}$$

where

$$v=u-F_\infty x\quad(u=F_\infty x+v) \tag{26}$$

and get the following theorem.

Theorem 2 (SF problem): All K(s) satisfying $\|G\|_\infty<1$ as well as internal stability are given by

$$K(s)=F_\infty+[I+W(s)\Pi B_2+N(s)B_1{}^+B_2]^{-1}[W(s)\Pi\Phi(s)+N(s)B_1{}^+\Phi(s)-N(s)B_1{}^TX] \tag{27a}$$

where

$$N(s)\in BH_2(p\times r),\qquad W(s)\in RH_2(p\times\overline{n-r}) \tag{27b}$$

are the free parameters.

Proof From Theorem 1 and (25), the input of the SF system can be given by

$$u=K_1\begin{bmatrix}x\\ w\end{bmatrix}=LF\{\Sigma_I,(Q_x,\ Q_w)\}\begin{bmatrix}x\\ B_1^+(\Phi x-B_2 v)\end{bmatrix}$$

$$=F_\infty x-(I-Q_x\Phi^{-1}B_2)^{-1}[(Q_x+Q_wB_1^+\Phi)x-Q_wB_1^+B_2v] \qquad (28)$$

which is equivalent to the input of the FI system in sense of a tansfer-function equivalence and satisfies $\|G\|_\infty<1$. We will prove that this u assures the internal stability of the SF system in the last paragraph.

If (28) is rearranged using $u=F_\infty x+v$, it becomes

$$v=-(I-Q_x\Phi^{-1}B_2-Q_wB_1^+B_2)^{-1}(Q_x+Q_wB_1^+\Phi)x \qquad (29)$$

where

$$I-Q_x\Phi^{-1}B_2-Q_wB_1^+B_2=(I+NB_1^TX\Phi^{-1}B_2)^{-1}[I+NB_1^+B_2+(W_x\Phi^{-1}+W_wB_1^+)B_2] \qquad (30a)$$

$$Q_x+Q_wB_1^+\Phi=(I+NB_1^TX\Phi^{-1}B_2)^{-1}[NB_1^TX-NB_1^+\Phi-(W_x\Phi^{-1}+W_wB_1^+)\Phi] \qquad (30b)$$

Then it follows from (15d) and (24) that we can parametrize the term $W_x\Phi^{-1}+W_wB_1^+$ in (30) as

$$W_x\Phi^{-1}+W_wB_1^+=W_x\Phi^{-1}(I-B_1B_1^+)+(W_x\Phi^{-1}B_1+W_w)B_1^+=W_x\Phi^{-1}(I-B_1B_1^+)$$

$$=W_x\Phi^{-1}(\Pi^T,B_1)\begin{bmatrix}\Pi\\ B_1^+\end{bmatrix}(I-B_1B_1^+)=W_x\Phi^{-1}\Pi^T\Pi=W\Pi \qquad (31)$$

where W(s) is a new parameter defined by

$$W(s)=W_x(s)\Phi^{-1}(s)\Pi^T \qquad (32)$$

It is readily seen that $W\in RH_2$ iff $W_x\in RH_\infty$ (see APPENDIX 2).

Therefore (29) becomes

$$v=-(I+NB_1^+B_2+W\Pi B_2)^{-1}(NB_1^TX-NB_1^+\Phi-W\Pi\Phi)x \qquad (33)$$

and K(s) is written by

$$K(s)=F_\infty-(I+NB_1^+B_2+W\Pi B_2)^{-1}(NB_1^TX-NB_1^+\Phi-W\Pi\Phi) \qquad (34)$$

But to make K(s) proper we need N(s) to be strictly proper since $B_1^+\neq 0$ (see APPENDIX 3).

Finally if this K(s) is represented by the form $K=LF\{\Sigma,Q\}$ we get

$$Q(s)=(I+NB_1^TX\Phi^{-1}B_2)^{-1}[N(B_1^TX-B_1^+\Phi)-W\Pi\Phi] \qquad (35)$$

Since W(s) and N(s) are strictly proper and $(I+NB_1^TX\Phi^{-1}B_2)^{-1}$ is stable, we can see $Q\in RH_\infty$ and it follows from (I) of Lemma 1 that K(s) internally stabilizes the SF system. Q.E.D.

3. Applications

3.1 Constant gain state feedback

In this section we will parametrize dynamical N(s) and W(s) which make K(s) just a constant gain other than the central solution F_∞. That is; we want to characterize all K(s) satisfying

$$K(s)=F=\text{constant} \qquad (36)$$

This problem is called CSF (constant state feedback) problem.

It will readily follow from (27) that this can be satisfied if and only if

$$-NB_1{}^TX + W\Pi\Phi + NB_1{}^*\Phi = (I + W\Pi B_2 + NB_1{}^*B_2)\Delta F \qquad (37)$$

Solving this eqaution we get

$$W(s)=\Delta F[sI-(A_F+B_1B_1{}^TX)]\Pi^T, \quad N(s)=\Delta F[sI-(A_F+B_1B_1{}^TX)]^{-1}B_1 \qquad (38)$$

where

$$A_F=A+B_2F \ , \ \Delta F=F-F_\infty \qquad (39)$$

If $N\in BH_2$ then $K(s)$ is a solution of the CSF problem since stability of N automatically implies that of W. The condition $N\in BH_2$ is easily obtained using bounded real lemma.

<u>**Theorem 3**</u> [9] <u>1)</u> F is a solution of the CSF problem if and only if

$$YA_F+A_F{}^TY+YB_1B_1{}^TY+F^TF+C_1{}^TC_1=0 \qquad (40)$$

has a stabilizing solution $Y\geq 0$. If this condition is satisfied the free parameters $W(s)$ and $N(s)$ are determined by (38).

<u>2)</u> The stabilizing solution $X\geq 0$ of (13b) is the minimum semi-positive definite stabilizing solution among all stabilizing solutions of (40).

<u>3.2 H_∞/H_2 control</u>

We will employ W and N to minimize an H_2 control performance. So the system considered here is given by

$$\dot{x}=Ax+B_1w+B_q\xi+B_2u \qquad (41a)$$

$$z=C_1x+D_{12}u \qquad (41b)$$

$$z_q=C_qx+D_qu \qquad (41c)$$

$$y=x \qquad (41d)$$

where x, z, u and w are the same as those of (1), $\xi\in R^q (q\leq n)$ and $z_q\in R^s$ are the disturbance and the controlled output of an H_2 control loop. We will take account all the assumptions imposed to (1). In addition we assume (C_q,A) is detectable and

$$C_q{}^TD_q=0, \qquad D_q{}^TD_q=R>0 \qquad (42)$$

where R is a constant weighting matrix.

We define $G_q(s)$ as the closed loop transfer function from ξ to z_q to get

$$z_q=G_q(s)\xi \qquad (43)$$

when the control (4) is applied which assures an internal stability of the closed loop system composed of (4) and (41).

We will consider the following problem here.

<u>**Problem**</u>: Minimize $\|G_q(s)\|_2$ by unconstrained $W(s)$ and $N(s)$ and check whether they satisfy the condition (27b) so that $\|G(s)\|_\infty<1$ holds.

This problem is a sufficient one of the original H_∞/H_2 control problem [3]. Rotea et al [5] have parametrized unconstrained H_2 optimal K(s) and have found K(s) that satisfies $\|G(s)\|_\infty<1$ in their Problem B. Therefore the present problem will solve the Problem B in the reverse direction.

The system (41) together with K(s) in (27) can be represented by

$$\dot{x}=A_\infty x+B_1 w+B_q \xi+B_2 v \tag{44a}$$

$$\begin{bmatrix} z \\ z_q \end{bmatrix}=\begin{bmatrix} C_1+D_{12}F_\infty \\ C_q+D_q F_\infty \end{bmatrix}x+\begin{bmatrix} D_{12} \\ D_q \end{bmatrix}v \tag{44b}$$

$$\begin{bmatrix} \eta_1 \\ \eta_2 \end{bmatrix}=\begin{bmatrix} 0 \\ -B_1{}^T X \end{bmatrix}x+\begin{bmatrix} 0 \\ I \end{bmatrix}w+\begin{bmatrix} \Pi \\ B_1{}^+ \end{bmatrix}B_q \xi \tag{44c}$$

$$v=W(s)\eta_1+N(s)\eta_2 \tag{44d}$$

as a semi-closed loop system where

$$A_\infty=A+B_2 F_\infty \tag{44e}$$

Considering W and N new unconstrained controllers, we minimizes $\| G_q(s)\|_2$ by W and N to get [10]

$$W(s)=\Delta F[sI-(A_F+LB_1B_1{}^TX)]^{-1}L\Pi^T,\quad N(s)=\Delta F[sI-(A_F+LB_1B_1{}^TX)]^{-1}LB_1 \tag{45}$$

where

$$L=B_q B_q{}^+,\quad \Delta F=F-F_\infty,\quad A_F=A+B_2F \tag{46}$$

and F is the H_2 optimal state feedback gain such as

$$F=-R^{-1}B_2{}^TP,\quad PA+A^TP-PB_2R^{-1}B_2{}^TP+C_q{}^TC_q=0 \tag{47}$$

And the global H_2 minimum is attained by these W and N. Then the condition $N\in BH_2$ will lead to the following theorem.

<u>Theorem 4</u> [10] The necessary and sufficient conditions for the existence of a solution of the H_∞/H_2 control problem are as follows.

1. (13a) has a stabilizing solution $X\geq 0$.
2. $YA_F+A_F{}^TY+F^TF+C_1{}^TC_1+[X(I-L)+YL]B_1B_1{}^T[L^TY+(I-L^T)X]=0$ (48)

has a stabilizing solution $Y\geq 0$.

3. $Y\geq X$ (49)

4. Conclusion

We have obtained the parametrization of H_∞ controllers for the FI problem and the state feedback control problem and have expressed constant state feedback gain in terms of the free parameters derived in this parametrization. Further H_∞/H_2 control has also treated in this context.

References

[1] H.Kimura and R.Kawatani (1988). "Synthesis of H∞ Controllers Based on Conjugation", Proc. of 27th CDC, pp. 7-13

[2] J.Doyle, K.Glover, P.P.Khargonekar & B.A.Francis (1989). "State-space solutions to standard H_2 and H_∞ control problems", IEEE Trans. Auto. Control, **AC-34**, pp. 831-847.

[3] D.S. Bernstein and W.M.Haddad (1989). "LQG control with an H_∞ performance bound: A Riccati equation approach", IEEE Trans. Auto. Control, **AC-34**, pp.293-305.

[4] J.Doyle, K.Zhou & B.Bodenheimer (1989). "Optimal control with mixed H^2 and H^∞ performance objectives", Proc. 1989 ACC, pp. 2065-2070.

[5] M.A.Rotea & P.P.Khargonekar(1990). "Simultaneous H^2/H^∞ optimal control with state feedback", Proc. 1990 ACC, pp.2830-2834

[6] I.R.Perterson (1987). "Disturbance attenuation and H_∞ optimization: A design method based on the algebraic Riccati equation", IEEE Trans. Auto. Control, AC-32, pp. 427-429

[7] K.Zhou and P.P.Khargonekar (1988). "An algebraic Riccati equation approach to H_∞ optimization", Systems & Control Letters,11, pp.85-91

[8] K.Z.Liu, T.Mita & R.Kawatani (1990)."Parametrization of state feedback H_∞ controllers", Int. J. Control, 51, pp.535-551

[9] T.Mita (1991)."Complete parametrizations of H_∞ controllers in full informat ion problem and state feedback problem", Proc. SICE 13th DST, pp. 59-64

[10] T. Mita (1990). "H_∞/H_2 mixed control", J. Fac. Enrg. Chiba Univ., 42-2, pp.13-22, Proc. MTNS'91 Kobe

[11] B.A.Francis (1987). A course in H_∞ control theory, Springer-Veralag

APPENDIX 1: It is clear that $I+NB_1{}^TX\Phi^{-1}B_2\in RH_\infty$. Since

$$X(A+B_2F_\infty)+(A+B_2F_\infty)^TX+XB_2B_2{}^TX+XB_1B_1{}^TX=-C_1{}^TC_1\leq 0 \qquad \text{(A1)}$$

has a stabilizing solution $X\geq 0$, $B_1{}^TX\Phi^{-1}B_2\in BH_\infty$ which proves $(I+NB_1{}^TX\Phi^{-1}B_2)^{-1}\in RH_\infty$ by virtue of the small gain theorem as well as $N\in BH_2$.

APPENDIX 2: It follows from (15d) and (32) that

$$W_x(s)\Phi^{-1}(s)(\Pi^T,B_1)=[W(s),-W_w(s)] \qquad \text{(A2)}$$

It also can be written as

$$W_x(s)=[W(s),-W_w(s)]\begin{pmatrix}\Pi\\ B1+\end{pmatrix}\Phi(s) \qquad \text{(A3)}$$

Therefore $W_x(\infty)\in RH_\infty$ iff $W(\infty)=0$ and $W_w(\infty)=0$, i.e., $W(s)\in RH_2$ and $W_w(s)\in RH_2$.

APPENDIX 3: Since $W(\infty)=0$ and $N(s)\in BH_\infty$, when $s\uparrow\infty$

$$K(s)=F_\infty-[I+N(\infty)B_1{}^+B_2]^{-1}[-sN(\infty)B_1{}^+-\text{constant}] \qquad \text{(A4)}$$

This shows $N(s)$ must be in BH_2 to assure $K(\infty)=\text{constant}$ since $B^+{}_1\neq 0$.

THE MIXED H_2 AND H_∞ CONTROL PROBLEM

A.A. Stoorvogel [1]
Dept. of Electrical Engineering
University of Michigan
Ann Arbor, MI 48109-2122
U.S.A.

H.L. Trentelman
Mathematics Institute
University of Groningen
P.O. Box 800
9700 AV Groningen
The Netherlands

Abstract

This paper is concerned with robust stability in combination with nominal performance. We pose an H_∞ norm bound on one transfer matrix to guarantee robust stability. Under this constraint we minimize an upper bound for the H_2 norm of a second transfer matrix. This transfer matrix is chosen so that its H_2 norm is a good measure for performance. We extend earlier work on this problem. The intention is to reduce this problem to a convex optimization problem.

Keywords : The H_2 control problem, the H_∞ control problem, robust stability, auxiliary cost.

1 Introduction

In all classical design methods robustness, i.e. insensitivity to system perturbations, plays an important role. In the last decade, the quest for robust controllers has resulted in a widespread search for controllers which robustly stabilize a system (see e.g. [6, 8, 9]). The application of the small gain theorem yields constraints on the H_∞ norm of the transfer matrix. The problem of minimizing the H_∞ norm by state and output feedback was studied in many papers (see e.g. [3, 5, 12]).

Clearly, besides guaranteed stability, in practical design we are also very much concerned with performance. Since performance can very often be expressed in H_2 norm constraints, a possible new goal is: minimize the *nominal* H_2 norm (guarantees nominal performance) under the constraint of an H_∞ norm bound (guarantees robust stability). There is only one very limited result available for this setup which can be found in [10]. An alternative approach of robust stability <u>and</u> robust performance can be found in [2, 11]. Several recent papers investigate the minimization of an auxiliary cost function which guarantees an H_∞ norm bound while minimizing an upper bound for the nominal H_2 norm (see [1, 4, 7, 15]). In this paper we will extend the results of [7].

---

[1] On leave from the Dept. of Mathematics, Eindhoven University of Technology, the Netherlands, supported by the Netherlands Organization for Scientific Research (NWO).

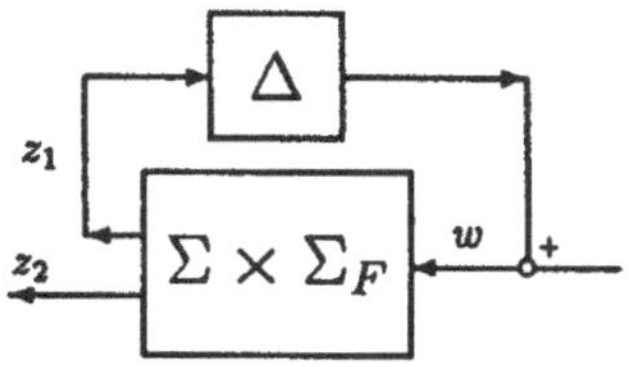

Figure 1: Closed loop system $\Sigma \times \Sigma_F$ with perturbation Δ

2 Problem formulation and main results

Let a linear, time-invariant system Σ be given by the following equations:

$$\Sigma : \begin{cases} \dot{x} = Ax + Bu + Ew, \\ y = C_0 x + D_0 w, \\ z_1 = C_1 x + D_1 u, \\ z_2 = C_2 x + D_2 u. \end{cases} \tag{2.1}$$

x, u, w, y, z_1 and z_2 take values in finite dimensional vector spaces: $x(t) \in \mathcal{R}^n$, $u(t) \in \mathcal{R}^m$, $w(t) \in \mathcal{R}^l$, $y(t) \in \mathcal{R}^q$, $z_1(t) \in \mathcal{R}^s$ and $z_2(t) \in \mathcal{R}^p$. The system parameters A, B, E, C_0, C_1, C_2, D_0, D_1 and D_2 are matrices of appropriate dimensions. We assume that the system is time-invariant, i.e. the system parameters are independent of time. The input w is the disturbance working on the system. We apply compensators Σ_F to Σ of the form:

$$\Sigma_F : \begin{cases} \dot{p} = Kp + Ly, \\ u = Mp + Ny, \end{cases} \tag{2.2}$$

The resulting closed loop system is of the form:

$$\Sigma \times \Sigma_F : \begin{cases} \dot{x}_e = A_e x_e + E_e w, \\ z_1 = C_{1,e} x_e + D_{1,e} w, \\ z_2 = C_{2,e} x_e + D_{2,e} w. \end{cases} \tag{2.3}$$

where $x_e = (x^T, p^T)^T$. We will call a controller internally stabilizing if the resulting matrix A_e in (2.3) is a stability matrix. The closed loop transfer matrix from w to z_1 and from w to z_2 will be denoted by G_{1,Σ_F} and G_{2,Σ_F} respectively. Our overall goal is to minimize the H_2 norm of the transfer matrix G_{2,Σ_F} over all internally stabilizing controllers of the form (2.2) under the constraint that the H_∞ norm of G_{1,Σ_F} is strictly less than 1. The motivation for this problem formulation is that the H_∞ norm constraint guarantees that the interconnection depicted in figure 1 is asymptotically stable for all, possibly time-varying or non-linear, Δ with gain less than or equal to 1, i.e. $\|\Delta z\|_2 \leq \|z\|_2$ for all $z \in \mathcal{L}_2$. Hence we guarantee robust stability. On the other hand, by minimizing the H_2 norm from w to z_2 we optimize nominal performance, i.e. performance for $\Delta = 0$, where we assume that performance is expressed in terms of the H_2 norm. Frequency-dependent weighting matrices can be taken into account, by incorporating them in the plant. Clearly the more general picture, where the output of the disturbance system enters the system Σ via a different input is even more attractive but very hard to handle.

In this paper we investigate an alternative cost-function which minimizes an upper bound for the H_2 norm. This bound has been introduced in [1] and further investigated in [4, 7, 15]. Define the following algebraic Riccati equation

$$A_e Y_e + Y_e A_e^T + (Y_e C_{1,e}^T + E_e D_{1,e}^T)(I - D_{1,e} D_{1,e}^T)^{-1}(C_{1,e} Y_e + D_{1,e} E_e^T) + E_e E_e^T = 0 \tag{2.4}$$

with stability condition

$$\sigma\left(A_e + (Y_e C_{1,e}^T + E_e D_{1,e}^T)(I - D_{1,e} D_{1,e}^T)^{-1} C_{1,e}\right) \subset C^- \tag{2.5}$$

where σ denotes the spectrum of a matrix and C^- the open left half plane. We have the following basic fact, see e.g. [1, 14, 16]

Lemma 2.1 : *Let Σ be described by (2.1). If Σ_F, described by (2.2) is such that the closed loop system, described by (2.3) satisfies $\sigma(A_e) \subset C^-$, then the following statements are equivalent:*

(*i*) $\|G_{1,\Sigma_F}\|_\infty < 1$

(*ii*) *There exists a real symmetric matrix Y_e that satisfies (2.4) and (2.5).*

If such Y_e exists then it is unique. In addition $\|G_{2,\Sigma_F}\|_2 < \infty$ if and only if $D_{2,e} = 0$ and in that case

$$\|G_{2,\Sigma_F}\|_2^2 \leq \text{Trace}\left(C_{2,e} Y_e C_{2,e}^T\right)$$

□

It can be shown that Trace $(C_{2,e} Y_e C_{2,e}^T)$ depends only on the transfer matrix from w to (z_1, z_2). We could therefore denote it by $\mathcal{J}_\Sigma(\Sigma_F)$, a value that only depends on Σ_F and not on a particular realization of Σ. $\mathcal{J}_\Sigma(\Sigma_F)$ is called the *auxiliary cost* associated with Σ_F. Our problem becomes:

> minimize $\mathcal{J}_\Sigma(\Sigma_F)$ over all compensators such that $\Sigma \times \Sigma_F$ is internally stable, $\|G_{1,\Sigma_F}\|_\infty < 1$ and $D_{2,e} = D_2 N D_0 = 0$.

Obviously, the first question to be posed is: does there exist a compensator Σ_F such that $\sigma(A_e) \subset C^-$, $\|G_{1,\Sigma_F}\|_\infty < 1$ and $D_2 N D_0 = 0$. In order to answer this question we need the following definitions. For $P \in \mathcal{R}^{n\times n}$ we define the following matrix:

$$F(P) := \begin{pmatrix} A^T P + PA + C_1^T C_1 + PEE^T P & PB + C_1^T D_1 \\ B^T P + D_1^T C_1 & D_1^T D_1 \end{pmatrix}.$$

If $F(P) \geq 0$, we say that P is a solution of the quadratic matrix inequality. We also define a dual version of this quadratic matrix inequality. For any matrix $Q \in \mathcal{R}^{n\times n}$ we define the following matrix:

$$G(Q) := \begin{pmatrix} AQ + QA^T + EE^T + QC_1^T C_1 Q & QC_0^T + ED_0^T \\ C_0 Q + D_0 E^T & D_0 D_0^T \end{pmatrix}.$$

If $G(Q) \geq 0$, we say that Q is a solution of the dual quadratic matrix inequality. In addition to these two matrices, we define two matrices pencils, which play dual roles:

$$L(P,s) := \begin{pmatrix} sI - A - EE^T P & -B \end{pmatrix},$$

$$M(Q,s) := \begin{pmatrix} sI - A - QC_1^T C_1 \\ -C_0 \end{pmatrix}.$$

Finally, we define the following two transfer matrices:

$$\begin{aligned} G_{ci}(s) &:= C_1 (sI - A)^{-1} B + D_1, \\ G_{di}(s) &:= C_0 (sI - A)^{-1} E + D_0, \end{aligned}$$

Let $\rho(M)$ denote the spectral radius of the matrix M. By $\text{rank}_{\mathcal{R}(s)} M$ we denote the rank of M as a matrix with elements in the field of real rational functions $\mathcal{R}(s)$. We define the invariant zeros of a system (A, B, C, D) as the points in the complex plane where the Rosenbrock system matrix loses rank. We are now in a position to recall the main result from [12]:

Theorem 2.2 : *Consider the system (2.1). Assume that both the system (A, B, C_1, D_1) as well as the system (A, E, C_0, D_0) have no invariant zeros on the imaginary axis. Then the following two statements are equivalent:*

(i) *For the system (2.1) there exists a controller Σ_F of the form (2.2) such that the resulting closed-loop system from w to z_1, with transfer matrix G_{1,Σ_F}, is internally stable and has H_∞ norm less than 1, i.e. $\|G_{1,\Sigma_F}\|_\infty < 1$.*

(ii) *There exist positive semi-definite solutions P, Q of the quadratic matrix inequalities $F(P) \geq 0$ and $G(Q) \geq 0$ satisfying $\rho(PQ) < 1$, such that the following rank conditions are satisfied*

(a) *rank* $F(P) = \text{rank}_{\mathcal{R}(s)} G_{ci}$,

(b) *rank* $G(Q) = \text{rank}_{\mathcal{R}(s)} G_{di}$,

(c) $\text{rank} \begin{pmatrix} L(P,s) \\ F(P) \end{pmatrix} = n + \text{rank}_{\mathcal{R}(s)} G_{ci} \qquad \forall\, s \in C^0 \cup C^+,$

(d) $\text{rank} \begin{pmatrix} M(Q,s) & G(Q) \end{pmatrix} = n + \text{rank}_{\mathcal{R}(s)} G_{di} \qquad \forall\, s \in C^0 \cup C^+.$ □

We note that the existence and determination of P and Q can be performed by investigating reduced order Riccati equations. Moreover, if condition (ii) holds then one can always find Σ_F with $N = 0$ so $D_{2,e} = D_2 N D_0 = 0$. We will assume throughout this paper that the assumption on the imaginary axis zeros as stated in the previous theorem holds. We will also assume that there exists matrices P and Q such that condition (ii) above is satisfied. This is needed to guarantee the existence of a controller Σ_F which makes $\mathcal{J}_\Sigma(\Sigma_F)$ finite. In the next section we show how to reduce our problem by a suitable system transformation.

3 A system transformation

We will use the idea of system transformations, introduced for this setting in [12]. We construct the following system,

$$\Sigma_Q : \begin{cases} \dot{x}_Q &= A_Q x_Q + B_Q u_Q + E_Q w_Q, \\ y_Q &= C_0 x_Q + D_0 w_Q, \\ z_{1,Q} &= C_1 x_Q + D_1 u_Q, \\ z_{2,Q} &= C_2 x_Q + D_2 u_Q. \end{cases} \tag{3.1}$$

where

$$G(Q) = \begin{pmatrix} E_Q \\ D_0 \end{pmatrix} \begin{pmatrix} E_Q^T & D_0^T \end{pmatrix}, \tag{3.2}$$

and

$$\begin{aligned} A_Q &:= A + QC_0^T C_0 \\ B_Q &:= B + QC_0^T D_0 \end{aligned}$$

It has been shown in [12] that the factorization in (3.2) is always possible. It has been shown in [12] that this new system is such that (A_Q, E_Q, C_0, D_0) is left invertible and minimum phase. It is interesting to see how this transformation affects our cost function $\mathcal{J}_\Sigma(\Sigma_F)$. The following theorem is an extension of a theorem in [12]:

Theorem 3.1 : *Let an arbitrary compensator Σ_F of the form (2.2) be given. The following two statements are equivalent:*

(i) *The compensator Σ_F applied to the original system Σ in (2.1) is internally stabilizing, yields $D_2 N D_0 = 0$ and the resulting closed loop transfer function from w to z_1 has H_∞ norm less than 1.*

(ii) *The compensator Σ_F applied to the new system Σ_Q in (3.1) is internally stabilizing, yields $D_2 N D_0 = 0$ and the resulting closed loop transfer function from w_Q to $z_{1,Q}$ has H_∞ norm less than 1.*

Moreover,

$$\mathcal{J}_\Sigma(\Sigma_F) = \text{Trace } [(C_2 + D_2 N C_0) Q (C_2 + D_2 N C_0)^T] + \mathcal{J}_{\Sigma_Q}(\Sigma_F) \tag{3.3}$$

□

In earlier papers (see [1, 4, 7, 15]) the assumption was made that D_0 is surjective. Then our constraint that $D_2 N D_0 = 0$ guarantees that $D_2 N C_0 = 0$ and therefore the first term on the right is independent of our specific choice for the compensator. The question is whether this is also true for the case that D_0 is not surjective. It is, as is stated in the following lemma:

Lemma 3.2 :*Let Q satisfying $G(Q) \geq 0$ be given. Then we have*

$$\text{Im}\, C_0 Q \subset \text{Im}\, D_0$$

Moreover, we have:

$$\mathcal{J}_\Sigma(\Sigma_F) = \text{Trace } C_2 Q C_2^T + \mathcal{J}_{\Sigma_Q}(\Sigma_F)$$

□

This enables us to follow the same lines as the analysis in [7]. We start by defining the following two classes of controllers:

- $\mathcal{A}(\Sigma)$ denotes the set of controllers Σ_F of the form (2.2) such that the closed loop system is internally stable, $D_2 N D_0 = 0$, and $\|G_{1,\Sigma_F}\|_\infty < 1$.
- $\mathcal{A}_m(\Sigma)$ denotes the set of state feedbacks Σ_F of the form $u = Fx$ such that the closed loop system is internally stable and $\|G_{1,\Sigma_F}\|_\infty < 1$.

We define the following system

$$\Sigma_{Q,sf}: \begin{cases} \dot{x}_Q &= A_Q x_Q + B_Q u_Q + E_Q w_Q, \\ y_Q &= x_Q, \\ z_{1,Q} &= C_1 x_Q + D_1 u_Q, \\ z_{2,Q} &= C_2 x_Q + D_2 u_Q. \end{cases} \tag{3.4}$$

$\Sigma_{Q,sf}$ is the "state feedback version" of Σ_Q. It turns out that Σ_Q has the nice property that the state feedback and the dynamic feedback cases are strongly correlated as is worked out in the following theorem (see [7]):

Theorem 3.3 : *Let Σ_Q be described by (3.1) and let $\Sigma_{Q,sf}$ be the state feedback version of Σ_Q. Then we have the following equalities:*

$$\inf_{\Sigma_F \in \mathcal{A}(\Sigma_Q)} \mathcal{J}_{\Sigma_Q}(\Sigma_F) = \inf_{\Sigma_F \in \mathcal{A}_m(\Sigma_Q)} \mathcal{J}_{\Sigma_{Q,sf}}(\Sigma_F)$$

□

Moreover, suppose we have a minimizing sequence for $\bar{\mathcal{J}}_{\Sigma_{Q,sf}}(\Sigma_F)$ over the set $\mathcal{A}_m(\Sigma_Q)$ (note that the infimum will in general not be attained). This will be a sequence of state feedbacks of the form $u_Q = F_\varepsilon x_Q$ ($\varepsilon \downarrow 0$).
It has been shown in [12] that (A_Q, E_Q, C_0, D_0) is left invertible and minimum phase. Therefore (see [13]) there exists a sequence of observer gains K_ε such that the observer:

$$\Sigma_{obs}: \quad \dot{p} = A_Q p + K_\varepsilon(C_0 p - y_Q), \tag{3.5}$$

for the system Σ_Q has the following property:

- Both the H_2 norm and the H_∞ norm of the closed loop transfer matrix from w to $x - p$ go to 0 as $\varepsilon \downarrow 0$.

Then it is easy to show that the following sequence of controllers yields a minimizing sequence for $\mathcal{J}_{\Sigma_Q}(\Sigma_F)$ over the set $\mathcal{A}(\Sigma_Q)$.

$$\Sigma_{F,\varepsilon}: \begin{cases} \dot{p} &= A_Q p + B_Q u_Q + K_\varepsilon(C_0 p - y_Q), \\ u_Q &= F_\varepsilon p. \end{cases} \tag{3.6}$$

By theorem 3.1, this sequence of controllers (with y_Q, u_Q replaced by y, u) is also a minimizing sequence for $\mathcal{J}_\Sigma(\Sigma_F)$ over the set $\mathcal{A}(\Sigma_Q)$. Hence our problem has been solved, except for the construction of this minimizing sequence of static feedbacks. Hence the problem has been reduced to a finite-dimensional optimization problem (an optimization over F). However, using the techniques from [7], it can be shown that we can reformulate this problem as a convex optimization problem.
Assume that the closed loop system after applying the feedback $u_Q = F x_Q$ to $\Sigma_{Q,sf}$ is of the form (2.3) (note that $D_{1,e} = 0$ and $D_{2,e} = 0$). We first use a different characterization of $\mathcal{J}_{\Sigma_{Q,sf}}(\Sigma_F)$. Define the following algebraic Riccati inequality

$$A_e Y + Y A_e^{\mathrm{T}} + (Y C_{1,e}^{\mathrm{T}} + E_e D_{1,e}^{\mathrm{T}})(C_{1,e} Y + D_{1,e} E_e^{\mathrm{T}}) + E_e E_e^{\mathrm{T}} < 0 \tag{3.7}$$

Note that (because of the strict inequality) the stability requirement (2.5) is equivalent to the requirement $Y > 0$. We get

$$\mathcal{J}_{\Sigma_Q}(\Sigma_F) = \inf\left\{ \text{Trace } C_{2,e} Y C_{2,e}^{\mathrm{T}} \mid Y > 0 \text{ satisfies (3.7)} \right\}$$

The above introduced overparametrization will allow us to transform the problem into a *convex* optimization problem.
Using the above, we can reformulate our problem:

$$\inf_{\Sigma_F \in \mathcal{A}_m(\Sigma_Q)} \mathcal{J}_{\Sigma_{Q,st}}(\Sigma_F) = \inf_{F,Y} \text{ Trace } (C_2 + D_2F)Y(C_2 + D_2F)^T$$

under the following constraints:

(*i*) $Y > 0$,

(*ii*) $A_Q + B_Q F$ is asymptotically stable,

(*iii*) Y satisfies (3.7).

Note that $A_e = A_Q + B_Q F$. Therefore, since $Y > 0$ and $Y(A_Q + B_Q F)^T + (A_Q + B_Q F)Y < 0$ we know that $A_Q + B_Q F$ is asymptotically stable. This makes constraint (ii) superfluous. Because $Y > 0$ we can instead optimize over $Y, W := FY^{-1}$. We get the following constraints:

(*a*) $Y > 0$

(*b*) $A_Q Y + Y A_Q^T + B_Q W + W^T B_Q^T + E_Q E_Q^T - (Y C_1^T + W^T D_1^T)(C_1 Y + D_1 W) < 0.$

with cost function:

$$\inf_{W,Y} \text{ Trace } C_2 Y C_2^T + 2 C_2 W D_2^T + D_2 W Y^{-1} W^T D_2^T$$

As shown in [7] this is a convex optimization. Note that since we only use the matrix Q from theorem 2.2 it can be shown that the above derivation holds under one single assumption: the system (A, E, C_0, D_0) has no invariant zeros on the imaginary axis.

4 Conclusion

In this paper we have shown that one of the assumptions made in [7] can be removed. We obtain a sequence of observers which, combined with a sequence of static feedbacks, results in a minimizing sequence. The resulting static feedback problem can be solved via a finite-dimensional convex optimization problem. We have an extra difficulty, introduced because D_0 is not necessarily surjective: we can not find an optimal observer and have to resort to a sequence of observers which in the limit has the desired property. Since we are investigating the suboptimal case this is not essential. A key point in the above analysis is that the resulting convex optimization problem is completely decoupled from the fact whether D_0 is surjective or not.

Clearly, the open problems remain: we would like to use the H_2 norm instead of the auxiliary cost. Moreover, we would like the output of the disturbance system Δ to enter the system arbitrarily and not the special case of figure 1.

References

[1] D.S. Bernstein and W.M. Haddad, "LQG control with an H_∞ performance bound: a Riccati equation approach", IEEE Trans. Aut. Contr., 34 (1989), pp. 293–305.

[2] ——, "Robust stability and performance analysis for state-space systems via quadratic Lyapunov bounds", SIAM J. Matrix. Anal. & Appl., 11 (1990), pp. 239–271.

[3] J. DOYLE, K. GLOVER, P.P. KHARGONEKAR, AND B.A. FRANCIS, "State space solutions to standard H_2 and H_∞ control problems", IEEE Trans. Aut. Contr., 34 (1989), pp. 831–847.

[4] J.C. DOYLE, K. ZHOU, AND B. BODENHEIMER, "Optimal control with mixed H_2 and H_∞ performance objectives", in Proc. ACC, Pittsburgh, PA, 1989, pp. 2065–2070.

[5] B.A. FRANCIS, *A course in H_∞ control theory*, vol. 88 of Lecture notes in control and information sciences, Springer-Verlag, 1987.

[6] P.P. KHARGONEKAR, I.R. PETERSEN, AND K. ZHOU, "Robust stabilization of uncertain linear systems: quadratic stabilizability and H_∞ control theory", IEEE Trans. Aut. Contr., 35 (1990), pp. 356–361.

[7] P.P. KHARGONEKAR AND M.A. ROTEA, "Mixed H_2/H_∞ control: a convex optimization approach", To appear in IEEE Trans. Aut. Contr., 1990.

[8] I.R. PETERSEN, "A stabilization algorithm for a class of uncertain linear systems", Syst. & Contr. Letters, 8 (1987), pp. 351–357.

[9] I.R. PETERSEN AND C.V. HOLLOT, "A Riccati equation approach to the stabilization of uncertain linear systems", Automatica, 22 (1986), pp. 397–411.

[10] M.A. ROTEA AND P.P. KHARGONEKAR, "H_2 optimal control with an H_∞ constraint: the state feedback case", Automatica, 27 (1991), pp. 307–316.

[11] A.A. STOORVOGEL, "The robust H_2 control problem: a worst case design", Submitted for publication.

[12] ——, "The singular H_∞ control problem with dynamic measurement feedback", SIAM J. Contr. & Opt., 29 (1991), pp. 160–184.

[13] H.L. TRENTELMAN, *Almost invariant subspaces and high gain feedback*, vol. 29 of CWI Tracts, Amsterdam, 1986.

[14] J.C. WILLEMS, "Least squares stationary optimal control and the algebraic Riccati equation", IEEE Trans. Aut. Contr., 16 (1971), pp. 621–634.

[15] K. ZHOU, J. DOYLE, K. GLOVER, AND B. BODENHEIMER, "Mixed H_2 and H_∞ control", in Proc. ACC, San Diego, CA, 1990, pp. 2502–2507.

[16] K. ZHOU AND P.P. KHARGONEKAR, "An algebraic Riccati equation approach to H_∞ optimization", Syst. & Contr. Letters, 11 (1988), pp. 85–91.

Nash Games and mixed H_2/H_∞ Control

[1]D.J.N. Limebeer [2]B.D.O. Anderson [1] B. Hendel

Department of Electrical Engineering, [1]
Imperial College,
Exhibition Rd.,
London.

Department of Systems Engineering, [2]
Australian National University,
Canberra,
Australia.

Abstract

The established theory of non-zero sum games is used to solve a mixed H_2/H_∞ control problem. Our idea is to use the two pay-off functions associated with a two player Nash game to represent the H_2 and H_∞ criteria separately. We treat the state feedback problem, and we find necessary and sufficient conditions for the existence of a solution. A full stability analysis is available in the infinite horizon case [13], and the resulting controller is a constant state feedback law which is characterised by the solution to a pair of cross-coupled Riccati differential equations.

1 Introduction

It is a well-known fact that the solution to multivariable H_∞ control problems is hardly ever unique. If the solution is not optimal it is never unique, even in the scalar case. With these comments in mind, the question arises as to what one can sensibly do with the remaining degrees of freedom. Some authors have suggested recovering uniqueness by strengthening the optimality criterion and solving a "super-optimal" problem [10, 11, 14, 21, 22]. Another possibility is entropy minimisation which was introduced into the literature by Arov and Krein [2], with other contributions coming from [7, 8, 9, 15, 18]. Entropy minimisation is of particular interest in the present context because entropy provides an upper bound on the H_2 norm of the input-output operator [17]. One may therefore think of entropy minimisation as minimising an upper bound on the H_2 cost. This gives entropy minimisation an H_2/H_∞ interpretation. As an additional bonus we mention that the solution to the minimum entropy problem is particularly simple. In the case of most state-space representation formulae for all solutions, all one need do is set the free parameter to zero [9, 15, 18].

Another approach to mixed H_2/H_∞ problems is due to Bernstein and Haddad [5]. Their solution turns out to be equivalent to entropy minimisation [16]. The work of Bernstein and Haddad is extended in [6, 23], where a mixed H_2/H_∞ problem with output feedback is considered. The outcome of this work is a formula for a mixed H_2/H_∞ controller which is parameterised in terms of coupled algebraic Riccati equations. The central drawback of this approach is concerned with solution procedures for the Riccati equations; currently an expensive method based on homotopy algorithms is all that is on offer. Khargonekar and Rotea examine multiple-objective control problems which include mixed H_2/H_∞ type problems. In contrast to the other work in this area, they give an algorithmic solution based on convex optimisation [19].

In the present paper we seek to solve a mixed H_2/H_∞ problem via the solution of an associated Nash game. As is well known [4, 20], two player non-zero sum games have two performance criteria, and the idea is to use one performance index to reflect an H_∞ criterion, while the second reflects an H_2 optimality criterion. Following the precise problem statement given in section 2.1, we supply necessary and sufficient conditions for the existence of a Nash equilibrium in Section 2.2. The aim of section 2.3 is to provide our previous work with a stochastic interpretation which resembles classical LQG control. Section 2.4 provides a reconciliation with pure H_2 and H_∞ control. In particular, we show that H_2, H_∞ and H_2/H_∞ control may all be captured as special cases of a two player Nash game. Brief conclusions appear in section 3.

Our notation and conventions are standard. For $A \in \mathcal{C}^{n\times m}$, A' denotes the complex conjugate transpose. $\mathcal{E}\{\cdot\}$ is the expectation operator. We denote $||R||_{2i}$ as the operator norm induced by the usual 2-norm on functions of time. That is :

$$||R||_{2i} = \sup_{u\in\mathcal{L}_2[t_0,t_f]} \frac{||Ru||_2}{||u||_2}.$$

2 The H_2/H_∞ Control Problem

2.1 Problem Statement

We begin by considering an H_2/H_∞ problem in which there is a single exogenous input. Given a linear system described by

$$\dot{x}(t) = A(t)x(t) + B_1(t)w(t) + B_2(t)u(t) \qquad x(t_0) = x_0 \qquad (2.1)$$

$$z(t) = \begin{bmatrix} C(t)x(t) \\ D(t)u(t) \end{bmatrix} \qquad (2.2)$$

in which the entries of $A(t), B_1(t), B_2(t), C(t)$ and $D(t)$ are continuous functions of time. We suppose also that $D'(t)D(t) = I$. In the remainder of our analysis we will take it as read that the problem data is time varying.

We wish to

(1) find a control law $u^*(t)$ such that

$$||z||_2^2 \le \gamma^2 ||w||_2^2 \qquad \forall w(t) \in \mathcal{L}_2[t_0,t_f]. \qquad (2.3)$$

This condition can be interpreted as an $\mathcal{L}_\infty$ type norm constraint of the form

$$||\mathcal{R}_{zw}||_{2i} \le \gamma \qquad (2.4)$$

where the operator $\mathcal{R}_{zw}$ maps the disturbance signal $w(t)$ to the output $z(t)$ when the optimal control law $u^*(t)$ is invoked.

(2) In addition, we require the control $u^*(t)$ to regulate the state $x(t)$ in such a way as to minimise the output energy when the worst case disturbance is applied to the system.

As we will now show, this problem may be formulated as an LQ, non-zero sum, differential game between two opposing players $u(t,x)$ and $w(t,x)$. We begin by defining the strategy sets for each player. To avoid difficulties with non-unique global Nash solutions with possibly differing Nash costs, we force both players to use linear, memoryless feedback controls. This restriction also results in a simple controller implementation which is easily compared with the corresponding linear quadratic and H_∞ controllers. The difficulties associated with feedback strategies involving memory are beautifully illustrated by example in Basar [3]. We introduce the cost functions to be associated with the H_2 and H_∞ criteria as

$$J_1(u,w) = \int_{t_0}^{t_f} \left(\gamma^2 w'(t,x)w(t,x) - z'(t)z(t)\right) dt \tag{2.5}$$

and

$$J_2(u,w) = \int_{t_0}^{t_f} z'(t)z(t)\, dt, \tag{2.6}$$

and we seek equilibrium strategies $u^*(t,x)$ and $w^*(t,x)$ which satisfy the Nash equilibria defined by

$$J_1(u^*,w^*) \leq J_1(u^*,w) \tag{2.7}$$

$$J_2(u^*,w^*) \leq J_2(u,w^*). \tag{2.8}$$

It turns out that the equilibrium values of $J_1(\cdot,\cdot)$ and $J_2(\cdot,\cdot)$ are quadratic in x_0. Consequently, if $x_0 = 0$, we have that $J_1(u^*,w^*) = 0$ and $J_2(u^*,w^*) = 0$. This observation motivates our particular choice of $J_1(\cdot,\cdot)$ as follows : Since $J_1(u^*,w^*) = 0$, we must have $\|z\|_2^2 \leq \gamma^2\|w\|_2^2$ for $u(t,x) = u^*(t,x)$ and all $w \in \mathcal{L}_2[t_0,t_f]$, which ensures $\|\mathcal{R}_{zw}\|_{2i} \leq \gamma$. The second Nash inequality shows that u^* regulates the state to zero with minimum control energy when the input disturbance is at its worst.

2.2 The necessary and sufficient conditions for the existence of linear controls

The aim of this section is to give necessary and sufficient conditions for the existence of linear, memoryless Nash equilibrium controls. When controllers exist, we show that they are unique and parameterised by a pair of cross-coupled Riccati differential equations. We use Ω to denote the set of all linear and memoryless state feedback controls on $[t_0,t_f]$.

Theorem 2.1 *Given the system described by :*

$$\dot{x}(t) = Ax(t) + B_1 w(t,x) + B_2 u(t,x) \qquad x(t_0) = x_0 \tag{2.9}$$

$$z(t) = \begin{bmatrix} Cx(t) \\ Du(t,x) \end{bmatrix} \qquad D'D = I, \tag{2.10}$$

there exist Nash equilibrium strategies $u^*(t,x) \in \Omega$ and $w^*(t,x) \in \Omega$ such that

$$\begin{aligned} J_1(u^*,w^*) &\leq J_1(u^*,w) \quad \forall w(t,x) \in \Omega \\ J_2(u^*,w^*) &\leq J_2(u,w^*) \quad \forall u(t,x) \in \Omega \end{aligned}$$

where

$$J_1(u,w) = \int_{t_0}^{t_f} \left[\gamma^2 w'(t,x)w(t,x) - z'(t)z(t)\right] dt \qquad (2.11)$$

$$J_2(u,w) = \int_{t_0}^{t_f} z'(t)z(t)dt \qquad (2.12)$$

if and only if the coupled Riccati differential equations

$$\begin{aligned} -\dot{P}_1(t) &= A'P_1(t) + P_1(t)A - C'C \\ &- \begin{bmatrix} P_1(t) & P_2(t) \end{bmatrix} \begin{bmatrix} \gamma^{-2}B_1B_1' & B_2B_2' \\ B_2B_2' & B_2B_2' \end{bmatrix} \begin{bmatrix} P_1(t) \\ P_2(t) \end{bmatrix} \qquad (2.13) \\ -\dot{P}_2(t) &= A'P_2(t) + P_2(t)A + C'C \\ &- \begin{bmatrix} P_1(t) & P_2(t) \end{bmatrix} \begin{bmatrix} 0 & \gamma^{-2}B_1B_1' \\ \gamma^{-2}B_1B_1' & B_2B_2' \end{bmatrix} \begin{bmatrix} P_1(t) \\ P_2(t) \end{bmatrix} \qquad (2.14) \end{aligned}$$

with $P_1(t_f) = 0$ and $P_2(t_f) = 0$ have solutions $P_1(t) \leq 0$ and $P_2(t) \geq 0$ on $[t_0, t_f]$. If solutions exist, we have that

(i) the Nash equilibrium strategies are uniquely specified by

$$u^*(t,x) = -B_2'P_2(t)x(t) \qquad (2.15)$$

$$w^*(t,x) = -\gamma^{-2}B_1'P_1(t)x(t) \qquad (2.16)$$

$$(ii)\, J_1(u^*,w^*) = x_0'P_1(t_0)x_0 \qquad (2.17)$$

$$J_2(u^*,w^*) = x_0'P_2(t_0)x_0. \qquad (2.18)$$

(iii) In the case that $u(t,x) = u^*(t,x)$ with $x_0 = 0$,

$$\|\mathcal{R}_{zw}\|_{2i} < \gamma \quad \forall w \in \mathcal{L}_2[t_0, t_f] \qquad (2.19)$$

where the operator $\mathcal{R}_{zw}$ is defined by

$$\dot{x}(t) = (A - B_2B_2'P_2(t))x(t) + B_1w(t,x) \qquad (2.20)$$

$$z(t) = \begin{bmatrix} C \\ DB_2'P_2(t) \end{bmatrix} x(t). \qquad (2.21)$$

Proof This is given in [13]

2.3 A Stochastic Interpretation

In the previous section we found a control law $u^*(t,x)$ which achieves $\|\mathcal{R}_{zw}\|_{2i} < \gamma$, and at the same time solves the deterministic regulator problem

$$\min_u \left\{ J_2(u, w^*) = \int_{t_0}^{t_f} z'(t)z(t)dt \right\}.$$

In this section we extend the analysis to the case of a second white noise disturbance input. In particular, we show that the control law $u^*(t,x)$ solves the stochastic linear regulator problem

$$\min_u \left\{ \bar{J}_2(u, w^*) = \mathcal{E}\{\int_{t_0}^{t_f} z'(t)z(t)dt\} \right\}$$

when the equation for the state dynamics is replaced by

$$\dot{x}(t) = Ax(t) + B_0 w_0(t) + B_1 w(t) + B_2 u(t)$$

in which $w_0(t)$ is a realisation of a white noise process.

Theorem 2.2 *Suppose*

$$\dot{x}(t) = Ax(t) + B_0 w_0(t) + B_1 w(t,x) + B_2 u(t,x) \quad x(t_0) = x_0 \qquad (2.22)$$

$$z(t) = \begin{bmatrix} Cx(t) \\ Du(t) \end{bmatrix} \qquad D'D = I \qquad (2.23)$$

with $\mathcal{E}\{x_0 x_0'\} = Q_0$ *and* $\mathcal{E}\{w_0(\tau)w_0(t)\} = I\delta(t-\tau)$.
Then the control law $u^*(t,x) = -B_2' P_2(t)x(t)$

(i) *results in* $\|\mathcal{R}_{zw}\|_{2i} < \gamma$, *where the operator* $\mathcal{R}_{zw}$ *is described by :*

$$\dot{x}(t) = (A - B_2' P_2(t))x(t) + B_1 w(t,x) \qquad x(t_0) = 0$$

$$z(t) = \begin{bmatrix} C \\ -DB_2' P_2(t) \end{bmatrix} x(t)$$

and

(ii) *solves the stochastic linear regulator problem*

$$\min_u \left\{ \bar{J}_2(u, w^*) = \mathcal{E}\{\int_{t_0}^{t_f} z'(t)z(t)dt\} \right\} \qquad (2.24)$$

with $w^*(t,x) = -\gamma^{-2} B_1' P_1(t)x(t)$. *In addition, we get*

$$\bar{J}_2(u^*, w^*) = tr\left[P_2(t_0)Q_0 + \int_{t_0}^{t_f} P_2(t)dt \right]. \qquad (2.25)$$

Remark
Implementing $u^*(t,x)$ and $w^*(t,x)$ gives

$$\begin{aligned}\dot{x} &= (A-\gamma^{-2}B_1B_1'P_1 - B_2B_2'P_2)x + B_0w_0 \qquad x(t_0)=0\\ w^* &= -\gamma^{-2}B_1'P_1x.\end{aligned}$$

It is immediate that

$$\begin{aligned}-(\dot{P}_1+\dot{P}_2) &= (A-\gamma^{-2}B_1B_1'P_1 - B_2B_2'P_2)'(P_1+P_2)\\ &+ (P_1+P_2)(A-\gamma^{-2}B_1B_1'P_1 - B_2B_2'P_2) + \gamma^{-2}P_1B_1B_1'P_1\end{aligned}$$

in which $(P_1+P_2)(t_f)=0$. It then follows from a standard result on stochastic processes, [12] Thm 1.54, that the energy in the worst-case disturbance is given by :

$$\begin{aligned}\mathcal{E}\left\{\gamma^2\int_{t_0}^{t_f} w^{*\prime}(t)w^*(t)dt\right\} &= \mathcal{E}\left\{\gamma^2\int_{t_0}^{t_f} x'(\gamma^{-2}P_1B_1B_1'P_1)xdt\right\}\\ &= tr\left\{\int_{t_0}^{t_f} B_0'(P_1+P_2)(t)B_0dt\right\}.\end{aligned}$$

2.4 Reconciliation of H_2, H_∞ and H_2/H_∞ theories

In this section we establish a link between linear quadratic control, H_∞ control and mixed H_2/H_∞ control problems. Each of the three problems may be generated as special cases of the following non-zero sum, two-player Nash differential game:

Given the system described by (2.9), find Nash equilibrium strategies $u^*(t,x)$ and $w^*(t,x)$ in Ω (the set of linear memoryless feedback laws) which satisfy $J_1(u^*,w^*)\leq J_1(u^*,w)$ and $J_2(u^*,w^*)\leq J_2(u,w^*)$, where

$$\begin{aligned}J_1(u,w) &= \int_{t_0}^{t_f}\left[\gamma^2w'(t,x)w(t,x) - z'(t)z(t)\right]dt\\ J_2(u,w) &= \int_{t_0}^{t_f}\left[z'(t)z(t) - \rho^2w'(t,x)w(t,x)\right]dt.\end{aligned}$$

The solution to this game is given by

$$u^*(t,x) = -B_2'S_2(t)x(t) \qquad \text{and} \qquad w^*(t,x) = -\gamma^{-2}B_1'S_1(t)x(t),$$

where $S_1(t)$ and $S_2(t)$ satisfy the coupled Riccati differential equations

$$\begin{aligned}-\dot{S}_1(t) &= A'S_1(t)+S_1(t)A - C'C\\ &- \begin{bmatrix} S_1(t) & S_2(t)\end{bmatrix}\begin{bmatrix}\gamma^{-2}B_1B_1' & B_2B_2'\\ B_2B_2' & B_2B_2'\end{bmatrix}\begin{bmatrix}S_1(t)\\ S_2(t)\end{bmatrix} \qquad S_1(t_f)=0\\ -\dot{S}_2(t) &= A'S_2(t)+S_2(t)A + C'C\\ &- \begin{bmatrix} S_1(t) & S_2(t)\end{bmatrix}\begin{bmatrix}\rho^2\gamma^{-4}B_1B_1' & \gamma^{-2}B_1B_1'\\ \gamma^{-2}B_1B_1' & B_2B_2'\end{bmatrix}\begin{bmatrix}S_1(t)\\ S_2(t)\end{bmatrix} \qquad S_2(t_f)=0\end{aligned}$$

(i) The standard LQ optimal control problem is recovered by setting $\rho=0$ and $\gamma=\infty$. In which case $-S_1(t)=S_2(t)=P(t)$ and $u^*(t,x)=-B_2'P(t)x(t)$ and $w^*(t,x)\equiv 0$.
(ii) The pure H_∞ control problem is recovered by setting $\rho=\gamma$. Again, $-S_1(t)=S_2(t)=P_\infty(t)$ and $u^*(t,x)=-B_2'P_\infty(t)x(t)$ and $w^*(t,x)=\gamma^{-2}B_1'P_\infty(t)x(t)$.
(iii) The mixed H_2/H_∞ control problem comes from $\rho=0$.

3 Conclusion

We have shown how to solve a mixed H_2/H_∞ problem by formulating it as a two player Nash game. The necessary and sufficient conditions for the existence of a solution to the problem are given in terms of the existence of solutions to a pair of cross-coupled Riccati differential equations. If the controller strategy sets are expanded to include memoryless nonlinear controls which are analytic in the state, the necessary and sufficient conditions for the existence of a solution are unchanged, as are the control laws themselves. We have also established a link between H_2, H_∞ and mixed H_2/H_∞ theories by generating each as a special case of another two-player LQ Nash game. In conclusion we mention that it is possible to obtain results for the infinite horizon problem. Under certain existence conditions, we show that the solutions of the cross-coupled Riccati equations approach limits P_1 and P_2, (1) which satisfy algebraic versions of (2.13) and (2.14), (2) $P_1 \leq 0$ and $P_2 \geq 0$, and (3) P_1 and P_2 have stability properties which are analogous to those associated with LQG and pure H_∞ problems.

References

[1] B. D. O. Anderson and J. B. Moore, "Optimal Control – Linear Quadratic Methods," Prentice-Hall International, 1989

[2] D. Arov and M. G. Krein, "On the evaluation of entropy functionals and their minima in generalized extension problems," *Acta Scienta Math.*, Vol.45, pp 33-50, 1983, (in Russian)

[3] T. Basar, "On the uniqueness of the Nash solution in linear-quadratic differential games," *Int. Journal of Game Theory*, Vol. 5, pp 65-90, 1977

[4] T. Basar and G. J. Olsder, "Dynamic noncooperative game theory," Academic Press, New York, 1982

[5] D. S. Bernstein and W. M. Haddad, "LQG control with an H_∞ performance bound: A Riccati equation approach," *IEEE Trans. Automat. Control*, Vol. AC-34, No. 3, pp 293-305, 1989

[6] J. C. Doyle, K. Zhou and B. Bodenheimer, "Optimal control with mixed H_2 and H_∞ performance objectives," *Proc. 1989 American Control Conf.*, Pittsburgh, Pennsylvania, pp 2065-2070, 1989

[7] H. Dym, "J-contractive matrix functions, reproducing kernel Hilbert spaces and interpolation," Vol. 71 of Regional Conference Series in Mathematics, Am. Math. Soc., 1989

[8] H. Dym and I. Gohberg, "A maximum entropy principle for contractive interpolants," *J. Funct. Analysis*, Vol.65, pp 83-125, 1986

[9] K. Glover and D. Mustafa, "Derivation of the maximum entropy H_∞ -controller and a state-space formula for its entropy," *Int. J. Control*, Vol.50, pp 899-916, 1989

[10] I. M. Jaimoukha and D. J. N. Limebeer, "State-space algorithm for the solution of the 2-block super-optimal distance problem," *IEEE Conference on Decision and Control*, Brighton, United Kingdom, 1991

[11] H. Kwakernaak, "A ploynomial approach to minimax frequency domain optimisation of multivariable feedback systems," *Int. J. Control*, Vol.44, No.1, pp 117-156, 1986

[12] H. Kwakernaak and R. Sivan, "Linear optimal control systems," J. Wiley & son, 1972

[13] D. J. N. Limebeer, B. D. O. Anderson and B. Hendel, "A Nash game approach to mixed H_2/H_∞ control," submitted for publication

[14] D. J. N. Limebeer, G. D. Halikias and K. Glover, "State-space algorithm for the computation of super-optimal matrix interpolating functions," *Int. J. Control*, Vol.50, No.6, pp 2431-2466, 1989

[15] D. J. N. Limebeer and Y. S. Hung, "An analysis of the pole-zero cancellations in H_∞ -optimal control problems of the first kind," *SIAM J. Control Optimiz.*, Vol.25, pp 1457-1493, 1987

[16] D. Mustafa, "Relations between maximum entropy H_∞ control and combined H_∞ /LQG control," *Systems and Control Letters*, Vol. 12, pp 193-203, 1989

[17] D. Mustafa and K. Glover, "Minimum entropy H_∞ control," Heidelberg, Springer-Verlag, 1990

[18] D. Mustafa, K. Glover and D. J. N. Limebeer, "Solutions to the H_∞ general distance problem which minimize an entropy integral," *Automatica*, Vol. 27, pp. 193-199, 1991

[19] M. A. Rotea and P. P. Khargonekar, "H_2-optimal control with an H_∞ -constraint - the state feedback case," Internal Report, Univ. of Minnesota

[20] A. W. Starr and Y. C. Ho, "Nonzero-sum differential games," *J. Optimization Theory and Applications*, Vol.3, No. 3, pp 184-206, 1967

[21] M. C. Tsai, D. W. Gu and I. Postlethwaite, "A state space approach to super-optimal H_∞ control problems," *IEEE Trans. Auto. Control*, AC-33, No.9, 1988

[22] N. J. Young, "The Nevanlinna-Pick problem for matrix-valued functions," *J. Operator Theory*, Vol.15, pp239-265, 1986

[23] K. Zhou, J. C. Doyle, K. Glover and B. Bodenheimer, "Mixed H_2 and H_∞ control," *Proc. 1990 American Control Conf.*, San Diego, pp 2502-2507, 1990

ROBUST l^1-OPTIMAL CONTROL

J.B. Pearson
Department of Electrical and Computer Engineering
Rice University, Houston, Texas 77251-1892 U.S.A.

1. Introduction

This paper presents a brief account of l^1-optimal control theory and its role in the robust performance problem, which is one of the most important problems in control system design.

The l^1-optimization problem was first formulated in 1986 by Vidyasagar [1] and solved in a series of papers [2-5] by Dahleh and Pearson. In its original form, the only source of uncertainty was in the description of the exogenous signals (reference and disturbance inputs). Further work by Dahleh and Ohta [6] determined a necessary and sufficient condition for robust stability of a system with respect to unstructured plant perturbations with bounded l^∞-induced norms. This result was generalized by Khammash and Pearson [7] to structured perturbations. It was also established in [7] that the robust performance problem and the robust stability problem are equivalent in a certain sense. This result makes it possible to solve the robust performance problem for the class of generalized plants having a linear, time invariant part plus l^∞-norm bounded, structured perturbations which can be nonlinear and time-varying.

This paper is organized as follows. In Section 2, the robust performance problem is stated and l^1 theory is briefly reviewed. In Section 3, we consider plant perturbations and review the current status of the robust stability problem and how it applies to the robust performance problem. Finally, in Section 4 we discuss the solution of the robust performance problem and some recent results on solutions of the L^1-optimization problem.

2. The Robust Performance Problem

Figure 1 shows a system in which the uncertain plant is described by the pair $\tilde{G} = (G, \Delta)$ where G is a fixed linear-time-invariant system and Δ is a diagonal operator with entries Δ_i which represents causal, norm-bounded perturbations of the nominal plant G. C is a linear-time-invariant controller, and w and z represent the exogenous inputs (reference and disturbance inputs) and the regulated outputs respectively.

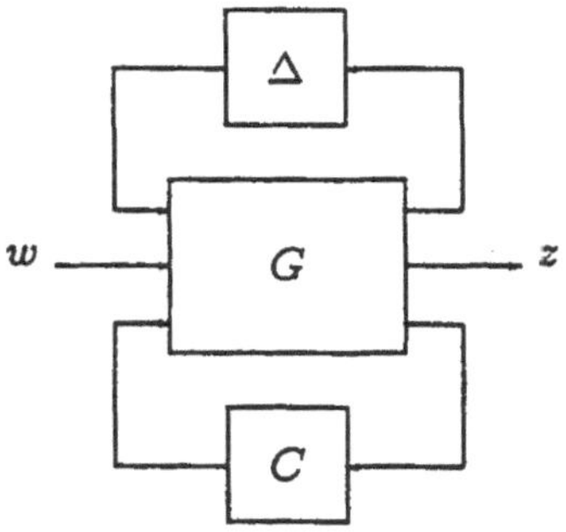

Figure 1

The class of admissible perturbations is given by the induced norms

$$\|\Delta_i\| = \sup_{x \neq 0} \frac{\|\Delta_i x\|_p}{\|x\|_p} \leq 1$$

where $p = 2$ or $p = \infty$. In the first case, if the Δ_i are linear-time-invariant, then the induced norm is the H^∞ norm. In the second case, the induced norm is the L^1 norm when the Δ_i are linear-time-invariant.

System performance is measured by the system "gain"

$$\|T_{zw}\| = \sup_{w \neq 0} \frac{\|z\|_p}{\|w\|_p} \qquad p = 2, \infty \ .$$

The problem considered in this paper is $p = \infty$ and the Δ_i represent causal norm-bounded perturbations which can be nonlinear and/or time-varying. In this case the system gain is the L^∞ induced norm. We will first consider discrete-time systems and l^∞ induced norms on the perturbations and system gain. G represents a linear-shift-invariant plant and the problem of interest is the

Robust Performance Problem (RPP).

Given the system shown in Figure 1, find a linear-shift-invariant controller C such that the system is internally stable and

$$\|T_{zw}\| < 1$$

for all admissible $\tilde{G}$.

This problem is the most fundamental of all control problems and is the underlying motivation for most of the introductory control systems textbooks that have been written in the last fifty years. Unfortunately, the problem was never posed in this manner until recently and as a consequence, our libraries are full of textbooks that never precisely state what the control problem is and represent merely a collection of various tools that may or may not apply. There are a few notable exceptions such as Bower and Schultheiss [8] and Horowitz [9]. Horowitz was the first to come to grips with the problem of plant uncertainty which is the primary motivation for feedback control. Almost all other textbooks treat plant uncertainty superficially, if at all, and then only by means of specifications such as gain and phase margins. We owe our current enlightment primarily to George Zames and John Doyle via the introduction of H^∞ theory and the structured-singular-value (SSV) or μ-methodology.

We will begin our discussion of l^1-optimal control theory with the **Nominal Performance Problem** (NPP) where $\Delta = 0$ and the objective is to minimize the system gain $\|T_{zw}\|$ over all stabilizing controllers C. In this case, there is no nonlinear-time-varying part so the system gain is the l^1-norm of the system impulse response.

Assume that the map T_{zw} is represented by the impulse response sequence $\Phi = \{\Phi_i\}$ which can be represented as

$$\Phi = H - U * Q * V$$

where H, U and V are matrix l^1 sequences, i.e.

$$\sum_{k=0}^{\infty} |H_{ij}(k)| < \infty , \qquad \sum_{k=0}^{\infty} |U_{ij}(k)| < \infty , \qquad \sum_{k=0}^{\infty} |V_{ij}(k)| < \infty ,$$

and $*$ represents convolution. Q is also a matrix l^1 sequence and is the well-known YJBK [10-12] parameter. The first part of our problem is to obtain a characterization of all l^1 sequences $\{K_{ij}(k)\}$ that can be represented as

$$K = U * Q * V$$

with $Q \in l^1$. This can be done and for a general result see McDonald and Pearson [13]. We will define

$$M = \{ K \in l^1 \mid K = U * Q * V *, \ Q \in l^1 \} .$$

Our optimization problem now becomes

$$\mu_0 = \inf_{K \in M} \|H - K\|_1$$

or

$$\mu_0 = \inf_{K \in M} \|\Phi\|_1$$

with $\Phi = H - K$.

This problem reduces to a finite linear programming (LP) problem in the case where the z-transforms of U and V, i.e. $\hat{U}$ and $\hat{V}$ have full row and full column ranks respectively. This is the so-called *one-block problem*. Other cases result in the two- or four-block problems [13].

It has been shown that if $\hat{U}$ and $\hat{V}$ have no zeros on the unit circle, there always exist solutions to the optimization problem. In the one block case, the solution is straightforward. Furthermore, the optimal impulse response sequence is of finite length and thus the optimal controller is always rational when the plant is rational.

In the multi-block case, the situation is quite different. Here we know that optimal solutions exist, and it is possible to calculate sub-optimal solutions that approach the optimal. The current method of solution is to set the problem up in such a way that finite impulse response solutions are feasible solutions of the LP problem. Then for a given length impulse response, an optimal solution is calculated which is sub-optimal for the original problem. These solutions converge to the optimal as the number of terms in the impulse response goes to infinity.

The advantage of this method of solution is that at any stage, we may stop the calculation and know that we have a stabilizing controller that furnishes a value of performance index equal to the value of the last LP problem.

3. The Robust Stability Problem

Consider the system shown in Figure 1. Part of the RPP is the

Robust Stability Problem (RSP).

Find a linear shift invariant C such that the system in Figure 1 is internally stable for all admissible $\tilde{G}$.

The first step is to consider the analysis problem; given G and C, when is the system robustly stable (i.e. stable for all admissible $\tilde{G}$)?

Consider Figure 2 where G and C are lumped into the linear-shift-invariant operator M.

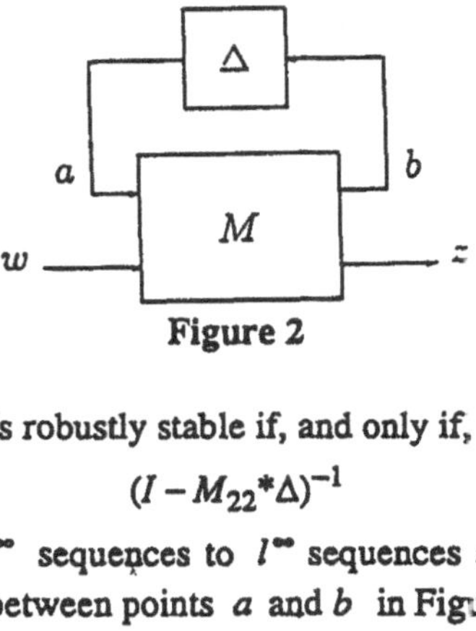

Figure 2

We say that this system is robustly stable if, and only if,

$$(I - M_{22} * \Delta)^{-1}$$

is l^∞ stable, i.e. it maps l^∞ sequences to l^∞ sequences and is bounded. M_{22} is the open loop impulse response between points a and b in Figure 2.

We can define robust performance for the system by stating that robust performance is achieved if, and only if, $(I - M_{22} * \Delta)^{-1}$ is l^∞ stable and

$$\|T_{zw}\| < 1 .$$

This can be converted into a RSP as shown in Figure 3.

Now define $\tilde{\Delta} = \begin{bmatrix} \Delta & 0 \\ 0 & \Delta_p \end{bmatrix}$ with $\|\Delta_p\| \leq 1$. Then the system in Figure 3 is

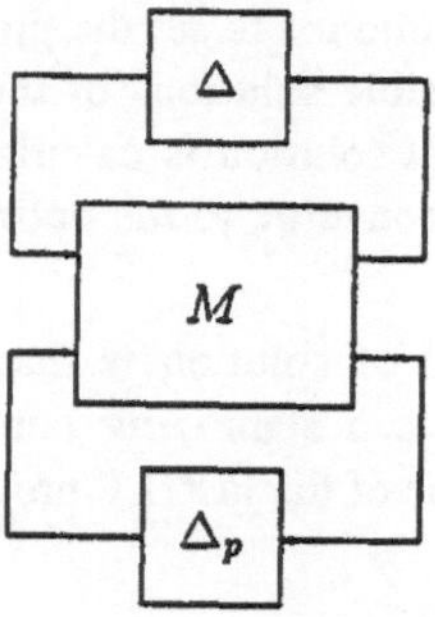

Figure 3

robustly stable if, and only if, $(I - M * \tilde{\Delta})^{-1}$ is l^∞ stable.

Recent results establish the following equivalence.

Theorem [7]

The system in Figure 2 achieves robust performance if, and only if, the system in Figure 3 achieves robust stability.

This is similar to the result established in [14] when the perturbations are linear-time-invariant and the norms are H^∞ norms. In our case, the result is true when $\tilde{\Delta}$ is nonlinear and/or time varying.

Now we will solve RPP if we can solve RSP for the system in Figure 3.

The solution to RSP is given as follows.

Define

$$M = \begin{bmatrix} M_{11} & \cdots & M_{1n} \\ \vdots & & \\ M_{n1} & \cdots & M_{nn} \end{bmatrix}$$

where the M_{ij} represent impulse response sequences between the jth input and i^{th} output and

$$\|M_{ij}\|_1 = \sum_{k=0}^{\infty} |M_{ij}(k)| \ .$$

Theorem [7]

The system in Figure 3 is robustly stable if, and only if, the system of inequalities

$$x_i \le \sum_{j=1}^{n} \|M_{ij}\|_1 x_j \qquad i = 1, \ldots, n$$

has no solutions in

$$(R^+)^n / \{0\} \ .$$

This result is quite easy to check. For example when $n=1$, we have

$$\|M\|_1 < 1$$

as the necessary and sufficient condition for robust stability. This is the result obtained by Dahleh and Ohta [6].

When $n=2$, we have

$$\|M_{11}\|_1 < 1$$

and

$$\|M_{22}\|_1 + \frac{\|M_{21}\|_1 \; \|M_{12}\|_1}{1 - \|M_{11}\|_1} < 1$$

and so forth.

These results are useful for stability analysis but not for synthesis. This problem is solved as follows.

Define

$$\overline{M} = \begin{bmatrix} \|M_{11}\|_1 & \cdots & \|M_{1n}\|_1 \\ \vdots & & \\ \|M_{n1}\|_1 & \cdots & \|M_{nn}\|_1 \end{bmatrix}$$

$$X = \mathrm{diag}\,(x_i) \quad x_i > 0 \qquad i=1,2,\ldots,n$$

$$\rho(\overline{M}) = \text{spectral radius of } \overline{M} \,.$$

Theorem [15]

The following are equivalent:

1) *The system in Figure 3 is robustly stable.*
2) $\rho(\overline{M}) < 1$.
3) $\min_X \|X^{-1}MX\| < 1$.

4. Solution of RPP

It follows from the Perron-Frobenius theory of positive matrices [16] that

$$\rho(\overline{M}) = \min_X \|X^{-1}MX\|_1$$

and the minimizing diagonal matrix X is made up of the entries of the eigenvector corresponding to $\rho(\overline{M})$. This gives us an approach to the synthesis problem, i.e. the problem of finding a controller that satisfies $\rho(\overline{M}) < 1$.

Consider the following:

Write

$$M = H - U*Q*V$$

and define

$$\mu_p = \min_{Q} \min_{X} \| X^{-1}(H - U^*Q^*V)X \|_1 .$$

An iteration can now be set up in order to determine if there exists a Q such that $\mu_p < 1$.

This is the same type of iteration that is used in μ-synthesis and is subject to some of the same problems. First, the minimization problem is non-convex, so only local minima can be found and, second, it is not clear how to choose the (in general) non-unique Q at each stage so as to decrease the spectral radius as much as possible.

An important question remaining is how do we handle continuous-time plants? It was shown in [3] that continuous-time L^1-optimal solutions consist of strings of delayed impulses. These are practically unrealizable and must be approximated by rational solutions. A recent paper on this type of approximation is that of Ohta, Maeda, and Kodama [17]. Another approach to the continuous time problem is, instead of trying to design a continuous-time controller, design a digital controller and, therefore, view the system as a sampled-data system in which the objective is to minimize the L^∞-induced system norm. This is a very active area of research, and new results are being obtained rapidly. The results directly applicable to the L^∞-induced norm are those of Sivashankar and Khargonekar [18], Dullerud and Francis [19], Bamieh et al. [20], and Khammash [21]. It is still too early to evaluate these results in terms of practical design algorithms, but it appears that we are on the right track, and that useful results will be forthcoming.

5. Acknowledgement

This research was sponsored under various grants by the National Science Foundation and the Air Force Office of Scientific Research.

6. References

[1] Vidyasagar, M., "Optimal rejection of persistent, bounded disturbances," *IEEE Trans. on Automatic Control*, AC-31(6), pp. 527-534 (Jun. 1986).

[2] Dahleh, M.A. and Pearson, J.B., "l^1-optimal feedback controllers for MIMO discrete-time systems," *IEEE Trans. on Automatic Control*, AC-32(4), pp. 314-322 (Apr. 1987).

[3] Dahleh, M.A. and Pearson, J.B., "L^1-optimal compensators for continuous-time systems," *IEEE Trans. on Automatic Control*, AC-32(10), pp. 889-895 (Oct. 1987).

[4] Dahleh, M.A. and Pearson, J.B., "Optimal rejection of persistent disturbances, robust stability, and mixed sensitivity minimization," *IEEE Trans. on Automatic Control*, AC-33(8), pp. 722-731 (Aug. 1988).

[5] Dahleh, M.A. and Pearson, J.B., "Minimization of a regulated response to a fixed input," *IEEE Trans. on Automatic Control*, AC-33(10), pp. 924-930 (Oct. 1988).

[6] Dahleh, M.A. and Ohta, Y., "A necessary and sufficient condition for robust BIBO stability," *Systems and Control Letters*, 11, pp. 271-275 (1988).

[7] Khammash, M. and Pearson, J.B., "Performance robustness of discrete-time systems with structured uncertainty," *IEEE Trans. on Automatic Control*, AC-36(4), pp. 398-412 (Apr. 1991).

[8] Bower, J.L. and Schultheiss, P.M., *Introduction to the Design of Servomechanisms*, New York: John Wiley & Sons (1958).

[9] Horowitz, I.M., *Synthesis of Feedback Systems*, New York: Academic Press (1963).

[10] Youla, D.C., Bongiorno, J.J., and Jabr, H.A., "Modern Wiener-Hopf design of optimal controllers – Part I: The single input-output case," *IEEE Trans. on Automatic Control*, AC-21, pp. 3-13 (1976).

[11] Youla, D.C., Bongiorno, J.J., and Jabr, H.A., "Modern Wiener-Hopf design of optimal controllers – Part II: The multivariable case," *IEEE Trans. on Automatic Control*, AC-21, pp. 319-338 (Jun. 1976).

[12] Kucera, V., *Discrete Linear Control: The Polynomial Equation Approach*, New York: John Wiley and Sons (1979).

[13] McDonald, J.S. and Pearson, J.B., "l^1-optimal control of multivariable systems with output norm constraints," *Automatica*, 27(2), pp. 317-329 (1991).

[14] Doyle, J.C., Wall, J.E., and Stein, G., "Performance and robustness analysis for structured uncertainty," Proceedings of the 21st IEEE Conference on Decision and Control, pp. 629-636 (Dec. 1982).

[15] Khammash, M. and Pearson, J.B., "Robustness synthesis for discrete-time systems with structured uncertainty," Proceedings of the 1991 Automatic Control Conference, Boston, MA, to appear (Jun. 1991).

[16] Gantmacher, F.R., *The Theory of Matrices, Vol. II*, New York: Chelsea Publishing (1959).

[17] Ohta, Y., Maeda, H., and Kodama, S., "Rational approximation of L_1-optimal controllers for SISO systems," Workshop on Robust Control, Tokyo, Japan (June 24-25, 1991).

[18] Sivashankar, N. and Khargonekar, P., "L_∞-induced norm of sampled-data systems," Proceedings of the 1991 Conference on Decision and Control, Brighton, England, to appear (Dec. 1991).

[19] Dullerud, G. and Francis, B., "L_1 performance in sampled-data systems," *IEEE Trans. on Automatic Control* (to appear).

[20] Bamieh, B., Dahleh, M.A., and Pearson, J.B., "Minimization of the L^∞-induced norm for sampled-data systems," Technical Report 9109, Dept. of Electrical and Computer Engineering, Rice University (1991).

[21] Khammash, M., "Necessary and sufficient conditions for the robustness of time-varying systems with applications to sampled-data systems," *IEEE Trans. on Automatic Control* (submitted).

Lecture Notes in Control and Information Sciences

Edited by M. Thoma and A. Wyner

Vol. 137: S. L. Shah, G. Dumont (Eds.)
Adaptive Control Strategies for Industrial Use
Proceedings of a Workshop
Kananaskis, Canada, 1988
VI, 360 pages. 1989

Vol. 138: D. C. McFarlane, K. Glover
Robust Controller Design Using Normalized Coprime Factor Plant Descriptions
X, 206 pages. 1990

Vol. 139: V. Hayward, O. Khatib (Eds.)
Experimental Robotics I
The First International Symposium
Montreal, June 19-21, 1989
XVII, 613 pages. 1990

Vol. 140: Z. Gajic, D. Petkovski, X. Shen
Singularly Perturbed and Weakly Coupled Linear Control Systems
A Recursive Approach
VII, 202 pages. 1990

Vol. 141: S. Gutman
Root Clustering in Parameter Space
VIII, 153 pages. 1990

Vol. 142: A. N. Gündeş, C. A. Desoer
Algebraic Theory of Linear Feedback Systems with Full and Decentralized Compensators
V, 176 pages. 1990

Vol. 143: H.-J. Sebastian, K. Tammer (Eds.)
System Modelling and Optimization
Proceedings of the 14th IFIP Conference
Leipzig, GDR, July 3-7, 1989
X, 960 pages. 1990

Vol. 144: A. Bensoussan, J. L. Lions (Eds.)
Analysis and Optimization of Systems
Proceedings of the 9th International Conference, Antibes, June 12-15, 1990
XII, 992 pages. 1990

Vol. 145: M. B. Subrahmanyam
Optimal Control with a Worst-Case Performance Criterion and Applications
X, 133 pages. 1990

Vol. 146: D. Mustafa, K. Glover
Minimum Entropy H_∞ Control
IX, 144 pages. 1990

Vol. 147: J. P. Zolésio (Ed.)
Stabilization of Flexible Structures
Third Working Conference,
Montpellier, France, January 1989
V, 327 pages, 1990

Vol. 148: In preparation

Vol. 149: K. H. Hoffmann, W. Krabs (Eds.)
Optimal Control of Partial Differential Equations
Proceedings of the IFIP WG 7.2 International Conference, Irsee, April 9-12, 1990
VI, 245 pages. 1991

Vol. 150: L. C. G. J. M. Habets
Robust Stabilization in the Gap-topology
IX, 126 pages. 1991

Vol. 151: J. M. Skowronski, H. Flashner, R. S. Guttalu (Eds.)
Mechanics and Control
Proceedings of the 3rd Workshop on Control Mechanics
in Honor of the 65th Birthday of George Leitmann
January 22-24, 1990, University of Southern California
IV, 497 pages. 1991

Vol. 152: J. D. Aplevich
Implicit Linear Systems
XI, 176 pages. 1991

Vol. 153: O. Hájek
Control Theory in the Plane
X, 272 pages. 1991

Vol. 154: A. Kurzhanski, I. Lasiecka (Eds.)
Modelling and Inverse Problems of Control for Distributed Parameter Systems
Proceedings of IFIP (W.G. 7.2)-IIASA Conference
Laxenburg, Austria, July 24-28, 1989
VII, 179 pages. 1991

Vol. 155: M. Bouvet, G. Bienvenu (Eds.)
High-Resolution Methods in Underwater Acoustics
V, 249 pages. 1991

Lecture Notes in Control and Information Sciences

Edited by M. Thoma and A. Wyner

Vol. 156: R. P. Hämäläinen, H. K. Ehtamo (Eds.)
Differential Games –
Developments in Modelling and Computation
Proceedings of the Fourth International Symposium on Differential Games and Applications
August 9-10, 1990, Helsinki University of Technology, Finland
XIII, 292 pages. 1991

Vol. 157: R. P. Hämäläinen, H. K. Ehtamo (Eds.)
Dynamic Games in Economic Analysis
Proceedings of the Fourth International Symposium on Differential Games and Applications
August 9-10, 1990, Helsinki University of Technology, Finland
XIII, 311 pages. 1991

Vol. 158: K. Warwick, M. Kárný, A. Halousková (Eds.)
Advanced Methods in Adaptive Control for Industrial Applications
X, 331 pages. 1991

Vol. 159: X. Li, J. Yong (Eds.)
Control Theory of Distributed Parameter Systems and Applications
Proceedings of the IFIP WG 7.2 Working Conference, Shanghai, China, May 6–9, 1990
VIII, 219 pages. 1991

Vol. 160: P. V. Kokotović (Ed.)
Foundations of Adaptive Control
IX, 525 pages. 1991

Vol. 161: L. Gerencsér, P. E. Caines (Eds.)
Topics in Stochastic Systems:
Modelling, Estimation and Adaptive Control
IV, 401 pages, 1991

Vol. 162: C. Canudas de Wit (Ed.)
Advanced Robot Control
IX, 314 pages, 1991

Vol. 163: V. L. Mehrmann
The Autonomous Linear Quadratic Control Problem
Theory and Numerical Solution
VI, 177 pages, 1991

Vol. 164: I. Lasiecka, R. Triggiani
Differential and Algebraic Riccati Equations with Application to Boundary/Point Control Problems: Continuous Theory and Approximation Theory
XI, 160 pages, 1991

Vol. 165: G. Jacob, F. Lamnabhi-Lagarrigue (Eds.)
Algebraic Computing in Control
Proceedings of the First European Conference
Paris, March 13–15, 1991
IX, 385 pages. 1991

Vol. 166: L. L. M. van der Wegen
Local Disturbance Decoupling with Stability for Nonlinear Systems
V, 135 pages, 1991

Vol. 167: M. Rao
Integrated System for Intelligent Control
VIII, 133 pages. 1992

Vol. 168: P. Dorato, L. Fortuna, G. Muscato
Robust Control for Unstructured Perturbations – An Introduction
VI, 117 pages. 1992

Vol. 169: V. M. Kuntzevich, M. Lychak
Guaranteed Estimates, Adaptation and Robustness in Control Systems
IV, 209 pages. 1992

Vol. 170: J. M. Skowronski, H. Flashner, R. S. Guttalu (Eds.)
Mechanics and Control
Proceedings of the 4th Workshop on Control Mechanics, January 21–23, 1991
University of Southern California, USA
IV, 301 pages. 1992

Vol. 171: P. Stefanidis, A. P. Papliński, M. J. Gibbard
Numerical Operations with Polynomial Matrices
Application to Multi-Variable Dynamic Compensator Design
VIII, 205 pages. 1992

Vol. 172: H. Tolle, E. Ersü
Neurocontrol
Learning Control Systems
Inspired by Neuronal Architectures and Human Problem Solving Strategies
X, 211 pages. 1992

Vol. 173: W. Krabs
On Moment Theory and Controllability of One-Dimensional Vibrating Systems and Heating Processes
VII, 174 pages. 1992

Lecture Notes in Control and Information Sciences

Edited by M. Thoma and A. Wyner

Vol. 174: A.J.M. Beulens, H.-J. Sebastian (Eds.)
Optimization-Based Computer-Aided
Modelling and Design
Proceedings of the First Working Conference
of the IFIP TC 7.6 Working Group,
The Hague, The Netherlands, 1991
VIII, 270 pages, 1992

Vol. 175: E. Rogers, D.H. Owens
Stability Analysis for Linear Repetitive Processes
VII, 197 pages, 1992

Vol. 176: B.L. Rozovskii, R.B. Sowers (Eds.)
Stochastic Partial Differential Equations
and Their Applications
Proceedings of IFIP WG 7/1 International Conference
University of North Carolina at Charlotte, NC
June 6 - 8, 1991
VIII, 251 pages, 1992

Vol. 177: I. Karatzas, D. Ocone (Eds.)
Applied Stochastic Analysis
Proceedings of a US-French Workshop,
Rutgers University, New Brunswick, N.J.
April 29 - May 2, 1991
X, 311 pages, 1992

Vol. 178: J.P. Zolésio (Ed.)
Boundary Control and Boundary Variation
Proceedings of IFIP WG 7.2 Conference,
Sophia Antipolis, France
October 15 - 17, 1990
VIII, 392 pages 1992

Vol. 179: Z.H. Jiang, W. Schaufelberger
Block Pulse Functions and Their Applications
in Control Systems
XII, 237 pages, 1992

Vol. 180: P. Kall (Ed.)
System Modelling and Optimization
Proceedings of the 15th IFIP Conference
Zurich, Switzerland, September 2-6, 1991
XIX, 969 pages, 1992

Vol. 181: C.R. Drane
Positioning Systems - A Unified Approach
X, 168 pages 1992

Vol. 182: J. Hagenauer (Ed.)
Advanced Methods for Sattellite
and Deep Space Communications
Proceedings of an International Seminar
Organized by Deutsche Forschungsanstalt
für Luft- und Raumfahrt (DLR)
Bonn, Germany, September 1992
VII, 196 pages 1992

Vol. 183: S. Hosoe (Ed.)
Robust Control
Proceedings of a Workshop held in Tokyo,
Japan, June 23 - 24, 1991
VII, 225 pages, 1992